introduction to calculus 2

The Houghton Mifflin GCMC Series
under the editorship of A. B. Willcox

Introduction to Calculus 1

A. B. Willcox · R. Creighton Buck · H. G. Jacob · D. W. Bailey

Introduction to Calculus 2

A. B. Willcox · R. Creighton Buck · H. G. Jacob · D. W. Bailey

Introduction to Calculus 1 and 2—combined edition

A. B. Willcox · R. Creighton Buck · H. G. Jacob · D. W. Bailey

Calculus of Several Variables

R. Creighton Buck · A. B. Willcox

Linear Algebra

H. G. Jacob · D. W. Bailey

Alfred B. Willcox
Executive Director
Mathematical Association of America

R. Creighton Buck
University of Wisconsin

Henry G. Jacob
University of Massachusetts

Duane W. Bailey
Amherst College

Houghton Mifflin Company · Boston

introduction to calculus 2

New York · Atlanta · Geneva, Illinois · Dallas · Palo Alto

Printed in the USA

Library of Congress Catalog Card Number: 71-141000

ISBN: 0-395-05539-3

To Our Wives

preface

Now that the book is written we come face-to-face with the inevitable question: Why did we do it? Every man who invests a part of his life in a pile of printed pages must answer that question, sooner or later, in his own mind. But if the fruit of his labor is a calculus book—*another* calculus book—an author has a special responsibility to answer the question in public. With that responsibility in mind, we offer several remarks . . .

. . . *about this series*. The recommendations for a *General Curriculum in Mathematics for Colleges* prepared by the Committee on the Undergraduate Program in Mathematics (CUPM) call for a basic sequence of five one-semester courses for the first two years of college. The sequence carries the student through beginning courses in calculus, probability and statistics, linear algebra, and multivariable calculus. It is not a sequence which most students can be expected to cover in two years—it is designed to accept students at a variety of levels of preparation and to provide a number of different tracks for students of a variety of interests and talents. *Introduction to Calculus 1 and 2* are the first and second books in a series designed to help bring these recommendations into the classroom.

In writing the books of this series, the authors have allowed their own experience to color freely the broad outlines of the committee. We have tried very hard to capture the spirit of the recommendations and hope that we have succeeded, at least in part. However, we also hope that in creating a concrete model we have improved upon the ideal. Most of all, we hope that our efforts will significantly help teachers to impart clarity and motivation to their courses which will challenge and excite their students.

. . . *about this book*. There are several features of this book which are novel. Aside from the overall adherence to the CUPM outline, the major focus is on readability and careful motivation. Our program for introducing continuity and limits is not the traditional one. Continuity comes first, via a highly geometrical approach.

Since this is the first significant idea of the calculus encountered by the student, considerable attention is devoted to its evolution from the most informal and intuitive image suggested by the words "continuous curve." Next, the idea of a function being limiting, or having a limit, at a point is defined as a sort of almost-continuity. This seems to us to be the natural progression from a simple intuitive idea to a more sophisticated variation. The discussion is informal—no ϵ's or δ's appear until the first chapter of *Introduction to Calculus 2*. The ideas of continuity and limits, which could hardly be more basic to the subject, must appeal to the intuition of the student if he is to take them at all seriously. We believe in rigor at this level only when it contributes to clarity, and we believe that an insistence on complete logical detail can often inhibit rather than enhance understanding. If the basic ideas of derivative and integrals can be transmitted by an intuitive approach, so can the ideas of continuity and limits. We have tried to do all these things.

The content of *Introduction to Calculus 1* (*C 1*) includes the necessary ingredients for a one-semester course covering both differential and integral calculus. Combining derivatives and integrals in a one-semester course along with the necessary basic techniques and applications to give these ideas flesh required some pruning. If you find that some of the topics which you consider absolutely essential for a first course are missing from *C 1*, you will probably find them in *Introduction to Calculus 2* (*C 2*), designed for the second semester. In *C 2* we return to each of the main ideas originally presented in *C 1*, sharpening techniques, extending concepts, and giving additional applications. This deeper examination of the basic concepts leads naturally into the discussions of series, elementary differential geometry of plane curves, and first order differential equations which conclude the second volume.

Thus, after the first semester the student should be ready to begin work in such subjects as probability, statistics, and physics, which are enhanced by some knowledge of the calculus. After two semesters he will have acquired a strong acquaintance with one-variable calculus and will be ready to branch out into higher dimensions.

In our writing we have had in mind the student who has had some introduction to the elementary functions and analytic geometry. This is in keeping with a trend to treat these subjects as precalculus ones which should properly be taught in high school and which in the ideal situation should already be familiar to the student when he begins his study of the calculus. However, reality dictates moderation of this view. Even those who espouse it zealously admit that its fulfillment lies in the future. Thus, our treatment of the elementary functions, although brief, is complete, and there is an appendix on analytic geometry, inequalities, and absolute value which can be used either for review or as a mini-text.

. . . for the teacher. This book is written for your student, not for you! It is designed to be read by your student, not to be a reference treatise. With this purpose in mind, we have tried to talk to the reader in much the same way that we have talked to our students in the classroom, attempting always to lead him on and to challenge him, but being careful not to outrun him. While frankly appealing to the reader's intuition, we have tried to keep our statements of mathematical definition and fact quite correct and expressed in the current idiom. There are many places for you to augment our presentation or to digress from it. We invite you also to disagree with it and alter it where you will. Treat the book as a companion in the classroom—one which has a point of view which need not always be your own but which can instruct by generating discussion and provoking thought. Calculus ex cathedra has not been our aim. The authors, who know the thrill of seeing a spark ignited in a student's eye, simply hope to help you bring the calculus to life for your students in your own way.

It has become almost a tradition in textbook writing to close a preface with some remarks to the teacher who anticipates difficulty in covering all of the offered material in the time available. We shall honor that tradition. However, the teacher is warned that this book is not a buffet from which meals can be selected for every appetite. We have striven for a connected narrative treating only those topics which most teachers consider essential for a first course. To shorten the treatment it is probably better to sample most topics lightly than to leave out many.

Exclusive of the sections numbered n.0, n = 1, 2, . . . , which are introductory and may be read rapidly, there are thirty-nine sections in the book. Generally, we would expect a teacher to assign one section per class meeting, but there must surely be exceptions to this rule of thumb. Many instructors will have fewer class meetings at their disposal, and most will wish to use some time at the beginning of the course for review and to bridge gaps between the starting point of this book and the student's previous experience. If the book is used in conjunction with *Introduction to Calculus 1* most instructors will wish to devote a few class periods to sections of the earlier book omitted for lack of time. Thus, the instructor may wish to double up some of the shorter sections, asking students to read some lightly and rapidly. It may be necessary—and it is entirely reasonable—to skip over some of the longer proofs, omitting most of the details. A few sections may be omitted entirely without leaving serious gaps. Chapters 5, 6, and 7 are relatively independent; any of these chapters may be omitted without serious loss of continuity.

There are so many variations to fit special interests and time restrictions that we are reluctant to write out any menus. However, as an example of what might be done in order to provide a course in thirty-seven class meetings, including time for review and bridging gaps, we offer the following one-semester diet. We hope no one will follow it without adapting it to his own tastes and the needs of his students.

Meeting	Sections	Topic
1.		Review and Bridging Gaps
2.		
3.		
4.		
5.	1.0, 1.1	Limits Revisited
6.	1.2	
7.	1.3	
8.	2.0, 2.1, 2.2	Derivatives Revisited
9.	2.3	
10.	2.4	
11.	2.5	
12.	3.0, 3.1	Integrals Revisited
13.	3.2	
14.	3.3	
15.	3.4	
16.	3.5	
17.	3.8	
18.	4.0, 4.1	Limits and Approximations
19.	4.2	
20.	4.3	
21.	5.0, 5.1	Infinite Series
22.	5.2	
23.	5.3	
24.	5.4	
25.	5.5	
26.	5.6	
27.	6.0, 6.1, 6.2	Smooth Curves; Elementary Differential Geometry
28.	6.3	
29.	6.4	
30.	6.5	
31.	6.6	
32.	6.7	
33.	7.0, 7.1	Simple Differential Equations
34.	7.2	
35.	7.3	
36.	7.4	
37.	7.5	

. . . about the help we have received. It would be impossible to list the names of all those who have influenced this book. Almost all of our teachers, colleagues, and students have been involved in shaping our ideas and attitudes. The writing project might never have been born if two of us had not been active participants in some of the deliberations of CUPM, and we value highly this opportunity for interaction with teachers representing the broadest spectrum of educational and pedagogical experience and thinking. However, several persons contributed so tangibly to the books that we must thank them individually and publicly. Mrs. Gloria Harris gave generously of her time typing draft after draft and learned more than she ever hoped to know about mathematicians and the foibles of their pens. Miss Veronica McLoud of Houghton Mifflin Company

contributed not only clarity to our prose but also accuracy to our mathematical presentation through her efficient editing. Without the help of these two women the books would still be lying in a dog-eared pile in a desk drawer. It was only through the enthusiasm and encouragement of Mr. Paul Kelly, college mathematics editor for Houghton Mifflin Company, that this writing project was born, and he has served as our efficient supporting team manager throughout. Finally, we are most grateful to Mr. Tom Hudson, a student at Amherst College, for the graphs in the *Solutions to Selected Exercises 1.*

A Note on Organization The book is divided into chapters, each of which is divided into sections. A notation "Section a.b" refers to Section b in Chapter a. Within a section, theorems, definitions, examples, and figures are numbered independently in serial order. Thus "Theorem a.b.c" refers to Theorem c in Section b of Chapter a. Lemmas and corollaries are considered theorems in the numbering system. Exercises occur in sets following each section. The reference "Exercise a.b.c" is to Exercise c in the set of exercises following Section a.b.

contents

Chapter 5 Infinite Series

Chapter 6 Smooth Curves; Elementary Differential Geometry

Chapter 7 Simple Differential Equations

Tables

limits revisited

chapter 1

1.0

Introduction

In *C 1*[1] we presented the elements of the calculus, developing its two central ideas, derivative and integral, and exploring a few of their many powerful implications. Our excursion has been admittedly rapid and sometimes superficial. Newly arrived tourists, we have completed a quick tour of the city for orientation and organization before exploring particular districts and streets in depth. We are now ready for a second look, this time in somewhat greater depth, with our appreciation of each alley and square enhanced by an overall understanding of how they all fit together in one metropolis.

Our tour (if we may continue the tourist analogy) began at the gates of the city, the idea of a continuous function and the related idea of a limit of a function. In one way or another, these ideas have underlain all that we have done. The important notion of a derivative was defined in terms of the limit of a certain constructed function. Although limits have not yet come into actual view in the definition of a definite integral, they are there and will emerge from the shadows soon (in Chapter 3). Moreover, in discussing integrals we did not often stray far from continuous functions, and the theorem that any continuous function is integrable is central to our discussion.

In fact, throughout our discussion of functions in *C 1* the concept of *continuity* was perhaps the strongest thread. Functions which have derivatives are necessarily continuous. Continuous functions have the intermediate value property, possess maximum and minimum values on closed intervals, have the inertia property,[2] and also are integrable and possess antiderivatives.

[1] *We shall refer to* Introduction to Calculus 1 *as* C 1 *and to* Introduction to Calculus 2 *as* C 2.

[2] *A function f has the inertia property if whenever $f(c) > 0$ (<0), then $f(x) > 0$ (<0) for all x in some neighborhood of c. See* C 1, *Chapter 1.*

Yet many of the functions with which we come into contact are inherently *not* continuous. For example, the function $(f(x_0 + h) - f(x_0))/h$ employed in the definition of the derivative is not continuous at $h = 0$. Fortunately, we find that we can often deal quite satisfactorily with functions which fail to be continuous if the failure is not too serious. A condition which often suffices and which does hold if the function is continuous is that the function have a limit at the appropriate point.

The notion of a function being *limiting* at a point and the related definition of the *limit*[3] of such a function were obtained by identifying a certain kind of "almost continuity." Although we have observed the basic role of the limit concept in the calculus, we have seen only the barest hint of the breadth of its uses. To appreciate this breadth one must see how the limit concept may be extended readily to situations in which our original conditions do not hold, some of them situations that do not even involve functions. In *C 1* we glanced briefly—and often casually—at some of these variations of the limit concept when we considered one-sided limits, infinite limits, and limits of sequences. There are many more such variations and generalizations, often at the very centers of the various districts and suburbs of the calculus.

[3] *A function is limiting at a point $x = c$ if it has a limit as $x \to c$. See* C 1, *Chapter 1.*

Although in *C 1* we used the concept of continuity in our definition of the limit of a function, we observed there that this was not a necessary ordering of the concepts. Our reasons for that approach were mainly pedagogical, and we saw in Theorem 1.7.1 of *C 1* that an independent definition of limits is possible. For purposes of future generalization we wish to focus on this independent definition here. (It will be restated in Section 1.1.)

However, we shall wish to make some changes in wording and style. The definitions employed in *C 1* were highly geometrical in their motivation. The use of such words as "neighborhood," "image," "mapping" was intended to be highly suggestive of position of points and placement of graphs. At that stage we made a definite effort to keep you thinking in geometric terms in order that the resulting picture of continuity should underlie and orient the more involved analytical considerations necessary in many of the later developments. But at the present level of penetration something new begins to happen. As we adapt and extend limit considerations to fit new situations the ties with the notion of continuity are weakened, and the geometrical point of view becomes more artificial and less enlightening. For purposes of maximum flexibility, therefore, we now find it expedient to enlarge our vocabulary somewhat by means of a different formulation of the limit definition. This reformulation of the definition of a limit and a more detailed discussion of some of the extensions of the limit concept which have been hinted at already are the subjects of this chapter.

1.1

Continuity and Limits

Let f be a function which is defined (at least) in a neighborhood of the number c.[1] Roughly speaking, when we say that f is *continuous* at $x = c$, we mean that the image $f(x)$ of x is near $f(c)$ whenever x is sufficiently near c. More precisely, we have expressed this idea (in Definition 1.4.1 of *C 1*) as follows:

(1.1.1) *The function f is continuous at c if every neighborhood V of $f(c)$ contains the image $f(U)$ of some neighborhood U of c.*

A function which either (a) is continuous at $x = c$ or (b) could be made continuous at $x = c$ by an appropriate definition (or redefinition) of the image $f(c)$ is called *limiting* at c. Roughly speaking, if a limiting function is not continuous at $x = c$, then at least the part of the graph for $x \neq c$ is "ready for continuity at c." The trouble lies entirely with $f(c)$.

If f is limiting at c, then the value L which must be given to $f(c)$ in order to make the function continuous at c is called the *limit* of f at c. It is written $L = \lim_{x\to c} f(x)$. Using the definition (1.1.1) we saw in Theorem 1.7.1 of *C 1* that

(1.1.2) *$L = \lim_{x\to c} f(x)$ if and only if every neighborhood V of L contains the image $f(U')$ of some deleted neighborhood U' of c.*[2]

Statement (1.1.2) expresses in precise terms what we mean intuitively when we say "$f(x)$ is arbitrarily close to L provided x is sufficiently close to (but not equal to) c." Lying in an arbitrary neighborhood V of L makes $f(x)$ "arbitrarily close to L"; lying in a corresponding (and sufficiently small) deleted neighborhood U' of c makes x "sufficiently near, but not equal to, c."

As you know, a neighborhood of a number x_0 is any open interval (u, v) such that $u < x_0 < v$. In a *symmetric* neighborhood of x_0, x_0 is the midpoint of the segment from u to v; that is, it is an interval of the form $(x_0 - r, x_0 + r)$, where r is a positive number. We might call r the *radius* of this symmetric neighborhood and adopt a standard notation for the symmetric neighborhood of x_0 with radius r:

$$N(x_0, r) = (x_0 - r, x_0 + r).$$

We shall call $N(x_0, r)$ the *r-neighborhood of x_0*. It is an easy exercise to show that

(1.1.3) *Every neighborhood of x_0 contains a symmetric neighborhood of x_0.*

For, if (u, v) is a neighborhood of x_0 and r is the distance from x_0 to the nearest endpoint, then $N(x_0, r)$ is contained in (u, v). Then

[1] *Recall that a neighborhood of c is just an open interval containing c. If, for example, the domain of definition of f is a closed interval, then the stated condition on f simply means that c should not be one of the endpoints of this interval.*

[2] *A deleted neighborhood of c is just a neighborhood of c with c removed. Some authors use the term* punctured *neighborhood.*

using (1.1.3) and (1.1.2) it is also easy to prove the following theorem. The proof and that of (1.1.3) will be left as exercises (Exercises 1.1.10, 1.1.11).

Theorem 1.1.1 *Let f be a function defined at least in a deleted neighborhood of c. Then* $\lim_{x\to c} f(x) = L$ *if and only if every symmetric neighborhood* $N(L, \epsilon)$ *of L contains the image of some deleted symmetric neighborhood* $N'(c, \delta)$ *of c.*

Theorem 1.1.1 simply tells us that in considering limits if we wish we may restrict our attention to *symmetric* neighborhoods. This flexibility often results in significant simplifications in proofs and calculations. It also makes possible another formulation of the limit definition. We first observe that once the numbers c and L are given, the symmetric neighborhoods $N(L, \epsilon)$ and $N'(c, \delta)$ are completely specified by the two positive numbers ϵ and δ. Thus, Theorem 1.1.1 could be rephrased as follows:

(1.1.4) $L = \lim_{x\to c}$ *if and only if for every* $\epsilon > 0$ *there is a* $\delta > 0$ *such that* $f(N'(c, \delta)) \subset N(L, \epsilon)$.

The final stage in "analytizing" or "de-geometrizing" the definition involves an attack on the statement $f(N'(c, \delta)) \subset N(L, \epsilon)$. What precisely does this statement mean? It means of course that if x is in $N'(c, \delta)$, then $f(x)$ is in $N(L, \epsilon)$. But x is in $N'(c, \delta)$ if and only if the distance between x and c is less than δ (that is, $|x - c| < \delta$) and x is *not* equal to c (that is, $|x - c| \neq 0$). Thus x is in $N'(c, \delta)$ if and only if $0 < |x - c| < \delta$. Similarly, $f(x)$ is in $N(L, \epsilon)$ if and only if $|f(x) - L| < \epsilon$. Therefore the statement $f(N'(c, \delta)) \subset N(L, \epsilon)$ is equivalent to the statement

$$|f(x) - L| < \epsilon \quad \text{provided} \quad 0 < |x - c| < \delta.$$

We need only combine the above with (1.1.4) to obtain the form of the limit definition that we shall use throughout most of this volume.

Theorem 1.1.2 *Let f be a function defined at least in a deleted neighborhood of c. Then* $\lim_{x\to c} f(x) = L$ *if and only if:*

$$\textit{for every} \quad \epsilon > 0 \quad \textit{there is a} \quad \delta > 0 \quad \textit{such that}$$
$$|f(x) - L| < \epsilon \quad \textit{provided} \quad 0 < |x - c| < \delta.$$

We illustrate the application of Theorem 1.1.2 by using it to verify that

(1.1.5) $$\lim_{x\to 9} \sqrt{x} = 3.$$

In order to prove (1.1.5), we must verify the final statement in Theorem 1.1.2 with

$$L = 3, \quad c = 9, \quad f(x) = \sqrt{x}.$$

Suppose that ϵ is an arbitrary positive number. (*Every* proof involving the limit definition in Theorem 1.1.2 begins with this sentence!) We must find a corresponding number δ so that the condition

(1.1.6) $$0 < |x - 9| < \delta$$

will imply the condition

(1.1.7) $$|\sqrt{x} - 3| < \epsilon.$$

Before we can choose δ wisely, we must analyze condition (1.1.7) more carefully. The presence of the radical in the expression on the left suggests that we multiply and divide by the "conjugate expression" $|\sqrt{x} + 3|$. We obtain

$$|\sqrt{x} - 3| = \frac{|\sqrt{x} - 3|\,|\sqrt{x} + 3|}{|\sqrt{x} + 3|} = \frac{|x - 9|}{\sqrt{x} + 3}.$$

Since $\sqrt{x} \geq 0$, we have $\sqrt{x} + 3 \geq 3$; so

(1.1.8) $$|\sqrt{x} - 3| \leq \tfrac{1}{3}|x - 9|.$$

Now we can see that any choice of δ less than or equal to 3ϵ will do. For if

$$\delta \leq 3\epsilon,$$

then (1.1.6) implies that

$$\tfrac{1}{3}|x - 9| \leq \tfrac{1}{3} \cdot 3\epsilon = \epsilon.$$

Thus, (1.1.6) with this choice of δ implies condition (1.1.7), using (1.1.8).

To summarize: Given an arbitrary number $\epsilon > 0$, we have found a $\delta > 0$ (any $\delta \leq 3\epsilon$ will do—if you wish, let $\delta = 3\epsilon$) such that $0 < |x - 9| < \delta$ implies $|\sqrt{x} - 3| < \epsilon$. By Theorem 1.1.2, then, we have verified (1.1.5).

Exercises

1 (a) Show that if $\epsilon > 0$, then $|x| < \epsilon$ if and only if $-\epsilon < x < \epsilon$.

(b) Show that if $\epsilon > 0$, then $|x - a| < \epsilon$ if and only if $a - \epsilon < x < a + \epsilon$.

2 Each part of this exercise consists of a limit statement and an ϵ. For each part find a δ such that the conclusion of Theorem 1.1.2 is valid.

(a) $\lim_{x\to 3} 7 = 7$, $\epsilon = .01$; (b) $\lim_{x\to 2} \frac{1}{2}x - 1 = 0$, $\epsilon = .01$;

(c) $\lim_{x\to -2} 3x + 4 = -2$, $\epsilon = .0001$; (d) $\lim_{x\to 1} 7x - 2 = 5$, $\epsilon = 15$.

3 (a) Show that if $|x - 2| < 1$, then $|x + 4| < 7$. (*Hint:* Use Exercise 1.)

(b) Use Theorem 1.1.2 to prove that $\lim_{x\to 2} x^2 + 2x - 1 = 7$. (*Hint:* Use part (a). An appropriate choice of δ here is the smaller of the two numbers 1 and $\epsilon/7$.)

4 (a) Show that if $|x - 2| < 1$, then $5 < |x + 4|$.

(b) Use Theorem 1.1.2 to prove that $\lim_{x\to 2} \dfrac{1}{x + 4} = 1/6$.

5 Use Theorem 1.1.2 to prove each of the following limit statements.

(a) $\lim_{x\to 1} 3x = 3$, (b) $\lim_{x\to -1} 2x - 3 = -5$,

(c) $\lim_{x\to 0} -2x + 7 = 7$, (d) $\lim_{x\to 2} x^2 = 4$,

(e) $\lim_{x\to -3} x^2 - 6 = 3$, (f) $\lim_{x\to 1/2} 3x^2 + \frac{1}{4} = 1$,

(g) $\lim_{x\to 5} x^2 - 4x - 1 = 4$, (h) $\lim_{x\to -2} -3x^2 + x + 10 = -4$,

(i) $\lim_{x\to -1/2} \dfrac{1}{x} = -2$, (j) $\lim_{x\to 4} \dfrac{1}{2x - 3} = \dfrac{1}{5}$,

(k) $\lim_{x\to 2} \dfrac{1}{x^2 - 2} = \dfrac{1}{2}$, (l) $\lim_{x\to 0} \dfrac{1}{x^2 - x + 1} = 1$,

(m) $\lim_{x\to 4} \dfrac{1}{\sqrt{x}} = \dfrac{1}{2}$, (n) $\lim_{x\to 8} \sqrt[3]{x} = 2$.

6 Prove that $\lim_{x\to 0} x \sin x = 0$.

7 Prove that if $\lim_{x\to a} f(x) = L_1$ and $\lim_{x\to a} f(x) = L_2$, then $L_1 = L_2$. (*Hint:* Suppose $L_1 \neq L_2$ and let $\epsilon = \frac{1}{3}|L_1 - L_2|$. Find δ_1 and δ_2 corresponding to the two limit statements and let δ be the smaller of the two. Choose $|x - a| < \delta$ and obtain a contradiction.)

8 What does it mean, expressed in terms of an ϵ-δ criterion, for $\lim_{x\to a} f(x)$ *not* to equal L?

9 Use Exercises 4 and 7 to show that $\lim_{x\to 2} 1/(x + 4) \neq \frac{1}{2}$.

10 Prove (1.1.3).

11 Prove Theorem 1.1.1, using (1.1.2).

1.2

Variations and Extensions of the Limit Concept

At several stages in *C 1* we made use of variants of the limit concept. We treated these variations informally, because a careful treatment would have distracted us from our principal goal in each instance.

We shall recall these variations in this section, indicating how the conditions in Theorem 1.1.2 may be altered to provide precise definitions.

Several times in *C 1* we required a concept of a *one-sided limit* of a function. If f is defined for all $x > c$ in some neighborhood of c, then we may consider the *right-hand limit* $\lim_{x\to c+} f(x)$. Such a limit is defined simply by adding the condition $x > c$ to the statement in Theorem 1.1.2. *Left-hand* limits are defined similarly.

Definition 1.2.1 We say that $\lim_{x\to c+} f(x) = L$ if and only if:

for every $\epsilon > 0$ there is a $\delta > 0$ such that
$|f(x) - L| < \epsilon$ provided $0 < |x - c| < \delta$ and $x > c$.

Definition 1.2.2 We say that $\lim_{x\to c-} f(x) = L$ if and only if:

for every $\epsilon > 0$ there is a $\delta > 0$ such that
$|f(x) - L| < \epsilon$ provided $0 < |x - c| < \delta$ and $x < c$.

Thus, if

(1.2.1) $$\beta(x) = [x], \text{ the greatest integer in } x,$$

then

(1.2.2) $$\lim_{x\to 2+} \beta(x) = 2, \quad \lim_{x\to 2-} \beta(x) = 1, \quad \lim_{x\to \frac{3}{2}+} \beta(x) = 1, \quad \lim_{x\to \frac{3}{2}-} \beta(x) = 1.$$

These results are suggested in Figure 1.2.1.

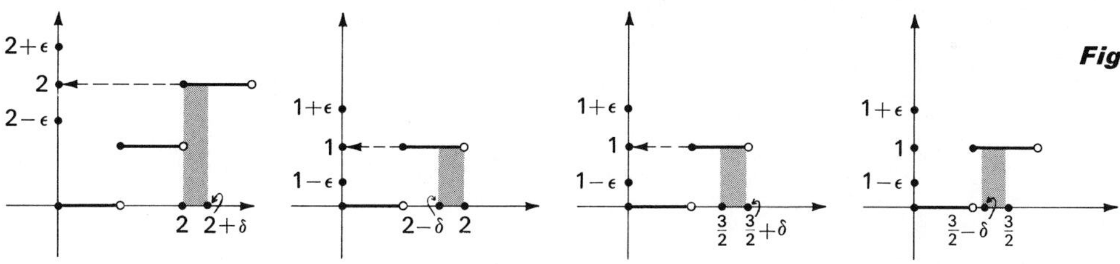

Figure 1.2.1

If we combine Definitions 1.2.1 and 1.2.2 and compare with Theorem 1.1.2, it is not difficult to see that

(1.2.3) $$\lim_{x\to c} f(x) = L \quad \text{if and only if} \quad \lim_{x\to c+} f(x) = \lim_{x\to c-} f(x) = L.$$

The proof of (1.2.3) is left as an exercise. Compare this statement with (1.2.2) and the previously established facts that the function β has no limit at $x = 2$ but that $\lim_{x\to 3/2} \beta(x) = 1$.

A second variation concerns *limits at* $\pm\infty$. In *C 1* in connection with asymptotes of the graph of a function we considered such limit statements as

(1.2.4) $$\lim_{x\to +\infty} f(x) = L.$$

In order to formulate a precise definition of statement (1.2.4) we must specify exactly what we mean by "being near $+\infty$." A natural method is suggested by the discussions leading to the ϵ-δ statement in Theorem 1.1.2. The condition $0 < |x - 2| < \delta$ expresses the idea that x is "δ-near 2." In order for x to be δ-near to 2, the smaller δ is, the nearer x must be to 2. Correspondingly, we might say that "x is δ-near $+\infty$" if $x > 1/\delta$. Here δ does not represent a distance from x to $+\infty$ (such an idea is clearly nonsense); but this scheme does preserve the feature that if x is δ-near $+\infty$ then the smaller δ is, the "nearer" x must be to $+\infty$. Moreover, this condition may be used to require x to be as close to $+\infty$ (that is, as *large*) as we please, simply by choosing δ small enough.

Thus, making the appropriate substitution in the ϵ-δ definition (Theorem 1.1.2), we obtain

Definition 1.2.3 We say that $\lim_{x\to+\infty} f(x) = L$ if and only if:

for every $\epsilon > 0$ there is a $\delta > 0$ such that $|f(x) - L| < \epsilon$ provided that $x > 1/\delta$.

Since "$x > 1/\delta$" (δ small) suggests "x near $+\infty$," then "$x < -1/\delta$" should suggest "x near $-\infty$." Thus, the corresponding definition for x approaching $-\infty$ is

Definition 1.2.4 We say that $\lim_{x\to-\infty} f(x) = L$ if and only if:

for every $\epsilon > 0$ there is a $\delta > 0$ such that $|f(x) - L| < \epsilon$ provided that $x < -1/\delta$.

From the above, it should be clear how to define such limit statements as $\lim_{x\to c} f(x) = \pm\infty$. We include these definitions without discussion. You should read them carefully and compare them with both Theorem 1.1.2 and Definitions 1.2.3, 1.2.4.

Definition 1.2.5 We say that $\lim_{x\to c} f(x) = +\infty$ if and only if:

for every $\epsilon > 0$ there is a $\delta > 0$ such that $f(x) > 1/\epsilon$ provided that $0 < |x - c| < \delta$.

Definition 1.2.6 We say that $\lim_{x\to c} f(x) = -\infty$ if and only if:

for every $\epsilon > 0$ there exists a $\delta > 0$ such that $f(x) < -1/\epsilon$ provided that $0 < |x - c| < \delta$.

Precise definitions of such statements as

$$\lim_{x\to+\infty} f(x) = +\infty, \quad \lim_{x\to+\infty} f(x) = -\infty, \quad \lim_{x\to c+} \quad = -\infty,$$

and others will be left to the exercises.

We have obtained several variations of the limit concept by shifting our attention to new regions of both the domain and the range of a function. We could obtain further variations by giving our imagination more rein to contemplate functions whose domains and ranges are not sets of real numbers. There are wide possibilities here, but exploring them would be more of a digression than we can afford. In the next section we shall consider a very modest move in this direction by considering functions whose domains contain no neighborhoods. We shall consider limits of sequences. A sequence is a function defined only on the set of positive integers.

Still further generalization is possible if we allow ourselves to contemplate "limits" of objects other than functions. What sorts of objects other than functions might we fruitfully consider? This question has been a basis for much research and many interesting and important results during this century. We cannot attempt an answer here. General theories of limits have been developed which combine a startlingly large and diverse collection of ideas encompassing almost all of the variety of modern analysis.

A slight hint at these possibilities comes from the following speculation. Suppose A and B are sets in each of which an appropriate definition of ϵ-near is possible (perhaps ϵ could mean something other than a positive number here!). Suppose further that F is a correspondence by means of which each element α of A is related to one *or more* elements β of B. We might use the symbol $F(\alpha)$ for *any* β in B which is so related by F to α. Then it might be reasonable to write

$$\lim_{\alpha\to\alpha_0} F(\alpha) = \beta_0$$

if for any ϵ there exists a δ such that *any* $F(\alpha)$ is ϵ-near β_0 provided that α is δ-near α_0. As to what A, B, F actually are, there is almost no limitation other than the imagination of the mathematician. We shall take one very modest step in this direction when we investigate a limit approach to the definite integral in Chapter 3.

We conclude with a remark and two examples concerning the evaluation of one-sided and infinite limits. We first remark that all of the theorems on limits of algebraic combinations and composites of functions *which make sense* are true for each of these new kinds of limits. The proofs are easy adaptations of the standard proofs for the original theorems. Thus, from the fact that $\lim_{x\to\pm\infty} 1/x = 0$ (Exercise 1.2.4) we can evaluate a wide variety of limits.

Example 1 To find

$$\lim_{x\to+\infty} \frac{x^3 - 4x}{2x^3 + x^2 + 1},$$

we first factor out the highest power of x in both the numerator and in the denominator, thus obtaining

(1.2.5)
$$\frac{x^3 - 4x}{2x^3 + x^2 + 1} = \frac{1 - 4/x^2}{2 + 1/x + 1/x^3}.$$

By the theorem on limits of products,

$$\lim_{x\to+\infty} \frac{4}{x^2} = \lim_{x\to+\infty} \frac{1}{x} = \lim_{x\to+\infty} \frac{1}{x^3} = 0.$$

Thus, using the limit theorems for sums and quotients, we see by way of (1.2.5) that

$$\lim_{x\to+\infty} \frac{x^3 - 4x}{2x^3 + x^2 + 1} = \frac{1 - 0}{2 + 0 + 0} = \frac{1}{2}.$$

Example 2 To evaluate

(1.2.6)
$$\lim_{x\to+\infty} \frac{x^4 + 2x + 1}{x^2 - 3x},$$

we use the same trick. Factoring out x^4 in the numerator and x^2 in the denominator, we have

(1.2.7)
$$\frac{x^4 + 2x + 1}{x^2 - 3x} = x^2\left(\frac{1 + 2/x^3 + 1/x^4}{1 - 3/x}\right).$$

Since $\lim_{x\to+\infty} x^2 = +\infty$, an application of Exercise 1.2.5 to (1.2.7) shows that the limit (1.2.6) is $+\infty$.

Exercises

1 Give a reasonable definition for each of the following limit statements.

(a) $\lim_{x\to c+} f(x) = +\infty$, (b) $\lim_{x\to c-} f(x) = -\infty$,

(c) $\lim_{x\to+\infty} f(x) = +\infty$, (d) $\lim_{x\to-\infty} f(x) = +\infty$.

2 Evaluate each of the following limits.

(a) $\lim_{x\to 0+} \frac{x}{|x|}$, (b) $\lim_{x\to 3+} \frac{|x-3|}{x-3}$,

(c) $\lim_{x\to 4+} [x] - x$, (d) $\lim_{x\to 2+} \frac{1}{x-2}$,

(e) $\lim_{x\to 0+} |x|x^2$, (f) $\lim_{x\to 1+} \frac{x-1}{\sqrt{x^2-1}}$,

(g) $\lim_{x\to\frac{3}{2}+} 2x - [2x]$,[1] (h) $\lim_{x\to\frac{3}{2}+} 3x - [2x]$,

(i) $\lim_{x\to 5+} (x - [x])^2$, (j) $\lim_{x\to -3+} x^{[x]}$.

[1] Here the square brackets refer to the greatest-integer function. See (1.2.1).

3 Evaluate each of the limits in Exercise 2 as a left-hand limit; i.e., in (a) evaluate $\lim_{x\to 0-} x/|x|$.

4 Use Definition 1.2.3 to prove that $\lim_{x\to+\infty} 1/x = 0$. Also prove that $\lim_{x\to-\infty} 1/x = 0$.

5 If $\lim_{x\to c} f(x) = +\infty$ and $\lim_{x\to c} g(x) = L > 0$, prove that $\lim_{x\to c} f(x)g(x) = +\infty$. Prove the same theorem for $x \to \pm\infty$.

6 Evaluate each of thè following limits.

(a) $\displaystyle\lim_{x\to+\infty} \frac{2x+3}{3x+4}$, (b) $\displaystyle\lim_{x\to+\infty} \frac{2x^2-4x+5}{7x^2+9x-2}$,

(c) $\displaystyle\lim_{x\to+\infty} \frac{3x-4}{5x^2+7x-11}$, (d) $\displaystyle\lim_{x\to+\infty} \frac{\sqrt{x^2+1}}{x+1}$,

(e) $\displaystyle\lim_{x\to+\infty} (x-\sqrt{x^2-2})$, (f) $\displaystyle\lim_{x\to+\infty} (x+\sqrt{x^2-2})$,

(g) $\displaystyle\lim_{x\to+\infty} (x^2-2x)$, (h) $\displaystyle\lim_{x\to+\infty} \left(\frac{1}{x}-\frac{1}{x^2}\right)$,

(i) $\displaystyle\lim_{x\to+\infty} \frac{x^2}{x^3+2}$, (j) $\displaystyle\lim_{x\to+\infty} \frac{x^4+x^2-1}{x^3+3x+1}$.

7 Evaluate each of the limits in Exercise 6 for x tending to $-\infty$. Thus, for example, in (a) evaluate $\lim_{x\to-\infty} \dfrac{2x+3}{3x+4}$.

8 Prove that if $\lim_{x\to+\infty} f(x) = a$, and g is continuous at a, then $\lim_{x\to+\infty} g(f(x)) = g(a)$.

9 Use Exercise 8 to evaluate

(a) $\displaystyle\lim_{x\to+\infty} e^{1/x}$, (b) $\displaystyle\lim_{x\to+\infty} \log\left(\frac{x+1}{x-1}\right)$,

(c) $\displaystyle\lim_{x\to+\infty} \sin\left(\frac{1}{x-2}\right)$, (d) $\displaystyle\lim_{x\to+\infty} e^{\sin(1/x)}$.

10 What theorems analogous to that stated in Exercise 8 can you find for other kinds of limits considered in this section?

11 Prove (1.2.3).

12 Show by examples that no conclusion can be made concerning $\lim_{x\to c} f(x)g(x)$ if $\lim_{x\to c} f(x) = \pm\infty$ and $\lim_{x\to c} g(x) = 0$.

1.3

Limits of Sequences

A function whose domain is the set of positive integers is called a sequence. If f is such a function and n is a positive integer, then the value $f(n)$ is usually written with "n" appearing as a subscript. Thus, we write

$$f(n) = a_n$$

and denote the sequence by $\{a_n\}$. That is,

$$\{a_n\} = \{a_1, a_2, a_3, \ldots\},$$

where not only the set of numbers $a_1, a_2, a_3, \ldots$ but also the order in which they occur determine the sequence. Thus,

(1.3.1) $$1, 2, 3, 4, 5, 6, \ldots$$

and

(1.3.2) $$2,\ 1,\ 4,\ 3,\ 6,\ 5,\ \ldots$$

involve the same numbers, but as written here represent different sequences The sequence (1.3.1) is

$$\{a_n\},\ a_n = n;$$

whereas the sequence (1.3.2) is

$$\{a_n\},\ a_n = \begin{cases} n+1 & \text{if } n \text{ is odd,} \\ n-1 & \text{if } n \text{ is even.} \end{cases}$$

Further examples of sequences are

(1.3.3) $$1, \frac{3}{2}, \frac{5}{3}, \frac{7}{4}, \ldots, 2 - \frac{1}{n}, \ldots,$$

(1.3.4) $$-1, \frac{3}{2}, -\frac{5}{3}, \frac{7}{4}, \ldots, (-1)^n\left(2 - \frac{1}{n}\right), \ldots,$$

(1.3.5) $$1, 2, 6, 24, \ldots, n!, \ldots,$$

(1.3.6) $$1, \frac{1}{2}, \frac{2}{9}, \ldots, \frac{n!}{n^n}, \ldots,$$

$$0, \frac{\log 2}{2}, \frac{\log 3}{3}, \ldots, \frac{\log n}{n}, \ldots,$$

$$2, \frac{9}{4}, \frac{64}{27}, \ldots, \left(1 + \frac{1}{n}\right)^n, \ldots,$$

(1.3.7) $$\sin x, \frac{\sin 2x}{\sqrt{2}}, \frac{\sin 3x}{\sqrt{3}}, \ldots, \frac{\sin nx}{\sqrt{n}}, \ldots.$$

Since a sequence is a function whose values (we call them "terms" in this context) are defined only for positive integers, the only kind of limit for a sequence that we can reasonably consider is a limit at ∞.[1] The appropriate definition is readily suggested in the light of the discussion in the last section.

[1] *When we write "∞" we always mean "$+\infty$."*

Definition 1.3.1 If $\{a_n\}$ is a sequence, we say that $\lim_{n\to\infty} a_n = l$ if and only if:
for every $\epsilon > 0$ there exists an $n_\epsilon > 0$ such that
$|a_n - l| < \epsilon$ for all $n \geq n_\epsilon$.

Note that Definition 1.3.1 is an ϵ-n_ϵ statement rather than an ϵ-δ statement. A choice of $\delta > 0$ and the condition $n > 1/\delta$ is clearly equivalent to the choice of some n_ϵ and the condition $n \geq n_\epsilon$.

Let us try out the definition on three of the sequences above.

Example 1 Let $a_n = 2 - 1/n$, the general term of sequence (1.3.3). Intuitively, it should be obvious that $\lim a_n = 2$. (Since the only kind of limit we consider is a limit as $n \to \infty$, we need not always write the full symbol $\lim_{n\to\infty} a_n$.) Let us test this statement against Definition 1.3.1.

If ϵ is an arbitrary positive number, then how large does n need to be in order to have

$$|a_n - 2| < \epsilon?$$

Since $|a_n - 2| = 1/n$, the above inequality will be true if

$$\frac{1}{n} < \epsilon, \quad \text{or} \quad n > \frac{1}{\epsilon}.$$

Thus, if we choose n_ϵ to be any integer greater than $1/\epsilon$, the conditions of Definition 1.3.1 are fulfilled. We therefore conclude that

$$\lim a_n = \lim \left(2 - \frac{1}{n}\right) = 2.$$

Example 2 It is an easy exercise to show that if $\lim a_n = l$ then for every $\epsilon > 0$ there exists an n_ϵ such that

$$|a_n - a_m| < 2\epsilon \tag{1.3.8}$$

for all $n \geq n_\epsilon$, $m \geq n_\epsilon$. (One need only observe that $|a_n - a_m| \leq |a_n - l| + |l - a_m|$ and apply the definition.)

Suppose now that $a_n = (-1)^n(2 - 1/n)$, the general term of sequence (1.3.4). Then

$$a_n - a_{n+1} = (-1)^n \left(4 - \frac{2n+1}{n(n+1)}\right).$$

Since $2n + 1 < 2(n + 1)$, we have

$$\frac{2n+1}{n(n+1)} < \frac{2}{n} \leq 2$$

for all n, so that

$$|a_n - a_{n+1}| = \left|4 - \frac{2n+1}{n(n+1)}\right| \geq 2.$$

Thus, we cannot have $|a_n - a_{n+1}| < 2\epsilon$ for any n if ϵ is smaller than 1. By our observation at the beginning of this example, then, this sequence cannot have *any* limit. (This is intuitively clear from a glance at the sequence, which seems to be undecided whether to converge to 2 or -2.)

Example 3 The sequence (1.3.5) also has no limit in the sense of Definition 1.3.1, for $|n! - (n + 1)!| = (n)n!$, and this certainly cannot be small for large n. However, it should be clear that with the proper definition of an infinite limit for a sequence, $\lim (n!) = +\infty$.

We may add, subtract, multiply, or divide sequences term-by-term to produce new sequences. These operations correspond exactly to adding, subtracting, multiplying, and dividing functions, and the same theorems about limits are true here as in the case of functions. Moreover, most of the other basic theorems from the function case hold also for sequences. We state several of these theorems here for future use. The proofs are exactly like the proofs of the earlier corresponding theorems.

In these theorems we shall write

$$\lim (ca_n), \quad \lim (a_n + b_n), \quad \lim (a_n b_n), \quad \text{etc.},$$

for the limits of the sequences $\{ca_n\}$, $\{a_n + b_n\}$, $\{a_n b_n\}$, etc., formed by multiplying the terms of the sequence $\{a_n\}$ by the fixed number c, adding corresponding terms of two sequences $\{a_n\}$ and $\{b_n\}$, multiplying corresponding terms of $\{a_n\}$ and $\{b_n\}$, etc.

Theorem 1.3.1 *If c is a real number and $\{a_n\}$ and $\{b_n\}$ are sequences with limits, then*
(i) $\lim (ca_n) = c \lim a_n$,
(ii) $\lim (a_n \pm b_n) = \lim a_n \pm \lim b_n$,
(iii) $\lim (a_n b_n) = \lim a_n \cdot \lim b_n$,
(iv) *if* $b_n \neq 0$ *for all* n *and* $\lim b_n \neq 0$, *then*

$$\lim \left(\frac{a_n}{b_n}\right) = \frac{\lim a_n}{\lim b_n},$$

(v) *if* $a_n \geq 0$ *and* u *is a real number, then*

$$\lim (a_n)^u = (\lim a_n)^u.$$

Theorem 1.3.2 *If* $a_n \leq c_n \leq b_n$ *for all* n *and* $\lim a_n = \lim b_n = l$, *then also*

$$\lim c_n = l.$$

If f is a function which is defined for all the numbers a_n, then we may form a new sequence $\{b_n\}$ by defining $b_n = f(a_n)$. In this context a theorem which corresponds to an earlier theorem about limits of composite functions (Theorem 1.7.3 in *C 1*) can easily be proved.

Theorem 1.3.3 *If* $\lim_{x \to a} f(x) = L$ *and if* $\lim a_n = a$ *but* $a_n \neq a$ *for all* n, *then*

$$\lim f(a_n) = L.$$

If f is continuous at $x = a$, then the restriction $a_n \neq a$ may be removed.

We come now to a new kind of limit theorem, for which we shall have use later. It is an existence theorem, a theorem which provides a criterion for the existence of the limit of a sequence. It will enable us to conclude that certain sequences have limits even though we may not know exactly what number the limit is. There are a number of other theorems of this nature, many of them significantly stronger than the one we shall state. However, this is one of the easiest such theorems to prove and we shall find it sufficient to meet our needs. First, a definition.

Definition 1.3.2 An *increasing* (*decreasing*) sequence $\{a_n\}$ is one for which $a_n \leq a_{n+1}$ ($a_n \geq a_{n+1}$) for every $n = 1, 2, 3, \ldots$. A *monotone* sequence is one which is either increasing or decreasing. A *bounded* sequence $\{a_n\}$ is one for which there exist constants m and M such that $m \leq a_n \leq M$ for all $n = 1, 2, 3, \ldots$.

Theorem 1.3.4 *Any bounded monotone sequence has a limit.*

Proof The proof will be based on the so-called *completeness property* of the real number system presented as an axiom in Chapter 6 of *C 1*. This property can be stated as follows:

If A and B are two nonempty sets of real numbers which have the property that every element of A is less than or equal to every element of B, then there must exist at least one number c such that $x \leq c \leq y$ for all x in A and all y in B.

Roughly speaking, this property asserts that any two sets one of which lies entirely "below" the other must be separated by some number or numbers. Note that the separating number c *may* be an element of one or both sets (as in the case where A and B are adjacent closed intervals).

We use the completeness property in the following way: Suppose that $\{a_n\}$ is an increasing sequence and

$$a_1 \leq a_2 \leq a_3 \leq \cdots \leq M.$$

Construct two sets A and B as follows:

$A = \{a_1, a_2, a_3, \ldots\}$, the numbers in the sequence.

$B = \{$all upper bounds (such as M) for the sequence $\{a_n\}\}$.

Obviously, every element of A is less than or equal to every element of B. This is ensured by the definition of B. If c is a separating number between A and B (whose existence is asserted by the completeness property), then two things can be said about c.

First, we see that c is itself an upper bound for the sequence $\{a_n\}$ (Why?) and it is therefore an element of B.

Second, c is less than or equal to every element of B (Why?), so c is the *least element* of B.

We shall show that $c = \lim a_n$. Let ϵ be an arbitrary positive number. Then since c is the *least* upper bound for the sequence $\{a_n\}$ and $c - \epsilon < c$, we see that $c - \epsilon$ is not an upper bound. Therefore some term of the sequence, let us call it a_{n_ϵ}, must be larger than $c - \epsilon$. That is,

$$c - \epsilon < a_{n_\epsilon} \leq c.$$

Since $\{a_n\}$ is an increasing sequence, we have, for all $n \geq n_\epsilon$,

$$c - \epsilon < a_{n_\epsilon} \leq a_n \leq c.$$

Thus, a_n is closer to c than is $c - \epsilon$, so the distance $|a_n - c|$ is less than ϵ. This holds for all $n \geq n_\epsilon$, so we have

$$|a_n - c| < \epsilon \quad \text{provided} \quad n \geq n_\epsilon.$$

The conditions of Definition 1.3.1 are therefore fulfilled and we conclude that $c = \lim a_n$. The case $\{a_n\}$ decreasing is covered by an analogous proof, and we omit the details.

We conclude with three examples illustrating the use of the above theorem in finding limits.

Example 4 Consider the sequence (1.3.6),

$$\{a_n\} = \left\{\frac{n!}{n^n}\right\}.$$

If we write

$$a_n = \frac{n}{n} \cdot \frac{n-1}{n} \cdot \frac{n-2}{n} \cdot \cdots \cdot \frac{1}{n},$$

we easily see that

$$0 < a_n \leq \frac{1}{n}.$$

We can also easily see, either intuitively or by an application of Definition 1.3.1, that

$$\lim (0) = \lim \left(\frac{1}{n}\right) = 0.$$

Thus, by Theorem 1.3.2, we must have

$$\lim a_n = \lim \left(\frac{n!}{n^n}\right) = 0.$$

Example 5 The general term of the sequence (1.3.7) is

$$a_n = \frac{\sin nx}{\sqrt{n}}.$$

Here we interpret x as a fixed but arbitrary real number. Since for all n

$$-1 \leq \sin nx \leq 1,$$

we see that

$$-\frac{1}{\sqrt{n}} \leq \frac{\sin nx}{\sqrt{n}} \leq \frac{1}{\sqrt{n}},$$

or equivalently

(1.3.9) $$-\sqrt{\frac{1}{n}} \leq \frac{\sin nx}{\sqrt{n}} \leq \sqrt{\frac{1}{n}}.$$

By an application of Theorem 1.3.1(v) we see that

(1.3.10) $$\lim \left(-\sqrt{\frac{1}{n}}\right) = \lim \sqrt{\frac{1}{n}} = 0,$$

and by another application of Theorem 1.3.2 we conclude from (1.3.9) and (1.3.10) that

$$\lim \frac{\sin nx}{\sqrt{n}} = 0.$$

Example 6 As a final example, consider

$$a_n = \left(1 + \frac{1}{n}\right)^n.$$

It is possible to show by purely algebraic manipulation that the sequence $\{a_n\}$ is increasing and that

$$2 \leq \left(1 + \frac{1}{n}\right)^n \leq 3.$$

It therefore follows from Theorems 1.3.2 and 1.3.4 that the sequence has a limit l, where l is a number between 2 and 3.

However, because of developments in *C 1*, we are in a position to reach this conclusion—and more than this—by an easy application of Theorem 1.3.3. Consider the new sequence

(1.3.11) $$\{b_n\} = \{\log a_n\} = \left\{n \log \left(1 + \frac{1}{n}\right)\right\}.$$

We can write

$$b_n = \frac{\log (1 + 1/n)}{1/n},$$

and then observe that if

(1.3.12) $$F(h) = \frac{\log (1 + h)}{h},$$

the sequence (1.3.11) is given by

(1.3.13) $$\{b_n\} = \left\{F\left(\frac{1}{n}\right)\right\}.$$

The function log was discussed in Chapter 7 of *C 1*. We found that this function is differentiable and that its derivative is the function with value $1/x$ for all $x > 0$. By the definition of a derivative

(1.3.14) $$\frac{1}{x} = \frac{d}{dx}\log x = \lim_{h\to 0}\frac{\log(x+h) - \log x}{h}.$$

Thus, from (1.3.12) and (1.3.14) with $x = 1$, we have

(1.3.15) $$\lim_{h\to 0} F(h) = \frac{1}{1} = 1.$$

We also know that

$$\frac{1}{n} \neq 0 \quad\text{and}\quad \lim\left(\frac{1}{n}\right) = 0,$$

so by Theorem 1.3.3 and (1.3.15) we have

$$\lim F\left(\frac{1}{n}\right) = 1.$$

Thus, by (1.3.13)

(1.3.16) $$\lim b_n = 1.$$

But

(1.3.17) $$a_n = e^{b_n} = \exp(b_n),$$

because of (1.3.11) and the known relationship between the functions log and exp. Moreover, exp is a continuous function. Thus we can apply Theorem 1.3.3 again and conclude from (1.3.16) and (1.3.17) that

(1.3.18) $$\lim\left(1 + \frac{1}{n}\right)^n = \lim a_n = e^{(\lim b_n)} = e.$$

The statement (1.3.18) is often taken as the definition of the irrational number *e*. The existence argument mentioned at the beginning of this example is then the proof of the existence of *e* and gives a first estimate of its size.

Exercises

1 Prove formula (1.3.8).

2 Give a reasonable definition for $\lim a_n = +\infty$ and for $\lim a_n = -\infty$.

3 Prove Theorem 1.3.2.

4 Prove Theorem 1.3.1.

5 Prove Theorem 1.3.3.

6 Evaluate the following limits.

(a) $\lim \dfrac{1}{n+4}$, (b) $\lim \dfrac{n}{n^2+2}$,

(c) $\lim \dfrac{3n^2+2n+1}{n^2}$, (d) $\lim \dfrac{3n^2+2n+1}{n^2}$,

(e) $\lim \dfrac{\sqrt{n}}{n+1}$, (f) $\lim \dfrac{n}{\sqrt{n+2}}$,

(g) $\lim (\sqrt{n^2+1}-n)$, (h) $\lim (\sqrt{n^2+n+1}-n)$,

(i) $\lim (\log(n-1)-\log(n+1))$, (j) $\lim \dfrac{\cos nx}{n}$,

(k) $\lim \dfrac{n!}{(2n)!}$, (l) $\lim \left(\dfrac{(n+3)!-n!}{(n+4)!}\right)$.

7 Repeat the argument in this section beginning with (1.3.11) to prove that $\lim_{h\to 0}(1+h)^{1/h} = e$ and $\lim_{x\to+\infty}(1+1/x)^x = e$.

8 Evaluate the following limits.

(a) $\lim \left(1+\dfrac{1}{2n}\right)^n$, (b) $\lim \left(\dfrac{n}{n+1}\right)^n$,

(c) $\lim \left(\dfrac{n+3}{n}\right)^n$, (d) $\lim \left(1+\dfrac{x}{n}\right)^n$.

9 Prove Theorem 1.3.4 for bounded decreasing sequences.

10 Given that $a_n = \dfrac{\alpha_0+\alpha_1 n+\cdots+\alpha_j n^j}{\beta_0+\beta_1 n+\cdots+\beta_k n^k}$, prove that

$$\lim a_n = \begin{cases} 0 & \text{if } j<k, \\ \pm\infty & \text{if } j>k, \\ \dfrac{\alpha_j}{\beta_k} & \text{if } j=k. \end{cases}$$

11 Show that $\lim \sqrt{1/n^p} = 1$ if $p>0$. (*Hint:* $n^{p/n} = e^{(p\log n)/n}$.)

12 Show that if $a>1$, $\lim a^n = +\infty$.

derivatives revisited

chapter 2

2.0

Introduction

In Chapter 2 of *C 1* we defined the *derivative* of a function as a certain limit. In that volume we explored some of the properties of derivatives, some shortcuts for calculating derivatives of specific functions, and some of the many applications of derivatives to problems arising in mathematics and many other fields. We left many loose ends in those discussions. Facts about derivatives were sometimes supported only by plausibility arguments; techniques for finding derivatives were restricted to the simplest situations; at least one important class of differentiable functions was not even mentioned; and important generalizations of the notion of derivative to functions depending on more than one variable were also not mentioned.

It is the purpose of this chapter to tie down a few of these loose ends. In Section 2.1 we shall review briefly the definition, the most immediate geometric interpretations, and a few of the basic properties of derivatives. In Section 2.2 we shall discuss very briefly the possibility of defining (and interpreting) derivatives for functions of several variables. In Section 2.3 we shall consider the problem of finding derivatives for functions which are defined implicitly by equations. In Section 2.4 we shall complete the discussion of inverse functions which we had begun earlier in Chapter 7 of *C 1*, and finally in Section 2.5 we shall introduce an important class of functions, namely the inverse trigonometric functions.

2.1

The Derivative of a Function

In Chapter 2 of *C 1* we defined the *derivative* $f'(x_0)$ of a function f at a point $(x_0, f(x_0))$ as follows:

(2.1.1)
$$f'(x_0) = \lim_{h\to 0} \frac{f(x_0+h) - f(x_0)}{h}.$$

Implicit in this definition, of course, is the condition that f be defined at least on a neighborhood of x_0. Moreover, in order that the limit (2.1.1) exist it is necessary for f to be continuous at x_0. That is, as we proved in *C 2*, Chapter 2,

(2.1.2) *If f is differentiable at x_0 (i.e., if $f'(x_0)$ exists), then f is continuous at x_0.*

We arrived at the definition in (2.1.1) through consideration of several questions relating to tangent lines of graphs of functions. Accordingly, we were led to the following geometric "interpretation."

(2.1.3) *The graph of $y = f(x)$ has a tangent line at the point $(x_0, f(x_0))$ if and only if $f'(x_0)$ exists, and $f'(x_0)$ is the slope of this tangent line* (*see* Figure 2.1.1).

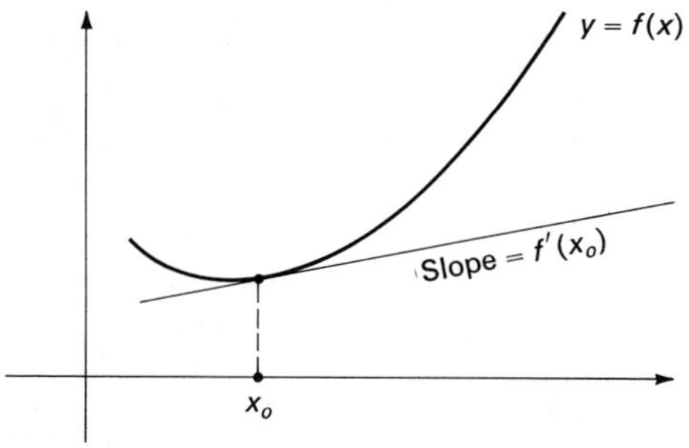

Figure 2.1.1

Statement (2.1.3) is, in fact, our *definition* of a tangent line. Another interpretation of a derivative, closely related to (2.1.3) but generally used in a slightly different context, is

(2.1.4) *The derivative $f'(x_0)$ is the rate of change of $f(x)$ with respect to x at the point where $x = x_0$.*

The most common use of (2.1.4) occurs when the function f designates the position coordinate s of an object moving on a coordinate line at any time t during a time interval of observation. If $s = f(t_0)$, we denote $f'(t_0)$ by v_0, and we call $f'(t_0) = v_0$ the *velocity* of the object at time $t = t_0$.

The converse of statement (2.1.2) is false, as we have seen. The function $f(x) = |x|$, which is continuous at $x = 0$ but not differentiable there, is the classical counterexample. If a function f is continuous at $x = x_0$ but not differentiable there, then if f is not too badly behaved we may be able to consider one or more replacements for the missing derivative. These are obtained by altering the limit statement (2.1.1) in accordance with some of the limit variants in Chapter 1.

For example, even though $f'(x_0)$ fails to exist, one or both of the following may exist:

(2.1.5) $$\text{the } \textit{left-derivative } f'_-(x_0) = \lim_{h\to 0-} \frac{f(x_0 + h) - f(x_0)}{h},$$

(2.1.6) $$\text{the } \textit{right-derivative } f'_+(x_0) = \lim_{h\to 0+} \frac{f(x_0 + h) - f(x_0)}{h}.$$

Clearly, $f'(x_0)$ exists if and only if both one-sided derivatives exist and are equal. If $f'_-(x_0) = f'_+(x_0)$, then $f'(x_0)$ exists and $f'(x_0) = f'_-(x_0) = f'_+(x_0)$. We may interpret $f'_-(x_0)$ as the slope of a *left-tangent* line to the graph of $y = f(x)$ at $(x_0, f(x_0))$, and $f'_+(x_0)$ as the slope of a *right-tangent* line.

Moreover, we may consider infinite derivatives of both the ordinary and the one-sided varieties. We need only apply to (2.1.1), (2.1.5), or (2.1.6) the notion of infinite limit mentioned in Chapter 1. This notion is of special geometrical importance, because we may say that if $f'(x_0) = +\infty$ or $f'(x_0) = -\infty$ (accordingly as the limit (2.1.1) is $+\infty$ or $-\infty$), then the graph of $y = f(x)$ has a vertical tangent line at $(x_0, f(x_0))$. If $f'(x_0)$ exists in the ordinary sense, then we may say that f has a *finite* derivative at x_0. Thus, the correct criterion for the existence of a tangent line for a graph is

The graph of $y = f(x)$ has a tangent line at $(x_0, f(x_0))$ if f has a finite or infinite derivative at $x = x_0$.

Example 1 Let

$$f(x) = |x^2 + 2x|.$$

An easy calculation shows that

$$\frac{f(0 + h) - f(0)}{h} = \begin{cases} |h + 2| & \text{if } h > 0, \\ -|h + 2| & \text{if } h < 0, \end{cases}$$

so

$$f'_-(0) = -2, \quad f'_+(0) = +2.$$

These one-sided derivatives are the slopes of the two "half-tangents" at (0, 0), which we see sketched in Figure 2.1.2.

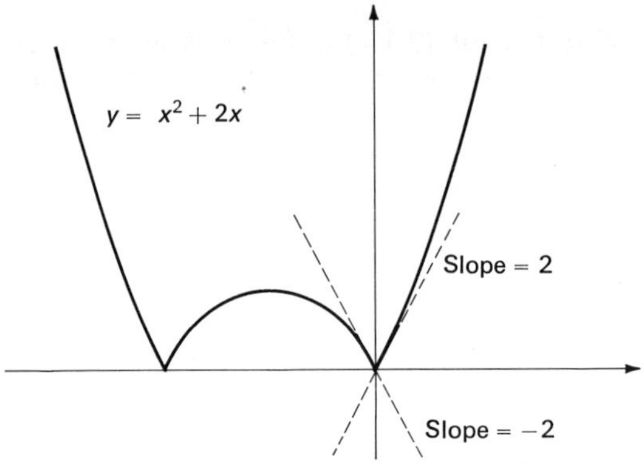

Figure 2.1.2

Example 2 Let

$$g(x) = x^{1/3}.$$

Then

$$\frac{g(0+h)-g(0)}{h} = h^{-2/3} = \left(\frac{1}{h^2}\right)^{1/3}, \ h \neq 0.$$

It is thus easy to see that

$$g'(0) = +\infty.$$

Accordingly, the graph of $y = g(x)$ (Figure 2.1.3) has a vertical tangent line at (0, 0).

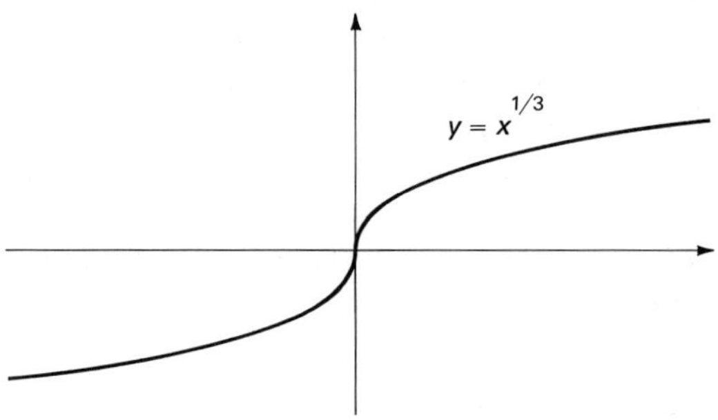

Figure 2.1.3

Example 3 If

$$f(x) = \tfrac{1}{2}\sqrt{4 - x^2}, \ -2 \leq x \leq 2,$$

then there is, of course, no possibility of considering a derivative at $x = 2$, for f is not defined throughout a neighborhood of $x = 2$. However, if h is negative, so that $|h| = -h$, then

(2.1.7) $$\frac{f(2+h)-f(2)}{h} = -\frac{1}{2}\left(\frac{4+h}{|h|}\right)^{1/2}.$$

(The derivation of (2.1.7) will be left as an exercise.)

Since $4 + h$ is near 4 and $|h|$ is near 0 whenever h is near 0, it follows (intuitively) that the limit as $h \to 0-$ of (2.1.7) is $-\infty$. Thus

$$f'_-(2) = -\infty$$

and the graph of $y = f(x)$ (Figure 2.1.4) has a vertical left-tangent line at (2, 0).

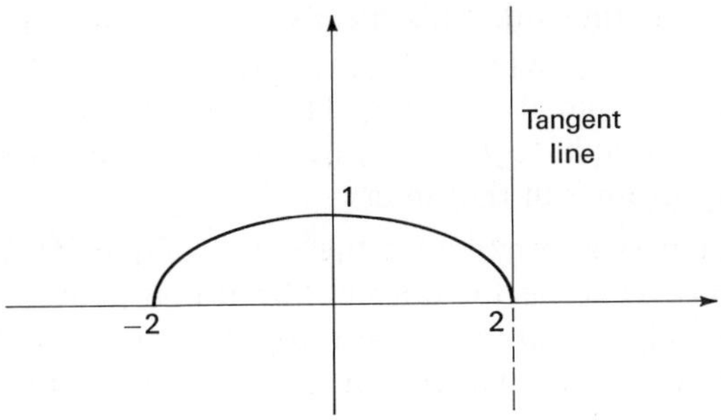

Figure 2.1.4

We proved a number of theorems in *C 1* concerning the problem of calculating derivatives of specific functions. We learned how to differentiate under appropriate circumstances all algebraic combinations of differentiable functions, e.g., $kf, f \pm g, fg, f/g, f^n$. We also learned how to differentiate the composite $f \circ g$ of two differentiable functions, provided that they "fit together" properly. To give these theorems some substance, we learned the derivatives of "elementary functions": polynomials, sine, cosine, tangent, log, exp. We shall not pause to recall these formulas here. They may be reviewed in *C 1* if necessary, and we shall recall them as necessary in future discussions.

Also in *C 1*, we explored a number of uses for the derivative in the study of functions, several of which will be extended or clarified in subsequent chapters. To refresh your memory and provide references for the future, we list here some of the important points of our previous discussions.

1. If $f'(x_0) > 0$ ($f'(x_0) < 0$), then the function f is *increasing* (*decreasing*) at x_0.

2. If f is defined in a neighborhood of c and $f(c)$ is a *relative extreme value* (maximum or minimum) of f, then if $f'(c)$ exists it must be 0.

3. The relationship expressed by the limit statement (2.1.1) is not the only link between the derivative of f and the difference quotient

$$\frac{f(x_0 + h) - f(x_0)}{h}.$$

If f is differentiable on an interval containing both x_0 and $x_0 + h$, then, for some (and perhaps several) $\bar{x}$ between x_0 and $x_0 + h$,

$$\frac{f(x_0 + h) - f(x_0)}{h} = f'(\bar{x}).$$

This is one form of the statement of the Mean Value Theorem.

4. From the Mean Value Theorem we were able to deduce a number of global theorems involving derivatives. One asserts that if $f'(x) \geq 0$ ($f'(x) \leq 0$) for all x in an interval, then f is increasing (decreasing) over that interval. This means that for any $x_1 < x_2$ in the interval we have $f(x_1) \leq f(x_2)$ ($f(x_1) \geq f(x_2)$). Another global theorem asserts that if $f''(x) \geq 0$ ($f''(x) \leq 0$) for all x in an interval, then the graph of $y = f(x)$ lies entirely above (below) any one of its tangent lines in that interval.

5. The line that best approximates the graph of $y = f(x)$ in a neighborhood of $(x_0, f(x_0))$ is the tangent line at that point. Equivalently, the linear polynomial which best approximates $f(x)$ for x near x_0 is the polynomial that has the same value and also the same derivative at x_0 as $f(x)$. This polynomial is

$$p(x) = f(x_0) + (x - x_0)f'(x_0).$$

Using the Mean Value Theorem we estimated the difference between $f(x)$ and the best linear approximation $p(x)$. If M is an upper bound of $|f''(x)|$ for x between x_0 and $x_0 + h$, then the distance between $f(x_0 + h)$ and $p(x_0 + h)$ is not greater than Mh^2.

In the following set of exercises you will have the opportunity to recall some of the basic facts about continuous and differentiable functions stated here and in Chapter 1, and to put them to use in deriving several new and interesting properties of derivatives. Since these facts close one or two gaps in past discussions and will be useful in several future ones, we outline them here very briefly.

If f is a function that is differentiable on an interval, then the derivative f' is a function, itself defined on that interval. What properties does f' have as a function? It is easy to see that f' does not itself have to be differentiable, or even continuous. For example, if

(2.1.8)
$$f(x) = \begin{cases} x^2 \sin \dfrac{\pi}{x}, & x \neq 0, \\ 0 & , x = 0, \end{cases}$$

then one can see that $f'(x)$ exists for all x. But $f'(x)$ is not continuous at $x = 0$ (Exercise 2.1.4).

However, $f'(x)$ does have many of the desirable properties which continuous functions have—in a manner of speaking, $f'(x)$ cannot be very badly discontinuous. For example,

(2.1.9) *If $f(x)$ is differentiable on an interval, then $f'(x)$ has the intermediate value property on that interval* (Exercise 2.1.5).

(2.1.10) *If* $f'(x)$ *exists for all* $x \neq x_0$, f *is continuous at* x_0, *and* $\lim_{x \to x_0+} f'(x) = L$, *then* $f'_+(x_0)$ *exists and equals* L. *If* $\lim_{x \to x_0-} f'(x_0) = L$, *then* $f'_-(x_0)$ *exists and equals* L (Exercise 2.1.14).

Properties (2.1.9), (2.1.10) say in different ways that if a derivative $f'(x)$ is defined throughout an interval, then it can have no "jumps" in that interval. From (2.1.9) follows a property of derivatives which we have used several times without proof in earlier discussions: If $f'(x) \neq 0$ on an interval, then $f'(x)$ must have constant sign on that interval. Thus, if $f'(x) \neq 0$ the function must be monotone over that interval, that is, either increasing or decreasing.

Another interesting property of derivatives states

(2.1.11) *If* $f'(x)$ *is monotone on an interval, then* $f'(x)$ *is continuous on that interval* (Exercise 2.1.7).

This property, too, states that a derivative cannot be discontinuous in just *any* way. It is, of course, closely related to (2.1.9), because the only way a monotone function can be discontinuous is to have a "jump."

Exercises

1 Each of the following functions fails to have a derivative at one or more points of its domain. Find the right- and left-hand derivatives, if they exist, at each of these points and sketch the corresponding tangent lines on a graph of the function.

(a) $f(x) = |x - 2|$, (b) $f(x) = |3x - x^2|$,
(c) $f(x) = |x^3 - x|$, (d) $g(x) = x^2 + [x]$,
(e) $g(t) = |\sin t|$, (f) $f(x) = |e^x - 1|$,
(g) $f(x) = \sqrt{|x| + 1}$, (h) $f(t) = \log(|t| + 1)$,
(i) $F(x) = x^{3/2}$, (j) $f(t) = (x + 2)^{3/2} + x$.

2 Sketch the graph of each of the following functions and indicate all vertical tangents which occur.

(a) $f(x) = \sqrt{x}$, (b) $f(x) = \sqrt[3]{x - 3}$,
(c) $f(x) = \dfrac{1}{x^2 - 2}$, (d) $g(t) = \sin \sqrt{t}$,
(e) $g(x) = \dfrac{1}{(x - 3)^2}$, (f) $f(x) = \log |x|$,
(g) $f(x) = \dfrac{x - 1}{(x - 2)^2}$, (h) $g(x) = \left|\dfrac{x - 3}{(x + 1)^2}\right|$.

3 Verify formula (2.1.7).

4 Compute f' for the function f defined in (2.1.8). Use established derivative formulas for $x \neq 0$, and (2.1.1) for $x = 0$. Show that f' is not continuous at $x = 0$.

5 Theorem 4.3.2 in *C 1* is a special case of (2.1.9). It asserts that if f is differentiable in $[a, b]$ and $f'(a) > 0$, $f'(b) < 0$, then there exists a number c between a and b such that $f'(c) = 0$. Use this theorem to prove (2.1.9). (*Hint:* If $f'(a) < c < f'(b)$, consider the function $g(x) = (cx - f(x)$.)

6 Prove that a monotone function f which has the intermediate value property on every interval $[a, b]$ (i.e., f assumes every value between $f(a)$ and $f(b)$) is continuous. (*Hint:* Suppose f is increasing. To show continuity at x_0, let $\epsilon > 0$ and choose $h_1, h_2 > 0$ so that $f(x_0 + h_1) = f(x_0) + \epsilon$ and $f(x_0 - h_2) = f(x_0) - \epsilon$. Choose δ to be the smaller of h_1, h_2.)

7 Use (2.1.9) and Exercise 6 to prove (2.1.11).

8 Can you find an increasing function f whose derivative is decreasing?

9 Show that a function need not be continuous at a point where it has an infinite derivative.

10 Is the derivative of a monotone function necessarily monotone?

11 Show that if f is a differentiable function at x_0, the sequence $\{a_n\}$,

$$a_n = n\left[f\left(x_0 + \frac{1}{n}\right) - f(x_0)\right],$$

converges to $f'(x_0)$.

12 Show by an example that convergence of the sequence defined in Exercise 11 is not sufficient for $f'(x_0)$ to exist. (Obviously we may define $f(x)$ arbitrarily if x is not one of the points x_0+1/n or x_0.)

13 Consider

$$f(x) = \begin{cases} x^3 + 2x^2 \text{ if } x \geq 0, \\ ax^2 + bx + c \text{ if } x < 0. \end{cases}$$

How can a, b, c be chosen to assure that f has a second derivative at 0?

14 Use the Mean Value Theorem to prove (2.1.10).

2.2

Functions of Several Variables; Partial Derivatives

So far we have considered only functions which depend on a single variable. In fact, it is our intention to adhere to this restriction throughout these two volumes, with the exception of a brief digression in this section and several applications in Chapter 7. We wish to comment on a very natural question: Can one extend the notion of derivative usefully to functions of several variables? In asking this question we open the lid of a large box. The calculus of functions of several variables is certainly a major branch of the calculus and will be treated in some detail in later volumes in this series. The box could indeed become a Pandora's box for us if we tried to release much of the contents. So we shall examine only one or two of the more elementary ideas.

Functions of several variables abound in nature and in all areas of scientific discourse. The volume of a rectangular box is a function of all three of its dimensions; the temperature on the earth is a function of both location and time; the mass of a solid sphere is a function of both its radius and its density. You can think of many examples for yourself; such functions far outnumber (a badly misused word, for we are dealing with infinite classes!) the functions of one variable that we have been considering.

As you might suspect, the big conceptual jump occurs in the transition from one independent variable to two, rather than from two to more than two. Most of the flavor of the calculus of functions of arbitrarily many variables can be found in the calculus of functions of two variables. Moreover, geometrical interpretations are more natural in the case of two independent variables—the expansion of our geometrical and algebraic horizons into higher dimensions is the purpose of another volume in this series, *Linear Algebra*. Thus, we shall stick to the two-variable path, where the view is at least adequate and the way more secure.

The domain of one of the functions we wish to consider will consist of a *set of ordered pairs* (x, y) *of real numbers*. For various sets of pairs, the following equations define *real-valued functions* F, G, H of the variables x, y:

(2.2.1) $$F(x, y) = 2 - 2x - y,$$

(2.2.2) $$G(x, y) = \sqrt{x^2 + y^2},$$

(2.2.3) $$H(x, y) = e^y \log x.$$

Functions of the above type may be represented geometrically by their *graphs*. It is not difficult to see what such a graph is. Just as the graph of a function f of one variable is a geometrical representation of the set of pairs $(x, f(x))$, the graph of a function F of two independent variables should be a geometrical representation of the set of all triples $(x, y, F(x, y))$. What is suggested, of course, is the use of rectangular coordinates in space analogous to the coordinates in a plane used to construct the graph of a function of one variable.

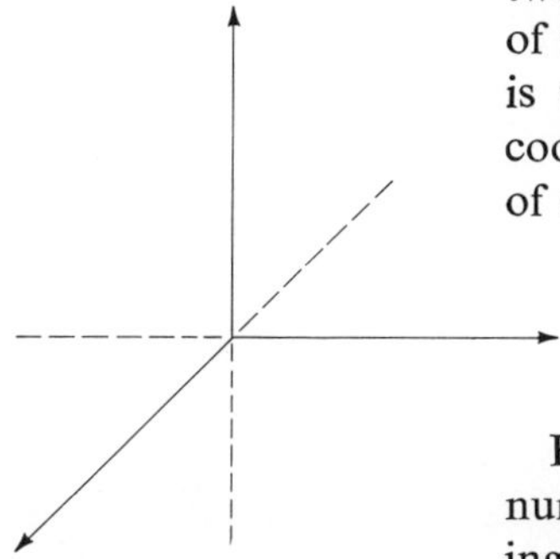

Figure 2.2.1

Briefly, we represent points in space by triples (x, y, z) of real numbers, "space-coordinates" of the points. We do this by establishing three intersecting mutually perpendicular oriented lines or *axes* (Figure 2.2.1); for convenience, we think of two as being horizontal and one vertical. Each axis is given a set of coordinates with the origin at the common point of intersection and the point 1 in the direction of the orientation from the origin. A point P in space is associated with a triple (x, y, z) if the perpendicular projections of P on the three axes have coordinates x, y, z respectively. We are presupposing an ordering of these axes here, of course. The usual convention is to consider the horizontal one perpendicular to the page first, the horizontal one in the page second, and the vertical third. Thus, x is the coordinate of the projection of P on the axis pointing at the reader, and so forth. For convenience we refer to the three axes in the indicated order as the x-, y-, and z-axes (Figure 2.2.2).

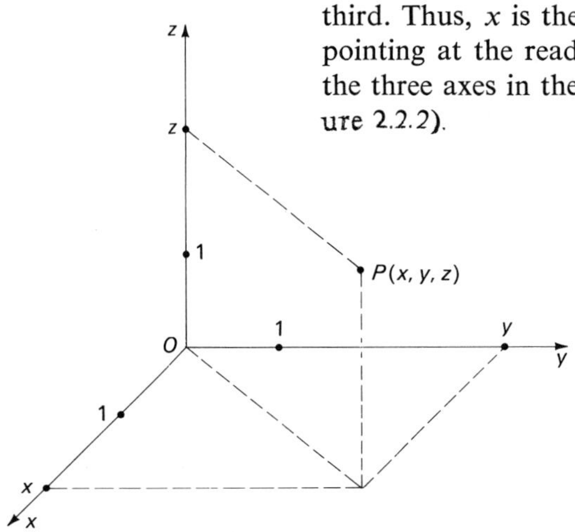

Figure 2.2.2

Now suppose that F is a function whose domain consists of a set R of ordered pairs (x, y) of real numbers. We may represent this domain R as a region in the plane of the x- and y-axes (we shall call this plane the xy-plane) by identifying the pair (x, y) with the special triple $(x, y, 0)$. For each such point $(x, y, 0)$ in R, there is one and only one point $(x, y, F(x, y))$, and we shall call the totality of such points $(x, y, F(x, y))$, with (x, y) in R, the *graph of F*. This graph forms a *surface* lying over the region R (Figure 2.2.3).

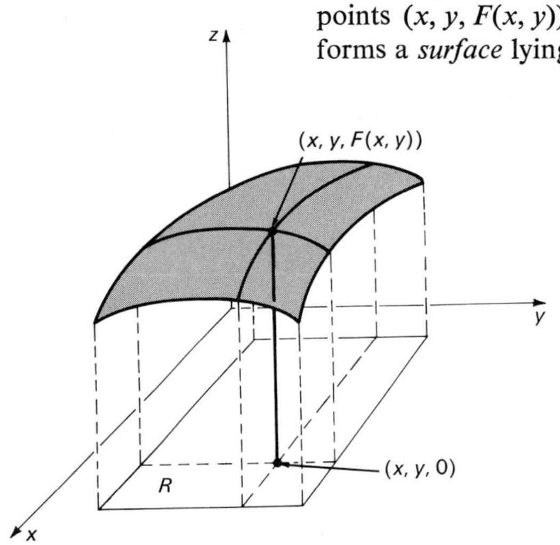

Graph of $z = F(x, y)$

Figure 2.2.3

Since we are here using $F(x, y)$ as a z-coordinate, it is natural to write $z = F(x, y)$. The graph of F could then also be called the *graph of the equation* $z = F(x, y)$. Graphs of the three functions in (2.2.1)–(2.2.3) are sketched in Figure 2.2.4. Thus, a function F of two variables determines a surface in space, the graph of F.

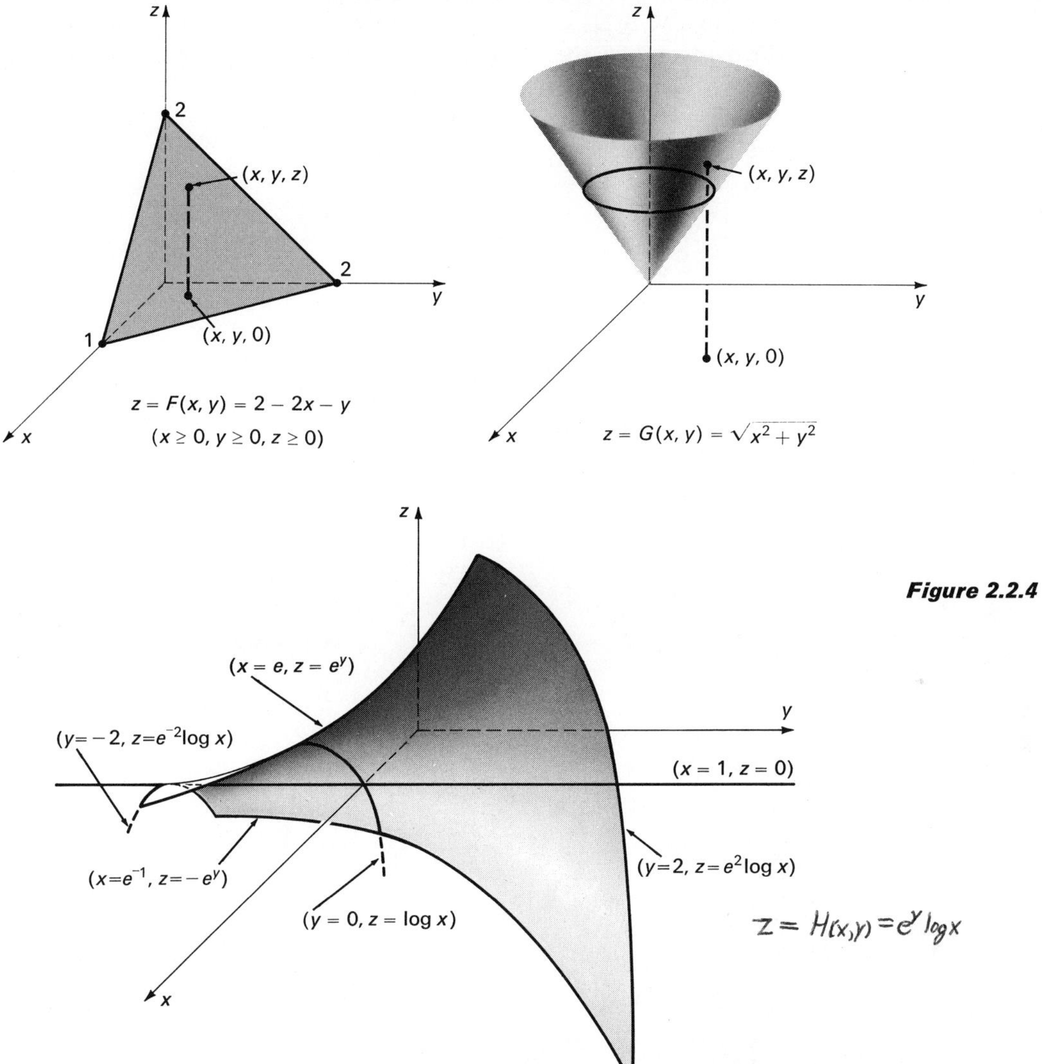

Figure 2.2.4

We can use this surface to motivate the extension of the notion of derivative. First note that if we consider a fixed value b of y and let x range over the set of values such that (x, b) is in R, then we obtain a function

$$f(x) = F(x, b)$$

of x alone. Note also that if $\Pi_{y=b}$ is the plane through $(0, b, 0)$

perpendicular to the y-axis, then $\Pi_{y=b}$ cuts the surface of F in a curve. We call this curve an *xz-section of the surface*. We thus have a function $F(x, b)$ of one variable, whose graph is a plane curve, an xz-section of the surface. It should be clear that the plane $\Pi_{y=b}$ contains axes (parallel to the x- and z-axes) relative to which the xz-section is the graph of the equation $z = F(x, b)$.

Similarly, if $\Pi_{x=a}$ is the plane through $(a, 0, 0)$ perpendicular to the x-axis, then the corresponding yz-section of the surface is the graph of the function

$$g(y) = F(a, y).$$

See Figure 2.2.5.

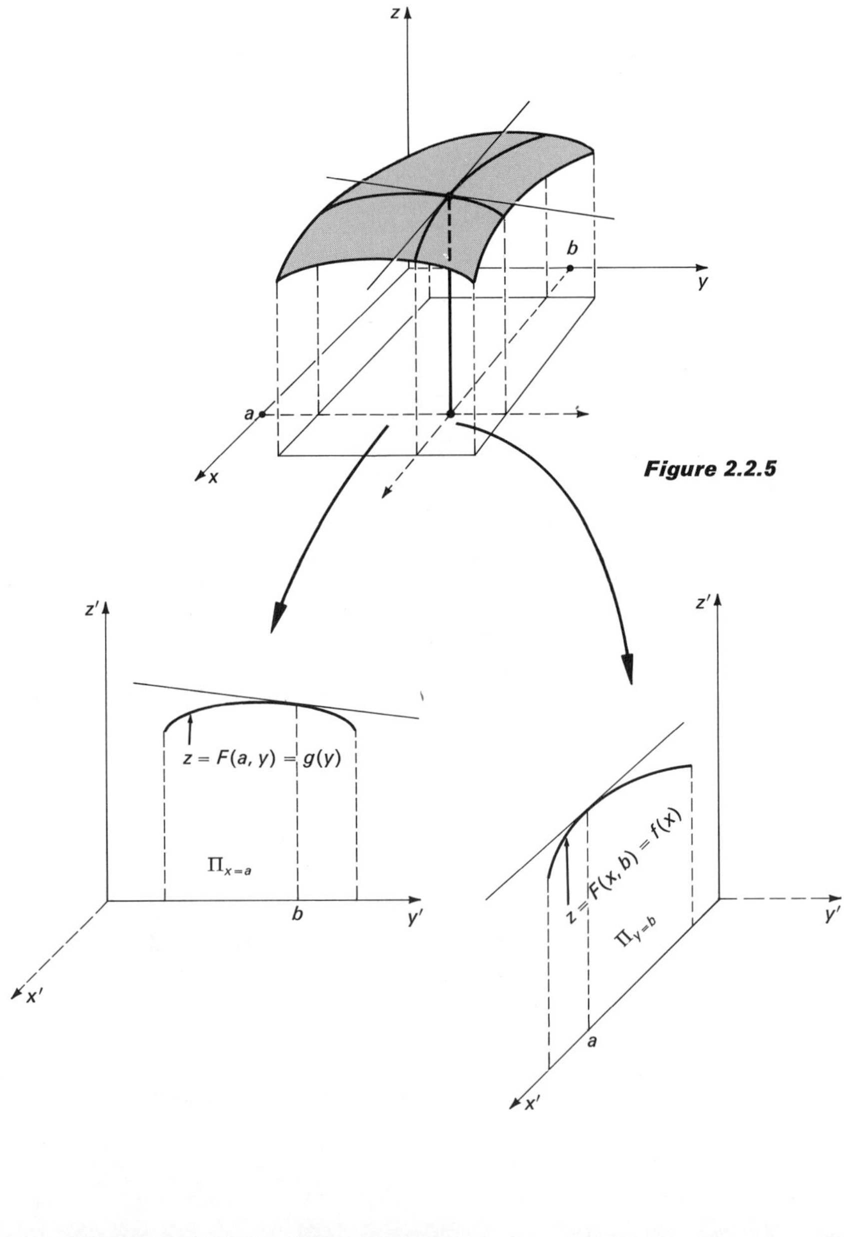

Figure 2.2.5

Since the functions (we might call them section-functions) $f(x) = F(x, b)$ and $g(y) = F(a, y)$ are ordinary functions of one variable, we may consider their derivatives

$$f'(x) = \frac{d}{dx}F(x, b), \; g'(y) = \frac{d}{dy}F(a, y).$$

These two derivatives (if they exist) are called the *partial derivatives* of F. We write

(2.2.4) $$f'(a) = F_x(a, b), \quad g'(b) = F_y(a, b).$$

In the light of the above, the geometrical interpretation of F_x and F_y is clear. $F_x(a, b)$ is the slope of the y-section, or, in other words, the slope of the surface in the x-direction at the point (a, b, c), $c = F(a, b)$. $F_y(a, b)$ is the slope of the x-section, that is, the slope of the surface in the y-direction at (a, b, c) (Figure 2.2.6).

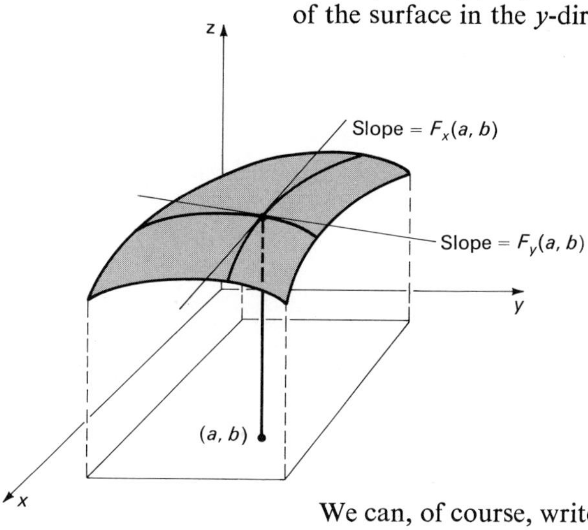

Figure 2.2.6

We can, of course, write out the definition (2.2.4) directly in terms of limits.

Definition 2.2.1 The partial derivatives of $F(x, y)$ with respect to x and y are defined as follows:

$$F_x(x, y) = \lim_{h \to 0} \frac{F(x + h, y) - F(x, y)}{h},$$

$$F_y(x, y) = \lim_{h \to 0} \frac{F(x, y + h) - F(x, y)}{h}.$$

In Definition 2.2.1 we have replaced the fixed point (a, b, c) by the point (x, y, z). This should cause no trouble if you clearly note one idiosyncrasy in the notation $F_x(x, y)$. Here, the "x" in the subscript refers to a direction—parallel to the x-axis—and the "(x, y)" refers to a particular point in the domain; the same care is required in the use of the symbol $F_y(x, y)$.

The above warning actually points out the mental process used to calculate partial derivatives. To find $F_x(x, y)$ one should think of "y" in the expression for $F(x, y)$ as representing, for the moment, a constant. The result is a function of x alone, and derivative formulas from *C 1* can, we hope, be used to calculate $F_x(x, y)$. Similarly, to find $F_y(x, y)$ simply think of x as a constant.

Referring to the functions F, G, H in (2.2.1)–(2.2.3), we easily see that

(2.2.5) $$F_x(x, y) = -2, \; F_y(x, y) = -1;$$

(2.2.6) $$G_x(x, y) = \frac{x}{\sqrt{x^2 + y^2}}, \; G_y(x, y) = \frac{y}{\sqrt{x^2 + y^2}};$$

(2.2.7) $$H_x(x, y) = \frac{e^y}{x}, \; H_y(x, y) = e^y \log x.$$

If F is any function of two variables for which at a point (a, b) both partial derivatives F_x and F_y exist, consider the tangents to the surface of F in the x- and y-directions. These are, of course, the lines in the planes $\Pi_{y=b}$, $\Pi_{x=a}$ through $(a, b, F(a, b))$ with slopes $F_x(a, b)$ and $F_y(a, b)$ respectively. Of special interest is the plane which contains these two lines; let us call the plane $\Pi(a, b)$. See Figure 2.2.7.

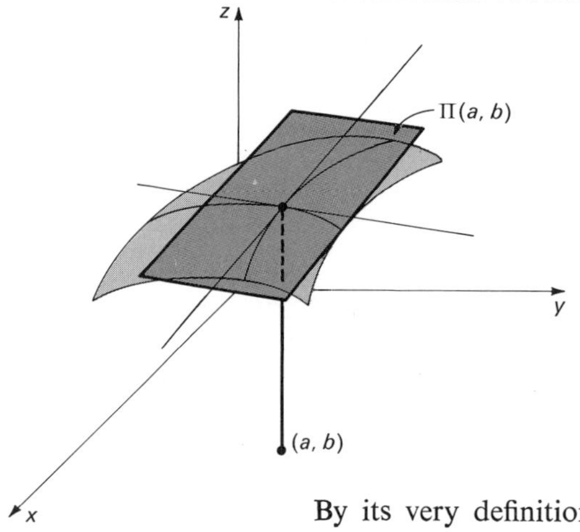

Figure 2.2.7

By its very definition we see that it is reasonable to call $\Pi(a, b)$ *tangent* to the surface in the x- and y-directions. But it does not follow that $\Pi(a, b)$ could reasonably be called a tangent plane to the surface. It may very well happen that $\Pi(a, b)$ does not "fit" the surface at all well in some direction other than the x- and y-directions. Examples for which this occurs do exist, but they are too complicated for presentation here. In such cases, the function F is too badly behaved for its surface to be approximated by a plane in the smooth way suggested by the word "tangent."

A condition on the function F can be formulated, though we shall not do it here, assuring us that the plane $\Pi(a, b)$ *does* fit the surface as we would expect a tangent plane to fit. A function which satisfies this condition at (a, b) is called *differentiable* at (a, b). Its graph is said to have a tangent plane at $(a, b, F(a, b))$.

What we want to point out here are two facts:

1. The existence of the partial derivatives F_x and F_y is not enough to ensure differentiability of F.

2. The condition of differentiability is *directly* related to the possibility of fitting a tangent plane to the surface exactly as differentiability was directly related to the existence of a tangent line in the one-variable, two-dimensional case.

We wish to take one more look into our two-variable box before closing the lid. Suppose $z = F(x, y)$ with both x and y functions of one or more additional variables $t_1, t_2, \ldots$. We might write $x = x(t_1, t_2, \ldots)$, $y = y(t_1, t_2, \ldots)$. Then z may be thought of as a function of $t_1, t_2, \ldots$. More precisely, we may consider a new composite function

$$f(t_1, t_2, \ldots) = F(x(t_1, t_2, \ldots), y(t_1, t_2, \ldots)).$$

Any theorem which gives the partial derivatives of f in terms of the partial derivatives of the other functions involved is called a *chain rule*. As you can imagine, there are infinitely many possible chain rules, although it is possible to include them all in one general form.

But this is material for another book. Let us look very briefly at only one situation, in which $x = x(t)$, $y = y(t)$ are functions of one variable t. Look at the composite functions $F(x(t), y(t))$, $G(x(t), y(t))$, $H(x(t), y(t))$, referring to (2.2.1)–(2.2.3). These can be written explicitly as

$$\begin{aligned} f(t) &= F(x(t), y(t)) = 2 - 2x(t) - y(t), \\ g(t) &= G(x(t), y(t)) = \sqrt{[x(t)]^2 + [y(t)]^2}, \\ h(t) &= H(x(t), y(t)) = e^{y(t)} \log x(t). \end{aligned}$$

Each can be differentiated using theorems from *C 1* (including the chain rule for functions of one variable) to obtain

(2.2.8) $$f'(t) = -2x'(t) - y'(t),$$

(2.2.9) $$g'(t) = \frac{x(t)x'(t) + y(t)y'(t)}{\sqrt{[x(t)]^2 + [y(t)]^2}},$$

(2.2.10) $$h'(t) = \frac{e^{y(t)}}{x(t)} x'(t) + [e^{y(t)} \log x(t)]y'(t).$$

If we compare (2.2.8)–(2.2.10) with (2.2.5)–(2.2.7), we quickly see that

(2.2.11) $$f'(t) = F_x(x(t), y(t))x'(t) + F_y(x(t), y(t))y'(t),$$

(2.2.12) $$g'(t) = G_x(x(t), y(t))x'(t) + G_y(x(t), y(t))y'(t),$$

(2.2.13) $$h'(t) = H_x(x(t), y(t))x'(t) + H_y(x(t), y(t))y'(t).$$

The similarity among (2.2.11)–(2.2.13) is not coincidence, of course, but a reflection of a general principle, the chain rule. This chain rule is not implausible, though a precise proof is fairly deep. The rate of change of z ($= F(x, y)$) produced by a simultaneous change in x and y (with rates $x'(t)$, $y'(t)$) can reasonably be expected to be the sum of the rate produced by a change in x alone ($F_x \cdot x'$) and in y alone ($F_y \cdot y'$).

We have here a good illustration of the danger of leaning too heavily on such plausibility discussions as the above (we are embarrassed to call it an argument). We would not even suspect from this discussion that the existence of the required partial derivatives F_x and F_y is *not* enough to assure the validity of this chain rule. There are, however, examples of functions for which the partial derivatives exist but for which the chain rule is not valid. To be certain that the rule applies, we must know that F is differentiable in the sense referred to above. Since we shall refer to this chain rule later in the chapter, we state it here as it can be proved, and is, in a subsequent volume in this series, *Calculus of Several Variables.*

Theorem 2.2.1 *If functions $F(x, y)$, $x(t)$, $y(t)$ are all differentiable, then so is the function*

$$f(t) = F(x(t), y(t)),$$

and

$$f'(t) = F_x(x(t), y(t))x'(t) + F_y(x(t), y(t))y'(t).$$

A common notation for partial derivatives is the following:

$$F_x = \frac{\partial F}{\partial x}, \quad F_y = \frac{\partial F}{\partial x},$$

the symbol "∂" resembling "d" in the df/dx notation for an ordinary derivative. In this notation the chain rule in Theorem 2.2.1 can be stated

$$\frac{dF}{dt} = \frac{\partial F}{\partial x}\frac{dx}{dt} + \frac{\partial F}{\partial y}\frac{dy}{dt}.$$

Exercises

1 Find the partial derivatives $\partial F/\partial x$ and $\partial F/\partial y$ if

(a) $F(x, y) = x^2 + y^2$, (b) $F(x,y) = 4x^2y^3$,

(c) $F(x, y) = x^2y + xy^2$, (d) $F(x, y) = \frac{x}{y} + \frac{y}{x}$,

(e) $F(x, y) = 2xy/(x^2 - y^2)$, (f) $F(x, y) = \sin (x + y^2)$,

(g) $F(x, y) = \tan \sqrt{xy}$, (h) $F(x, y) = x^{1/2}y^{1/3}$,

(i) $F(x, y) = e^{(x^2+y^2)}$, (j) $F(x, y) = \log \sqrt{x + y}$,

(k) $F(x, y) = e^x \sin y$, (l) $F(x, y) = (x^2 + xy + y^2)^{1/2}$,

(m) $F(x, y) = 2^x 3^y$, (n) $F(x, y) = x^y$,
(o) $F(x, y) = x^y y^x$, (p) $F(x, y) = \cos(1/(x + y^2))$.

2 Find $\dfrac{dF}{dt}$ if

(a) $F(x, y) = x^2 + y^2$, $x(t) = t$, $y(t) = t^2$;
(b) $F(x, y) = 2x^3y^4$, $x(t) = t^2$, $y(t) = \sin t$;
(c) $F(x, y) = x^2y + xy^2$, $x(t) = 2t + 4$, $y(t) = t^2 + 2t - 1$;
(d) $F(x, y) = x \cos y + y \sin x$, $x(t) = \sin 2t$, $y(t) = \cos 2t$.

In the same way that in *C 1*

$$f(x_0) + f'(x_0)(x - x_0)$$

was used to furnish an estimate of $f(x)$ for x close to x_0, we can use

$$F(x_0, y_0) + \frac{\partial F}{\partial x}(x_0, y_0)(x - x_0) + \frac{\partial F}{\partial y}(x_0, y_0)(y - y_0)$$

as an estimate of $F(x, y)$, where F is differentiable, x is close to x_0, and y is close to y_0.

3 Approximate the given function values.
(a) $F(1.1, 2.1)$ if $F(x, y) = x^2 + 4y^2$,
(b) $F(31°, 29°)$ if $F(x, y) = \sin x \cos y$,
(c) $F(0.1, 92°)$ if $F(x, y) = e^{x^2} \sin y$,
(d) $F(1.1, 0.9)$ if $F(x, y) = \log(xy^2)$.

4 Find the approximate change in the volume of a box whose dimensions are $3'' \times 4'' \times 5''$ if the sides are changed to $3'' \times 4\frac{1}{5}'' \times 5\frac{1}{2}''$.

5 Approximate the change in the hypotenuse of a right triangle with legs of length 6″ and 8″ when the shorter leg is lengthened by $\frac{1}{4}''$ and the longer leg is shortened by $\frac{1}{8}''$.

6 The altitude of a right circular cone is 15″ and is increasing at the rate of 0.2 in./min. The radius of the base is 10″ and is decreasing at the rate of 0.3 in./min. How fast is the volume changing?

7 The radius of a right circular cylinder is 12 in. and is increasing at the rate of 0.4 in./sec., while the altitude is 8″ and is decreasing at the rate of 0.4 in./sec. Find the rate at which the volume is changing.

2.3

Implicitly Defined Functions

So far, all the functions we have encountered have been defined by an explicit description of the rule or mapping. For example, a function f is defined *explicitly* by

$$f(x) = \sqrt{x^2 - 1}, \quad x \geq 1.$$

However, one is often introduced to a function not directly but indirectly, by means of some statement *about* the function. A statement which does not describe a function explicitly but which allows one to *infer* its existence and form is called an *implicit* definition. Most commonly, functions are defined implicitly by means of equations such as

(2.3.1) $$x^2 + 4y^2 = 4$$

or

(2.3.2) $$\log y - x = 0.$$

We say that equation (2.3.1) defines a function $f(x)$ implicitly, if

$$x^2 + 4[f(x)]^2 = 4$$

for all x in some domain. Similarly, (2.3.2) defines a function $g(x)$ if

$$\log (g(x)) - x = 0$$

for some values of x. It is clear that both of these equations do define functions of x; (2.3.2) defines the function exp: $\exp(x) = e^x$, for

$$\log e^x - x = 0$$

for all x. Moreover, exp is the *only* function for which this is true; we say that (2.3.2) *uniquely* defines the function exp.

Equation (2.3.1) implicitly defines more than one function. For if

$$f_1(x) = \tfrac{1}{2}\sqrt{4 - x^2}, \quad |x| \leq 2, \text{ and}$$
$$f_2(x) = -\tfrac{1}{2}\sqrt{4 - x^2}, \quad |x| \leq 2,$$

then

$$x^2 + 4[f_1(x)]^2 = 4 \quad \text{and} \quad x^2 + 4[f_2(x)]^2 = 4$$

both hold for all $|x| \leq 2$. Thus, either f_1 or f_2 is defined implicitly by equation (2.3.1). Still other functions defined implicitly by (2.3.1) for $-2 \leq x \leq 2$ can be constructed by tearing f_1 and f_2 apart and putting the pieces back together in different ways. If f_3 is a function which coincides with f_1 for $x \leq 0$ and with f_2 for $x > 0$, then f_3 is also defined implicitly by (2.3.1).

In each of these cases we have been able to obtain explicit expressions for the functions defined implicitly by the equations by solving the equations for y in terms of x. In one case the solution was unique; in the other, it was not. Such a solution is not always possible, and it is often decidedly not easy to find. Although the equation

(2.3.3) $$y^4 - xy^3 + (x^2 - 1)y = 9 - x$$

might be solved for y in terms of x, few (including the authors) would be eager to try to find such a solution. Yet, general theorems tell us that (2.3.3) does define a differentiable function for x in some neighborhood of, say, $x = 1$, and that there is only one such function $y = f(x)$ such that $f(1) = 2$. Whatever this function is, our *only* contact with it is the equation (2.3.3) that it satisfies.

We shall not take up here the general theorems just referred to that give conditions under which an equation necessarily defines a differentiable function implicitly. But if such an equation does define a function $f(x)$, we can ask whether it is possible to find the derivative $f'(x)$ *without actually calculating $f(x)$ explicitly*. The answer is Yes, and the procedure is simple.

Finding derivatives of implicitly defined functions might be called *an act of faith*. One must simply *believe* that an equation such as (2.3.3) defines y as a function of x. The equation itself then becomes an identity relating various combinations of x and that function, which we shall call $f(x)$. The two sides of the equation then define two new functions, different combinations of $f(x)$ $(=y)$ and x, which are the same for all x (in the domain of $f(x)$). Their derivatives must then be the same. When these derivatives, the derivatives with respect to x of the two sides of the equation, are calculated from whatever formulas apply, the results are expressions in x, *$f(x)$, and $f'(x)$*. When these expressions are set equal to each other—they *are* the same, as we have observed above—the resulting equation may be solved for $f'(x)$ in terms of x and $f(x)$.

Example 1 Let $f(x)$ be a function defined implicitly by (2.3.3) such that $f(1) = 2$. Note that $x = 1$, $y = 2$ *do* satisfy the equation. Then we have

(2.3.4) $$[f(x)]^4 - x[f(x)]^3 + (x^2 - 1)f(x) \equiv 9 - x.$$

Thus,

$$4[f(x)]^3 f'(x) - [f(x)]^3 - 3x[f(x)]^2 f'(x) + 2x\,f(x) + (x^2 - 1)f'(x) \equiv -1,$$

so that

(2.3.5) $$f'(x) = \frac{[f(x)]^3 - 2xf(x) - 1}{4[f(x)]^3 - 3x[f(x)]^2 + x^2 - 1}.$$

Although not an *explicit* expression for $f'(x)$ (the expression involves $f(x)$, which we do not know explicitly) (2.3.5) is an expression for $f'(x)$ which can be very useful. For example, we can say that

$$f'(1) = \frac{8 - 4 - 1}{32 - 12} = \frac{3}{20}.$$

Expressed in another way, the graph of (2.3.4) passes through the point (1, 2), and we have found that the graph has a tangent line with slope 3/20 at that point. This we have found by noting that a small portion of this graph near (1, 2) can be interpreted as the graph of a function $f(x)$ defined implicitly by the equation. We have found the derivative $f'(1)$, which by previous results we know to be the slope of the tangent.

The work of calculating the derivative of this implicitly defined function is a little easier if we use the dy/dx notation. We do not actually need to substitute $f(x)$ for each y which occurs in the equation. It is enough to *think* of y as representing the value $f(x)$ of a function, and hence think of the two sides as representing identical functions which may be differentiated and equated. Thus,

(2.3.6) $$y^4 - xy^3 + (x^2 - 1)y = 9 - x,$$

becomes, on differentiation with respect to x,

(2.3.7) $$4y^3 \frac{dy}{dx} - y^3 - 3xy^2 \frac{dy}{dx} + 2xy + (x^2 - 1)\frac{dy}{dx} = -1.$$

Solving (2.3.7) for dy/dx we have

(2.3.8) $$\frac{dy}{dx} = \frac{y^3 - 2xy - 1}{4y^3 - 3xy^2 + x^2 - 1}.$$

an expression which is valid for *any* function y of x defined by (2.3.6) (and there will be several such functions).

Expression (2.3.8) has another useful interpretation. We can assert that at any point (x, y) on the graph of (2.3.6), the slope of the tangent line is

$$\frac{y^3 - 2xy - 1}{4y^3 - 3xy^2 + x^2 - 1}.$$

Example 2 Equation (2.3.1) is the equation of an ellipse. If this equation defines y as a function of x, then

$$2x + 8y\frac{dy}{dx} = 0,$$

or

(2.3.9) $$\frac{dy}{dx} = -\frac{x}{4y}.$$

From (2.3.9) we can infer many things. The graph is rising (from left to right) in the second and fourth quadrants, falling in the first and third. The tangent is horizontal at points $(0, \pm 1)$ and vertical at $(\pm 2, 0)$. (It can be verified that with the proper interpretation, (2.3.9) is valid even when the denominator of the expression for dy/dx vanishes. This occurs only at an endpoint of the domain of one of the functions defined by (2.3.1), so the infinite derivative must be interpreted as a left or right derivative. We shall not attempt a full discussion of such isolated events!)

Moreover, even (2.3.9) may be interpreted as an identity between two functions of x. These functions must therefore have the same derivative. We therefore have

(2.3.10)
$$\frac{d^2y}{dx^2} = -\frac{4y - x(4\,(dy/dx))}{16y^2}.$$

Using the expression (2.3.9) in (2.3.10), we obtain

$$\frac{d^2y}{dx^2} = -\frac{4y + x^2/y}{16y^2},$$

or, using (2.3.1),

$$\frac{d^2y}{dx^2} = -\frac{4y^2 + x^2}{16y^3} = -\frac{1}{4y^3}.$$

Thus, whenever $y > 0$, $d^2y/dx^2 < 0$, and whenever $y < 0$, $d^2y/dx^2 > 0$. The graph of (2.3.1) is therefore concave down above the horizontal axis and concave up below the axis.

All this information combines to suggest the graph in Figure 2.3.1, which, of course, we recognize as an ellipse.

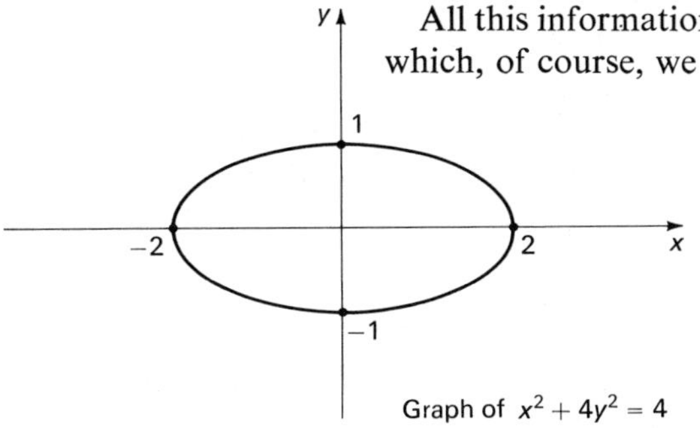

Graph of $x^2 + 4y^2 = 4$

Figure 2.3.1

In the above discussion of the procedure for finding the derivative of an implicit function, we have made several claims substantiated only by examples. We cannot make the discussion entirely rigorous at this stage, but the line of reasoning can be made much more cohesive by the application of some of the ideas from the last section.

An equation in x and y can always be written

(2.3.11)
$$F(x, y) = 0,$$

where F is some function of two variables. If f is a function of x defined implicitly by (2.3.11), then we have

(2.3.12)
$$F(x, f(x)) \equiv 0,$$

an identity which holds for all x in the domain of f. Looking at (2.3.12) we may consider the composite function $F(x, f(x))$ obtained by replacing y by $f(x)$ in $F(x, y)$. Here, x plays a dual role; it is both one of the original two variables and also the "new" variable. Using Theorem 2.2.1, we may find the derivative of this composite, a derivative which by (2.3.12) must be 0. We obtain

$$F_x(x, f(x)) + F_y(x, f(x))f'(x) = 0.$$

Thus

(2.3.13)
$$f'(x) = -\frac{F_x(x, f(x))}{F_y(x, f(x))}.$$

Of course, (2.3.13) can also be written

(2.3.14)
$$\frac{dy}{dx} = -\frac{F_x(x, y)}{F_y(x, y)}$$

if we interpret y here as a function defined implicitly by (2.3.11).

You should compare (2.3.14) with (2.3.9) and (2.3.8). Expression (2.3.14) gives perhaps the quickest way to write down the derivative of an implicit function provided the defining equation is given in form (2.3.11).

One final comment: In an equation of the form (2.3.11) there is no automatic hierarchy of variables. There is no reason besides custom to select x as the independent variable and let y denote a function. We may just as well ask whether (2.3.11) defines x implicitly as a function of y. The entire discussion then proceeds as above, with the roles of x and y reversed. If $x = g(y)$ is a function defined by (2.3.11), then

$$F(g(y), y) = 0,$$
$$F_x(g(y), y)g'(y) + F_y(g(y), y) = 0,$$
$$g'(y) = -\frac{F_y(g(y), y)}{F_x(g(y), y)}.$$

In another notation,

$$\text{If } F(x, y) = 0, \text{ then } \frac{dx}{dy} = -\frac{F_y(x, y)}{F_x(x, y)},$$

provided, of course, the given equation *does* define x as a differentiable function of y.

Exercises

1 (a) Find two differentiable functions $y = y(x)$ defined by the equation $y^2 - x - 1 = 0$.

(b) Sketch the graph of the equation $y^2 - x - 1 = 0$ and indicate on it the graphs of the two functions in (a).

(c) Find $y'(x)$ for each of the two functions in (a).

(d) Find dy/dx by differentiating the equation $y^2 - x - 1 = 0$ implicitly and compare the results with those in (c).

(e) Find the slopes of the tangent lines to the graph of $y^2 - x - 1 = 0$ at the points $(0, 1)$, $(0, -1)$, $(1, \sqrt{2})$, $(1, -\sqrt{2})$.

2 (a)–(d) Consider the questions for Exercise 1 when applied to the equation $x^2y^2 = 4$.

(e) Find the slopes of the tangent lines to the graph of $x^2y^2 = 4$ at the points $(1, 2)$, $(1, -2)$, $(-1, 2)$, $(-1, -2)$.

3 Assume that each of the given equations defines at least one differentiable function $y = y(x)$. Use the method of Example 1 to find the derivative dy/dx if

(a) $y^2 = 5x$, (b) $xy = 4$,
(c) $x^2 - 2y^2 = 3$, (d) $5x^2 + 4y^2 = 10$,
(e) $x^2 + y^2 = 16$, (f) $x + y + y^2 = 0$,
(g) $2x^2 + xy^2 - y^3 = 0$, (h) $x + \sqrt{xy} = 2y$,
(i) $\sin(x^2) + \cos(y^2) = 1$, (j) $x/y - 3y = 4x$,
(k) $x^2e^{y^2} = 1$, (l) $x^2 \log(xy)^{1/2} = y$.

4 Find the second derivatives d^2y/dx^2 of the functions in Exercise 3.

5 Assume that each of the given equations defines at least one differentiable function $y = y(x)$. Use formula (2.3.14) to find the derivative dy/dx.

(a) $\sin x + \cos y = 1$, (b) $x^2y + y^2x = 6$,
(c) $x^2 = \dfrac{x - y}{x + y}$, (d) $x^3 + xy + y^3 = 1$,
(e) $\sqrt{x} + \sqrt{y} = 10$, (f) $(x + y)^3 + (x - y)^3 = x^4 + y^4$,
(g) $\dfrac{1}{x} + \dfrac{1}{y} = 1$, (h) $y^2(x^2 + 1) = x^2 - 1$,
(i) $x^{2/3} + y^{2/3} = 1$, (j) $y = \sin x \cos y^2$.

6 Find an equation for the tangent line to each of the following curves at the indicated point.

(a) $x^2 - xy + y^2 = 7$, $(2, 3)$; (b) $x^2y^2 = 16$, $(2, 2)$;
(c) $y = \dfrac{x - y}{x + y}$, $(0, -1)$; (d) $(y - x)^2 = 2x + 4$, $(6, 2)$;
(e) $\sqrt{x} + \sqrt{y} = 2$, $(1, 1)$; (f) $x \sin y = 1$, $(2, \pi/6)$.

7 Find equations for the normal lines to the curves in Exercise 6 at the points indicated. (Recall that the *normal* line to a curve at a point is perpendicular to the tangent line at that point.)

2.4

Inverse Functions

In Chapter 7 of *C 1*, in introducing the function exp, we briefly discussed the notion of an inverse function. We shall complete that discussion in this section before introducing the important inverse trigonometric functions in the next section. If f is a function, then an *inverse function* f^{-1} of f is defined as any function which satisfies the conditions

(2.4.1)
$$f^{-1}(f(x)) = x,$$
$$f(f^{-1}(y)) = y,$$

for all x in some interval I and all y in $f(I)$.

It is not necessary to state both of the conditions in (2.4.1), for the two are equivalent. For example, in the second condition y is some number in the range of f, so $y = f(x)$ for some x. Thus $f^{-1}(y) = f^{-1}(f(x))$. But if we assume the first condition, then $f^{-1}(f(x)) = x$. Thus $f(f^{-1}(y)) = f(x) = y$, and the second condition follows.

A comment on the notation f^{-1} is in order. We adopt this widely used notation for an inverse function here so that you will become acquainted with the accepted language. The notation has obvious disadvantages, though, and we must be careful from this point on not to confuse the inverse function f^{-1} with the reciprocal $1/f$.

Recall from the discussions in C 1 that the graphs of a function f and any inverse function f^{-1} are simply related. For, if we write

$$y = f(x),$$

then by the first of the conditions (2.4.1)

$$x = f^{-1}(y).$$

Thus, if a point (x, y) is on the graph of f (so that $y = f(x)$), then the point (y, x) is on the graph f^{-1} (so that $x = f^{-1}(y)$).[1] Since the two points (x, y), (y, x) are mirror images of each other with respect to the line $y = x$, we conclude that the graph of f^{-1} is obtained by simply reflecting the graph of f through this line (see Figure 2.4.1).

[1] *If this seems confusing, recall that when dealing with the graph of a function we always consider values of the function as ordinates of points.*

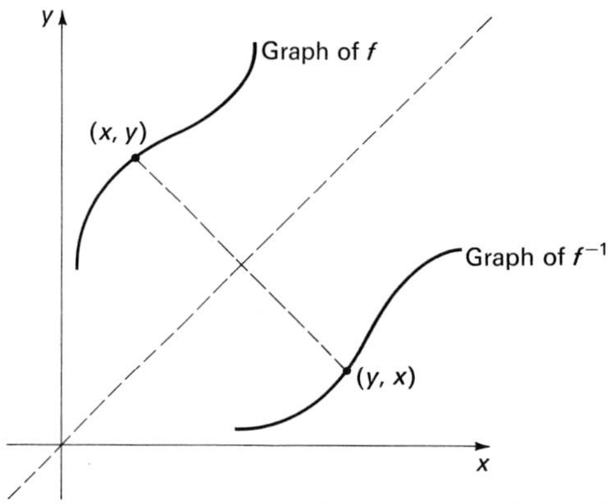

Figure 2.4.1

Another observation which we can make using (2.4.1) is that the equation

(2.4.2) $$F(x, y) = f(x) - y = 0$$

defines the function $f^{-1}(y)$ implicitly, i.e., $F(f^{-1}(y), y) \equiv 0$. Thus, the discussion at the end of the last section applies, and we conclude that

(2.4.3) $$[f^{-1}]'(y) = -\frac{F_y}{F_x} = -\frac{-1}{f'(x)} = \frac{1}{f'(x)}\cdot$$

Thus, if x and y are two numbers such that $y = f(x)$ and $x = f^{-1}(y)$, then the derivative of the inverse function f^{-1} evaluated at y and the derivative of f evaluated at x are reciprocals.

This, you may recall, was how we discovered the derivative formula for the function exp. For exp is the inverse function of log, that is, $\log(\exp(y)) = y$. Thus, by (2.4.3),

(2.4.4) $$\exp'(y) = \frac{1}{\log'(x)} = \frac{1}{1/x} = x.$$

But $x = \exp(y)$, because $y = \log(x)$, so by (2.4.4),

$$\exp'(y) = \exp(y),$$

the familiar formula.

The above development, like the one in the last section, is clouded by the fact that we had to *assume* that (2.4.2) defines a function x of y and to *assume* that this function is differentiable. We can dissolve this cloud. We shall obtain precise conditions under which (2.4.2) is guaranteed to define a differentiable function—and only *one* such function.

Let $f\colon A \to B$ be a function; that is, let $f(x)$ be defined for all x in A. Assume, further, that $B = f(A)$, that is, that for every y in B there is at least one x in A such that $f(x) = y$. We call f a *one-to-one function* if for every y in B there is *only* one x in A with $f(x) = y$. Clearly, a one-to-one function has an inverse function, and only one such inverse. For we may define $f^{-1}(y)$ for any y in B to be that unique x in A such that $f(x) = y$.

The simplest one-to-one functions we have encountered are the monotone functions, those which are either increasing ($x_1 < x_2$ implies $f(x_1) < f(x_2)$) or decreasing ($x_1 < x_2$ implies $f(x_1) > f(x_2)$). Notice that if f is monotone, then f^{-1} is also monotone. (The proof is left as an exercise.) We shall be content to discuss only monotone functions in this connection, and we have proved

Theorem 2.4.1 *Let f be a monotone function defined on A, and $B = f(A)$. Then f has one and only one inverse function f^{-1} defined on B, and f^{-1} is also monotone.*

Actually, we shall be content to discuss monotone functions that are defined on intervals. If A is an interval and f is monotone on A, then we might ask what kind of set of numbers $B = f(A)$ is. Here the Intermediate Value Theorem is suggested; for if f has the intermediate value property, then B is certainly also an interval. In fact, if $A = [a, b]$ and f is increasing on A, then B must be contained in the interval $[f(a), f(b)]$. If, in addition, f has the intermediate value property, then any number in $[f(a), f(b)]$, being between two values $f(a)$ and $f(b)$ of f, must be a value $f(x)$ and hence must be in B. Thus $B = [f(a), f(b)]$.

In Chapter 1 of *C 1* we noted that any continuous function has the intermediate value property. Thus, we have proved that *the inverse function of a continuous monotone function on an interval is also defined on an interval.*

But we can say more. If f is continuous and increasing on $[a, b]$ and $f^{-1}(y)$ is therefore defined on $[f(a), f(b)]$, then f^{-1} *also has the intermediate value property.* For suppose $x_1 = f^{-1}(y_1)$ and $x_2 = f^{-1}(y_2)$ are two values of f^{-1} and that x is a number between x_1 and x_2. If $x_1 < x < x_2$, then x is certainly also in $[a, b]$, so $f(x) = y$ is defined. Since f is increasing, $y_1 < y < y_2$. Thus f^{-1} takes on every value between x_1 and x_2 somewhere on the interval $[y_1, y_2]$. That is precisely the intermediate value property. The case of a decreasing function is handled similarly.

Thus, f^{-1} is also a monotone function with the intermediate value property. Exercise 2.1.6 shows that f^{-1} is then also continuous. We have therefore proved

Theorem 2.4.2 *Let f be a continuous monotone function defined on the closed interval $[a, b]$. Then its inverse function f^{-1} is a continuous monotone function defined on the closed interval between $f(a)$ and $f(b)$.*

Now let f be a differentiable function on $[a, b]$ such that $f'(x) \neq 0$ for x in $[a, b]$. Then, as we have observed before, f is monotone and continuous on $[a, b]$. Thus Theorem 2.4.2 applies, and we have a continuous monotone inverse function. To simplify our arguments let us assume that f, and hence also f^{-1}, is increasing. The case of decreasing f is handled similarly.

Let us investigate the possibility of a derivative for f^{-1}. We must examine the quotient

$$\frac{f^{-1}(y + k) - f^{-1}(y)}{k},$$

where k is any real number such that both y and $y + k$ lie in the domain $[f(a), f(b)]$ of f^{-1}. Since this domain is $f([a, b])$, there must exist two numbers x and $x + h$ in $[a, b]$ such that

(2.4.5) $$y = f(x), \quad y + k = f(x + h).$$

Since f is monotone (hence one-to-one), we see that $h \neq 0$ if (and only if) $k \neq 0$. The relations (2.4.5) can also be written

(2.4.6) $$f^{-1}(y) = x, \quad f^{-1}(y + k) = x + h.$$

Moreover, (2.4.5) implies that

(2.4.7) $$k = f(x + h) - f(x).$$

Using (2.4.6) and (2.4.7) we see that

(2.4.8)
$$\frac{f^{-1}(y+k)-f^{-1}(y)}{k}=\frac{h}{f(x+h)-f(x)}.$$

Since by (2.4.6) $h = f^{-1}(y + k) - f^{-1}(y)$ and since by Theorem 2.4.2 f^{-1} is continuous, we have

(2.4.9)
$$\lim_{k\to 0} h = 0.$$

The right side of (2.4.8) is a function F of h, which can also be considered a function of k. From the left side of (2.4.8), $\lim_{k\to 0} F(h) = [f^{-1}]'(y)$. But an easy corollary of Theorem 1.7.5 of *C 1* (trivial in the light of our original definition of a limit in *C 1*) shows that, by (2.4.9), $\lim_{k\to 0} F(h) = \lim_{h\to 0} F(h)$.[2] Thus,

$$\begin{aligned}[f^{-1}]'(y) &= \lim_{k\to 0}\frac{f^{-1}(y+k)-f^{-1}(y)}{k}\\ &= \lim_{h\to 0}\frac{h}{f(x+h)-f(h)}=\frac{1}{f'(x)}\end{aligned}$$

provided, of course, $f'(x) \neq 0$.

[2] *If this seems complicated, accept the result intuitively as a natural consequence of the fact (2.4.9) that as $k \to 0$ then also $h \to 0$.*

The above discussion establishes the main theorem of this section:

Theorem 2.4.3 *If f is a differentiable function with nonvanishing derivative on $[a, b]$, then f has a differentiable inverse function f^{-1} defined on the closed interval between $f(a)$ and $f(b)$. Moreover,*

$$[f^{-1}]'(y)=\frac{1}{f'(x)}=\frac{1}{f'(f^{-1}(y))}.$$

We have thus returned to (2.4.3), but with doubt concerning the existence of an inverse function removed. Recall that in the sole previous application of the above ideas, namely in the case $f(x) = \log(x)$, the derivative $f'(x) = 1/x$ is certainly nonvanishing. Thus, Theorem 2.4.3 provides our first conclusive proof of the *existence* of a differentiable function exp.

We shall see additional applications of these theorems in the next section.

Exercises

1 For each of the following functions f

(i) Sketch the graph of $y = f(x)$.

(ii) By considering the derivative f', find one or more intervals on which f has an inverse $x = f^{-1}(y)$.

(iii) Find a formula for the inverse function on each interval.

(iv) Use Theorem 2.4.3 to find a formula for df^{-1}/dy.

(a) $f(x) = 2x + 3$, (b) $f(x) = x^2 + 2$,

(c) $f(x) = 14$, (d) $f(x) = (x + 2)^3$,

(e) $f(x) = x^3 + 2$, (f) $f(x) = x^2 + 4x - 2$,
(g) $f(x) = \sqrt[3]{x + 4}$, (h) $f(x) = x + [x]$,
(i) $f(x) = \frac{1}{x}$, (j) $f(x) = \frac{x - 3}{x + 3}$,
(k) $f(x) = 3|x| + x$, (l) $f(x) = 2x/(x^2 + 1)$.

2 Prove that the inverse of an increasing (decreasing) function is an increasing (decreasing) function.

3 Suppose f and g are functions which have inverses f^{-1} and g^{-1} respectively. Show that $h = g \circ f$ has an inverse and that $h^{-1} = f^{-1} \circ g^{-1}$.

4 Prove that if f is an increasing differentiable function and the graph of f is concave upward at a point x_0, then the graph of f^{-1} is concave downward at $y_0 = f(x_0)$.

5 Can a function be both monotone increasing on $(-\infty, \infty)$ and bounded?

6 Find an equation for the tangent line to the graph of f^{-1} at the point (3, 1) if $f(x) = x^3 + 2x^2 - x + 1$.

7 Suppose the graphs of two functions f and g are orthogonal at a point (x_0, y_0) (i.e., their graphs intersect at (x_0, y_0) and have perpendicular tangents at that point). Prove that if f and g have inverses, the graphs of f^{-1} and g^{-1} are orthogonal at (y_0, x_0).

8 Let $f(x) = x^3$ for $-\infty < x < \infty$. Show that f has an inverse function but that it is not differentiable everywhere. Why does this not contradict Theorem 2.4.3?

9 Let $f(x) = \int_0^x e^{-t^2}\,dt$ for $x \geq 0$. Prove that f has an inverse function and find its derivative.

10 Suppose f and f^{-1} are twice differentiable functions. Prove that

$$(f^{-1})''(y) = \frac{-f''(x)}{[f'(x)]^3} = -\frac{f''(f^{-1}(y))}{[f'(f^{-1}(y))]^3}.$$

(*Hint:* Differentiate the formula in Theorem 2.4.3. Be certain you understand each step.)

2.5

The Inverse Trigonometric Functions

Inverse functions of the trigonometric functions play an important role in the calculus both because of their relationship to the geometry of a circle and because of the interesting properties of their derivatives. In this section we shall develop the inverses of three functions: sine, cosine, and tangent.

In looking for inverse functions, first observe that the trigonometric functions are not one-to-one. For a given number y between -1 and 1 there are many—in fact, infinitely many—values of x

such that $\sin x = y$. Thus, we cannot hope to construct a well-defined inverse for the function sine unless we first restrict the domain of definition sufficiently to obtain a one-to-one function. As we saw in the last section, it is sufficient to restrict the domain of $\sin x$ to an interval on which the derivative does not vanish, i.e., on which $\sin x$ is monotone. If we can find such an interval whose image is the entire range $-1 \leq y \leq 1$ of $y = \sin x$, so much the better. This will maximize the domain of the inverse function. We shall follow this plan for each of the functions: sine, cosine, and tangent.

First, a word about terminology. If $f(x) = \sin x$ and f^{-1} is an inverse function, then we can assign a geometrical meaning to $f^{-1}(y)$ for $-1 \leq y \leq 1$. Since by the definition of an inverse function

$$\sin (f^{-1}(y)) = y,$$

we can say that $f^{-1}(y)$ is (the radian measure of) an angle whose sine is y. Since radian measure is just the length of an arc of a circle of radius 1, we could also say that $f^{-1}(y)$ is the *arc* of a unit circle whose central angle has sine equal to y. Thus, it is suggestive to call this quantity arcsin (y), and the name *arcsine* for the inverse of sine has become commonly accepted. For similar reasons the inverses of the (suitably restricted) functions cosine, tangent have come to be called *arccosine* and *arctangent*. We shall follow this convention.

Arcsine Since $d/dx\,(\sin x) = \cos x$, we see that

$$\frac{d}{dx} \sin x > 0 \text{ for } -\frac{\pi}{2} < x < \frac{\pi}{2}.$$

Thus, sine is increasing on the closed interval $[-\pi/2, \pi/2]$. Since $\sin(-\pi/2) = -1$ and $\sin(\pi/2) = 1$, the image of $[-\pi/2, \pi/2]$ is the interval $[-1, 1]$, the entire range of sine. Thus, we shall consider only the *restricted sine function*

$$\sin x, \ -\frac{\pi}{2} \leq x \leq \frac{\pi}{2},$$

for the purposes of constructing an inverse.

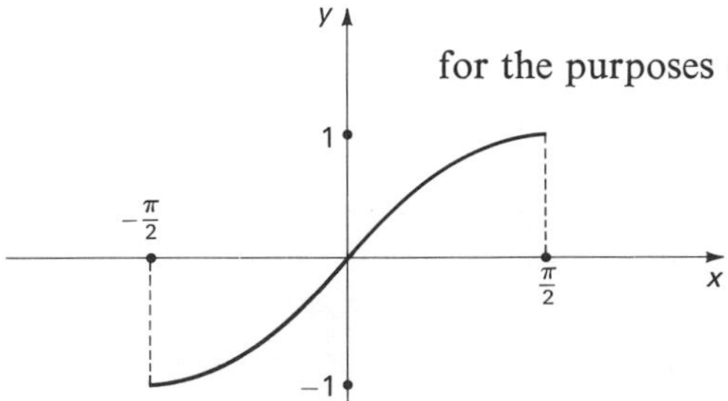

Figure 2.5.1

The restricted sine

Theorems 2.4.2 and 2.4.3 assert the existence of an increasing inverse function

$$\text{arcsine: } [-1, 1] \to \left[-\frac{\pi}{2}, \frac{\pi}{2}\right].$$

That is,

(2.5.1) $$-\frac{\pi}{2} \leq \arcsin(y) \leq \frac{\pi}{2} \text{ for all } -1 \leq y \leq 1.$$

The derivative of arcsine can be found from an application of Theorem 2.4.3. If $y = \sin x$ so that $x = \arcsin(y)$, we have

(2.5.2) $$\frac{d}{dy}\arcsin(y) = \frac{1}{d/dx\,(\sin x)} = \frac{1}{\cos x}.$$

A more useful expression for this derivative can be found if we note that, since $\cos x \geq 0$ for all x in this interval,

(2.5.3) $$\cos x = \sqrt{1 - \sin^2 x}.$$

But $\sin x = y$, so (2.5.3) can be written

(2.5.4) $$\cos x = \sqrt{1 - y^2},$$

and, by (2.5.2) and (2.5.4), we have

(2.5.5) $$\frac{d}{dy}\arcsin(y) = \frac{1}{\sqrt{1 - y^2}}.$$

If in dealing with arcsine we wish to observe our usual convention for function notation, then we must denote the independent variable by x. This notation convention is not a rule, of course, but when we consider the graph of the function arcsine, it *is* a rule that we use numbers in the domain as abscissas and numbers in the range as ordinates. By far the easiest way to conform to these conventions is to preserve our habit of denoting abscissas by x and ordinates by y, simply interchanging the x and y in the expression $x = \arcsin(y)$.

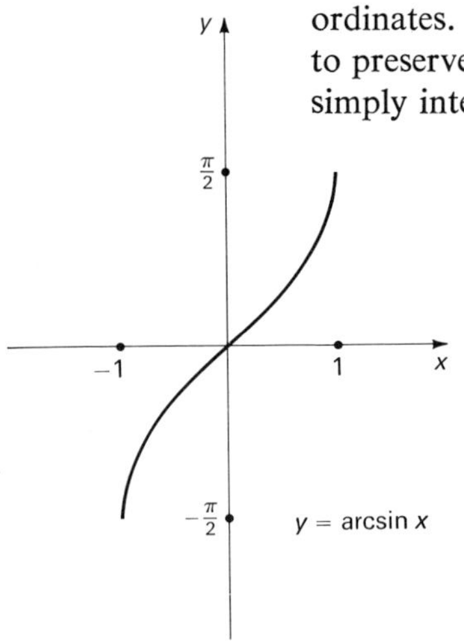

Figure 2.5.2

Then (2.5.1) and (2.5.5) become

(2.5.6)
$$-\frac{\pi}{2} \leq \arcsin(x) \leq \frac{\pi}{2} \text{ for all } -1 \leq x \leq 1,$$
$$\frac{d}{dx}\arcsin(x) = \frac{1}{\sqrt{1-x^2}}.$$

The graph of $y = \arcsin(x)$ is just the mirror image with respect to the line $y = x$ of the graph of $x = \arcsin(y)$. But the latter equation is the same as $y = \sin(x)$, whose graph is just the graph of the restricted sine function shown in Figure 2.5.1. Thus, the graph of arcsine is just the reflection of the graph of the restricted sine. It is shown in Figure 2.5.2.

Arccosine Since $d/dx(\cos x) = -\sin x$, we see that

$$\frac{d}{dx}\cos x < 0 \text{ for } 0 < x < \pi.$$

Thus, cosine is monotone (decreasing) on the closed interval $[0, \pi]$ and maps this interval onto the range $[-1, 1]$ of the cosine function. So we consider the *restricted cosine function*

$$\cos x,\ 0 \leq x \leq \pi,$$

whose graph is shown in Figure 2.5.3. We thus have a decreasing inverse function

$$\text{arccosine: } [-1, 1] \to \left[-\frac{\pi}{2}, \frac{\pi}{2}\right].$$

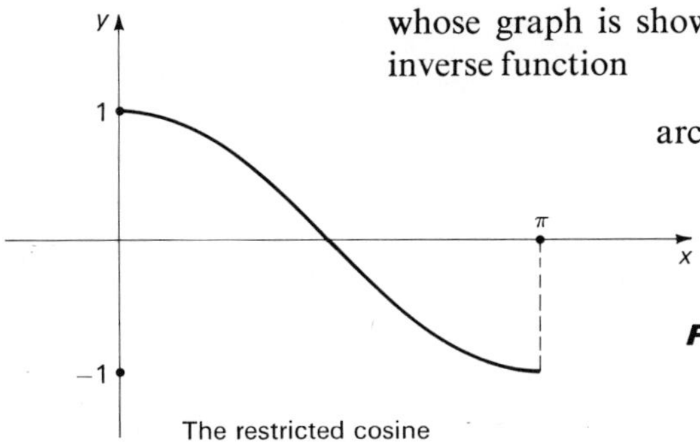

The restricted cosine

Figure 2.5.3

By Theorem 2.4.3 again we have

$$\frac{d}{dy}\arccos(y) = \frac{1}{d/dx(\cos x)} = \frac{-1}{\sin x}.$$

For $0 \leq x \leq \pi$ we have $\sin x \leq 0$, so

$$\sin x = -\sqrt{1-\cos^2 x} = -\sqrt{1-y^2};$$

and

$$\frac{d}{dy}\arccos(y) = \frac{-1}{\sqrt{1-y^2}}.$$

Reversing, as before, the roles of x and y, we then have for the function arccosine

(2.5.7)
$$0 \leq \arccos(x) \leq \pi \quad \text{for all} \quad -1 \leq x \leq 1,$$
$$\frac{d}{dx} \arccos(x) = \frac{-1}{\sqrt{1 - x^2}}.$$

The graph of arccosine, the mirror image with respect to the line $y = x$ of the graph of the restricted cosine, is shown in Figure 2.5.4.

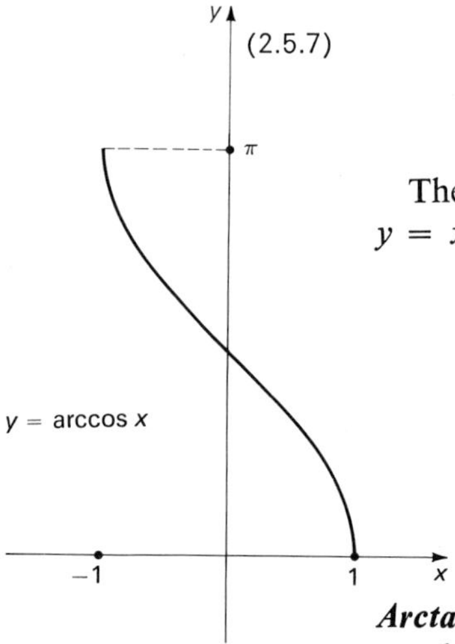

Figure 2.5.4

Arctangent The range of the tangent function is the set of all real numbers. The function is increasing on the open interval $(-\pi/2, \pi/2)$ and maps this open interval onto the entire range. Therefore, an appropriate *restricted tangent* function is

$$\tan x, \ -\frac{\pi}{2} < x < \frac{\pi}{2}.$$

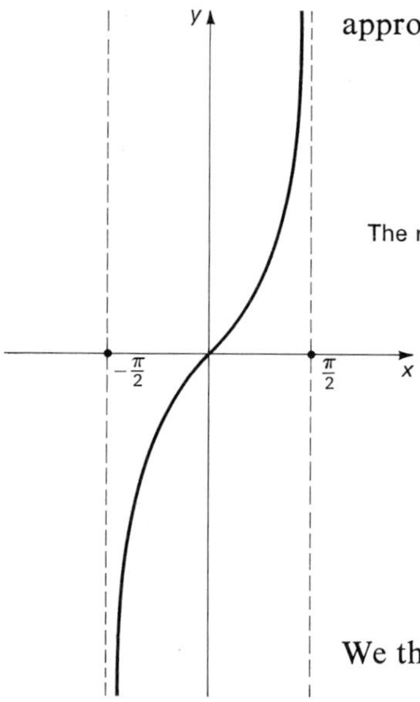

The restricted tangent

Figure 2.5.5

We thus have an increasing inverse function

(2.5.8)
$$\text{arctangent: } (-\infty, \infty) \to \left(-\frac{\pi}{2}, \frac{\pi}{2}\right).$$

The derivative can be obtained from the equations

$$\frac{d}{dy} \arctan(y) = \frac{1}{d/dx\,(\tan x)} = \frac{1}{\sec^2 x}.$$

But $\sec^2 x = 1 + \tan^2 x = 1 + y^2$, so we have

(2.5.9)
$$\frac{d}{dy}\arctan(y) = \frac{1}{1+y^2}.$$

Interchanging x and y and repeating (2.5.8) and (2.5.9), we have

(2.5.10)
$$-\frac{\pi}{2} < \arctan(x) < \frac{\pi}{2} \text{ for all } -\infty < x < \infty,$$
$$\frac{d}{dx}\arctan(x) = \frac{1}{1+x^2}.$$

The graph of arctangent, the mirror image of the graph in Figure 2.5.5, is shown in Figure 2.5.6.

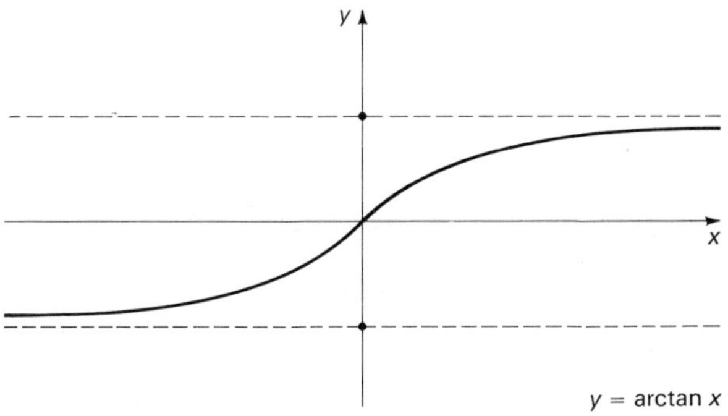

Figure 2.5.6

As we have said, part of the interest in these three new functions lies in the form of their derivatives. A glance at (2.5.6), (2.5.7), (2.5.10) reveals that we have three new integration formulas to add to our list. They are worth noting because the functions $(1 - x^2)^{-1/2}$ and $(1 + x^2)^{-1}$ are particularly simple functions which are closely related to a great many algebraic functions.

For convenient reference we list here in one place the new derivative and integral formulas.

$$\frac{d}{dx}\arcsin(x) = \frac{1}{\sqrt{1-x^2}},$$
$$\frac{d}{dx}\arccos(x) = \frac{-1}{\sqrt{1-x^2}},$$
$$\frac{d}{dx}\arctan(x) = \frac{1}{1+x^2},$$
$$\int \frac{dx}{\sqrt{1-x^2}} = \arcsin(x) = -\arccos(x),^1$$
$$\int \frac{dx}{1+x^2} = \arctan(x).$$

[1] *Following the convention established in* C 1, *$\int f(x)\,dx$ denotes* any *antiderivative of $f(x)$. Hence a formula such as this is to be understood to say that $-\arccos(x)$ is* an *antiderivative of $1/\sqrt{1-x^2}$.*

Exercises

1 Evaluate

(a) $\arcsin\left(\frac{\sqrt{3}}{2}\right)$, (b) $\arccos(0)$, (c) $\arctan(-1)$,
(d) $\arcsin(-\frac{1}{2})$, (e) $\arctan(\sqrt{3})$, (f) $\operatorname{arcsec}(2)$,
(g) $\operatorname{arccot}(1)$, (h) $\arcsin\left(-\frac{1}{\sqrt{2}}\right)$, (i) $\operatorname{arcsec}(-1)$,
(j) $\arccos(\frac{1}{2})$, (k) $\arccos(-\frac{1}{2})$, (l) $\arctan\left(-\frac{\sqrt{3}}{3}\right)$.

2 Suppose $y = \arcsin(\frac{1}{3})$. Find (a) $\cos y$, (b) $\tan y$, (c) $\sec y$, (d) $\cot y$, (e) $\csc y$.

3 Suppose $y = \arctan(-2)$. Find (a) $\sin y$, (b) $\cos y$, (c) $\cot y$, (d) $\sec y$, (e) $\csc y$.

4 Evaluate

(a) $\tan\left(\arcsin\left(\frac{1}{\sqrt{2}}\right)\right)$, (b) $\cot(\arccos(-\frac{1}{2}))$,
(c) $\csc(\arctan(-1))$, (d) $\sin(\arccos(x))$,
(e) $\cos\left(\arctan\left(\frac{1}{x}\right)\right)$, (f) $\tan(\operatorname{arcsec}(1+x))$.

5 Differentiate

(a) $f(x) = \arcsin(3x)$, (b) $f(x) = \arctan(-4x)$,
(c) $F(x) = \arccos(\sqrt{x})$, (d) $f(t) = t\arcsin(\sqrt{t} - 1)$,
(e) $f(x) = \arctan\left(\frac{x}{1-x^2}\right)$, (f) $s(t) = \arcsin\left(\frac{1}{t+2}\right)$,
(g) $F(y) = \frac{\arccos(\sqrt{y})}{y}$, (h) $f(x) = \log(\arctan(x))$,
(i) $f(x) = \arcsin(2x) + \arccos(-\frac{1}{2}x)$,
(j) $f(t) = e^t \arcsin(t^2)$.

6 Find $\partial F/\partial x$ and $\partial F/\partial y$ if

(a) $F(x, y) = x\arcsin(2y)$, (b) $F(x, y) = \arctan(x^2 + xy)$,
(c) $F(x, y) = \arccos(x\tan y)$, (d) $F(x, y) = \cos(x\arctan(y))$,
(e) $F(x, y) = \frac{\arcsin(xy)}{x^2 + y^2}$, (f) $F(x, y) = \sqrt{x}\arccos(\sqrt{xy})$.

7 Define the restricted secant function and derive a formula for $d/dx\,(\operatorname{arcsec}(x))$. (Choosing the proper restricted domain is not so routine here as in the earlier cases. Try using the union of $[0, \pi/2)$ and $[\pi, 3\pi/2)$. This is the range of arcsec. What is its domain?)

8 Define the restricted cotangent function and derive a formula for $d/dx\,(\operatorname{arccot}(x))$. (The range of arccot is $(0, \pi)$.)

9 Prove that $f(x) = \arcsin(x) + \arccos(x)$ is a constant function by considering its derivative. What is the constant?

10 Find dy/dx if $x\sin y + x^3 = \arctan(y)$.

11 Prove that $\arctan(x) + \arctan(1/x) = \begin{cases} \pi/2 & \text{if } x > 0, \\ -\pi/2 & \text{if } x < 0. \end{cases}$

12 A picture a feet high is placed on a wall with its base b feet above the level of an observer's eye. If the observer stands x feet from the wall, show that the angle of vision α subtended by the picture is

$$\alpha = \arctan\left(\frac{a+b}{x}\right) - \arctan\left(\frac{b}{x}\right).$$

How far from the wall should the observer stand in order to have a maximum angle of vision?

integrals revisited

chapter 3

3.0

Introduction

In *C 1* you were introduced to two concepts bearing the name "integral." The definitions of the definite integral and the indefinite integral are unrelated, yet the two concepts are very closely related, the connection being given by the Fundamental Theorem of the Calculus. In this section we shall extend our discussion of both of these concepts.

In sections 3.1, 3.2 we shall add to our collection of techniques for finding indefinite integrals. The trigonometric substitutions in Section 3.1 are suggested by the integral formulas found at the end of the last chapter. This discussion extends the one begun in Section 8.3 of *C 1*. The possibility of finding indefinite integrals for all rational functions, which is explored in Section 3.2, also depends in part on these formulas. Indefinite integrals form a theoretical bridge between differential calculus (calculus pertaining to derivatives) and integral calculus (pertaining to definite integrals), and also provide a bridge between the previous chapter and the present one.

In Section 3.3 we present a new way of looking at definite integrals: A definite integral is expressed as a certain kind of limit. In Section 3.4 we see how this method clears up a troublesome aspect of one of our applications of integrals in *C 1*.

In Section 3.5 we reexamine another application of the definite integral from a new point of view, and the discussion leads to still another representation of an integral as a kind of limit. This representation is discussed further in Section 3.6.

In Section 3.7 we show how these new representations of integrals lead to the solution of some unusual problems involving the physical quantity work.

Finally, in Section 3.8 we consider an extension of the notion of definite integral to certain functions which are not integrable in the strict sense of our present definition.

3.1

Trigonometric Substitutions

In Section 2.5 we discovered that the inverse trigonometric function arcsine has an algebraic derivative. Moving in the opposite direction we obtained the formula

(3.1.1) $$\int \frac{dx}{\sqrt{1 - x^2}} = \arcsin(x),$$

in which the inverse trigonometric function has surprisingly worked its way into an obviously nontrigonometric integral. Actually, we had stumbled onto this fact in *C 1*, Chapter 8, in connection with some examples of integration by substitution. At that time we could not explore very deeply this technique of integration because we kept encountering inverse trigonometric functions. We still didn't really know what they were or if they could be discussed in any systematic way.

Now that we are familiar with the needed inverse trigonometric functions, we can profitably return to the interesting possibility of reducing certain algebraic integrals to manageable form by means of trigonometric substitutions. For a start, look at (3.1.1) and let

$$\arcsin(x) = y,$$

which is to say,

(3.1.2) $$x = \sin y, \quad -\pi/2 \leq y \leq \pi/2.$$

We observe that formula (3.1.1) can be found directly by making the substitution (3.1.2) in the integral according to the methods presented in Chapter 8 of *C 1*. We write

$$x = \sin y \text{ and } dx = \cos y\, dy$$

and substitute in the integral in (3.1.1), remembering that since $-\pi/2 \leq y \leq \pi/2$, $\sqrt{1 - \sin^2 x} = \cos x$. Thus,

$$\int \frac{dx}{\sqrt{1 - x^2}} = \int \frac{\cos y\, dy}{\cos y} = \int dy = y.$$

Since $x = \sin y$, $y = \arcsin(x)$, and we obtain (3.1.1).
Consider the formula

$$\int \frac{dx}{1 + x^2} = \arctan(x),$$

from Section 2.5. If we set $y = \arctan(x)$, so that $-\pi/2 < y < \pi/2$, we see how the formula could have been derived by means of a trigonometric substitution. We would substitute in the integral

$$x = \tan y, \ -\pi/2 < y < \pi/2,$$

$$dx = \sec^2 y\, dy.$$

Recalling that $1 + x^2 = 1 + \tan^2 y = \sec^2 y$, we see that

$$\int \frac{dx}{1 + x^2} = \int \frac{\sec^2 y\, dy}{\sec^2 y} = \int dy = y = \arctan(x).$$

Substitutions of this type prove fruitful in many integrals more complicated than the above. A general rule of thumb is

1. In an integral containing $\sqrt{a^2 - x^2}$, substitute

$$x = a \sin y.$$

(3.1.3) 2. In an integral containing $\sqrt{x^2 - a^2}$, substitute

$$x = a \sec y.$$

3. In an integral containing $a^2 + x^2$, substitute

$$x = a \tan y.$$

The result is usually a purely trigonometric integral in the variable y. The substitution (3.1.3) is successful if this trigonometric integral can be calculated. A large number of ingenious techniques for evaluating trigonometric integrals issue from the rich supply of trigonometric identities. Thus, for the appropriate integrals, substitutions of the form (3.1.3) are usually successful. We shall not systematically discuss the common techniques for evaluating trigonometric integrals, but we shall illustrate a few of them in the following examples. You will find that most trigonometric integrals which result from such substitutions can be worked out by the use of the technique of integration by parts and the formulas for the integrals of the basic trigonometric functions.

Example 1 Consider the integral

(3.1.4) $$\int \frac{dx}{\sqrt{x^2 - 1}}.$$

We have suggested in (3.1.3)

(3.1.5) $$x = \sec y$$

as a favorable substitution. If we wish to calculate (3.1.4) for $x > 1$, then we may assume that $0 < y < \pi/2$. (Why?)

A simple mnemonic device, first mentioned in Section 8.3 of *C 1*, will help us keep track of most of the identities we shall need. The device is the right triangle in Figure 3.1.1. in which one acute angle has radian measure y and the sides are labeled so that (3.1.5) holds.

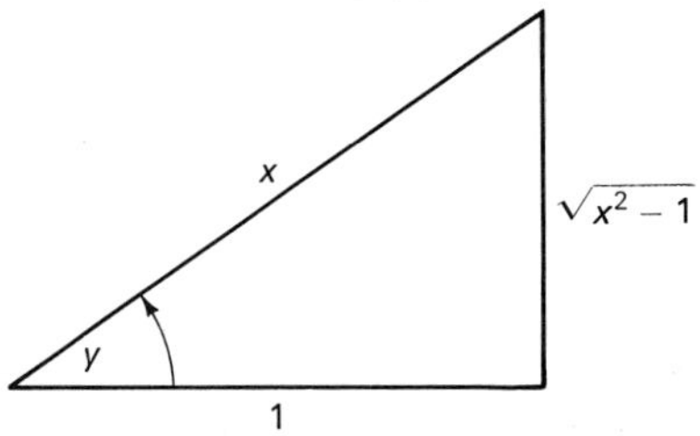

Figure 3.1.1

From (3.1.5) and the figure we see immediately that

(3.1.6) $$dx = \sec y \tan y \, dy, \qquad \sqrt{x^2 - 1} = \tan y,$$

so that

(3.1.7) $$\int \frac{dx}{\sqrt{x^2 - 1}} = \int \frac{\sec y \tan y \, dy}{\tan y} = \int \sec y \, dy.$$

We know from a previous formula that

(3.1.8) $$\int \sec y \, dy = \log (\sec y + \tan y),$$

so we have an expression for the integral (3.1.4).

We obviously desire an expression in x for the integral (3.1.4), and this can be found easily. In fact, we need only glance at (3.1.5) and (3.1.6) to see that

(3.1.9) $$\log (\sec y + \tan y) = \log (x + \sqrt{x^2 - 1}).$$

Thus, by (3.1.9), (3.1.8), (3.1.7),

$$\int \frac{dx}{\sqrt{x^2 - 1}} = \log (x + \sqrt{x^2 - 1}).$$

This result is easily checked by differentiation.

Example 2 Consider

$$\int \sqrt{4 - x^2}\, dx.$$

Here following (3.1.3) we try

(3.1.10) $$x = 2 \sin y, \quad -\frac{\pi}{2} \le y \le \frac{\pi}{2}.$$

The appropriate diagram for this substitution is given in Figure 3.1.2.

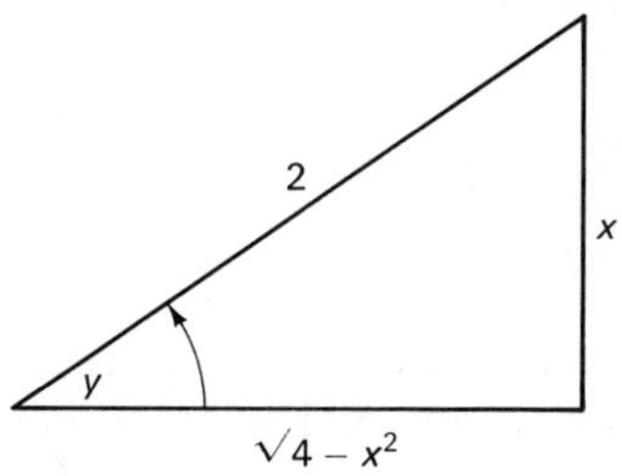

Figure 3.1.2

We see that

$$dx = 2 \cos y\, dy,$$
$$\sqrt{4 - x^2} = 2 \cos y,$$

so

(3.1.11) $$\int \sqrt{4 - x^2}\, dx = 4 \int \cos^2 y\, dy.$$

For an even power of sine or cosine, integration by parts always works. Here we split off one $\cos y$, combine it with dy to form $d(\sin y)$, and write

$$\int \cos^2 y\, dy = \int \cos y\, d(\sin y) = \sin y \cos y + \int \sin^2 y\, dy,$$

which, since $\sin^2 y = 1 - \cos^2 y$, may be written

(3.1.12) $$\int \cos^2 y\, dy = \sin y \cos y + y - \int \cos^2 y\, dy.$$

In *C 1* we saw that in an equation like (3.1.12) it is valid to add $\int \cos^2 y\, dy$ to both sides, writing

(3.1.13) $$2 \int \cos^2 y\, dy = \sin y \cos y + y$$

and finally,

$$\int \cos^2 y\, dy = \tfrac{1}{2}(\sin y \cos y + y).$$

By (3.1.10), we see that $y = \arcsin(x/2)$, and reading the expressions in terms of x for $\sin y$ and $\cos y$ from Figure 3.1.2, we have

$$\frac{1}{2}(\sin y \cos y + y) = \frac{1}{2}\left(\frac{x\sqrt{4 - x^2}}{4} + \arcsin\left(\frac{x}{2}\right)\right).$$

Thus, by (3.1.11),

$$\int \sqrt{4 - x^2}\, dx = \frac{x\sqrt{4 - x^2}}{2} + 2 \arcsin\left(\frac{x}{2}\right).$$

This, too, may be checked by differentiation.

Example 3 The integral

$$\int \frac{dx}{x\sqrt{1 - x^2}}$$

calls for the substitution $x = \sin y$ because of the presence of $\sqrt{1 - x^2}$. Hence we have $dx = \cos y\, dy$ and $\sqrt{1 - x^2} = \cos y$. Thus

$$\int \frac{dx}{x\sqrt{1 - x^2}} = \int \frac{\cos y\, dy}{\sin y \cdot \cos y} = \int \frac{dy}{\sin y} = \int \csc y\, dy.$$

In *C 1* we saw that

$$\int \csc y\, dy = -\log(\csc y + \cot y),$$

and we see from Figure 3.1.3 that

$$\log(\csc y + \cot y) = \log\left(\frac{1}{x} + \frac{\sqrt{1 - x^2}}{x}\right) = \log\left(\frac{1 + \sqrt{1 - x^2}}{x}\right).$$

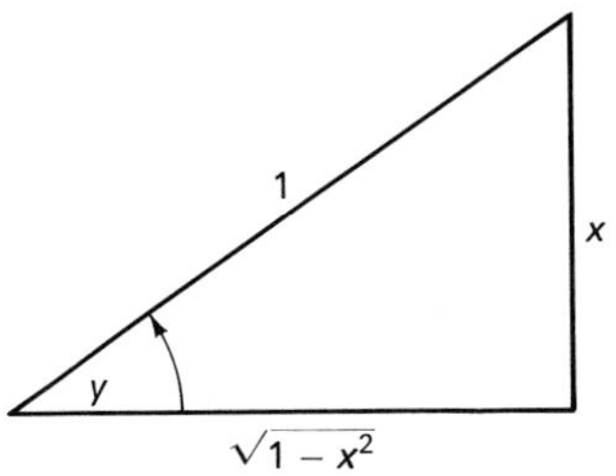

Figure 3.1.3

Thus

(3.1.14) $$\int \frac{dx}{x\sqrt{1 - x^2}} = -\log\left(\frac{1 + \sqrt{1 - x^2}}{x}\right) = \log\left(\frac{x}{1 + \sqrt{1 - x^2}}\right).$$

If in (3.1.14) we rationalize the denominator in $x/(1 + \sqrt{1 - x^2})$ by multiplying and dividing by $1 - \sqrt{1 - x^2}$, we obtain

$$\int \frac{dx}{x\sqrt{1-x^2}} = \log\left(\frac{1-\sqrt{1-x^2}}{x}\right).$$

Example 4 The area of the figure in the first quadrant bounded by the axes, the graph of $y^2 - x^2 = 1$, and the line $x = 1$ (Figure 3.1.4) is

(3.1.15) $$A = \int_0^1 \sqrt{1+x^2}\,dx.$$

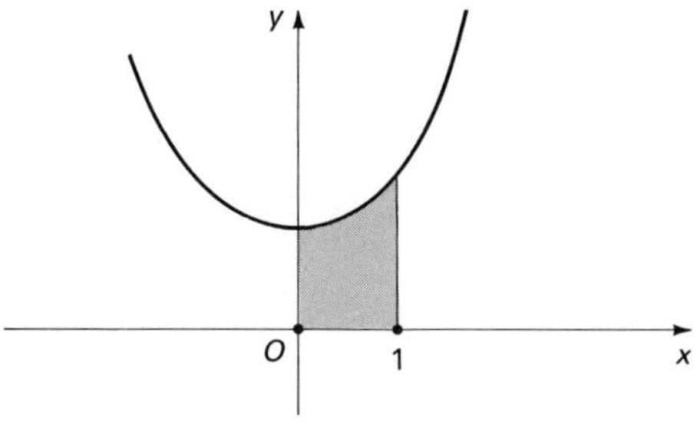

Figure 3.1.4

Let us calculate A. By (3.1.3) we are led to try the substitution

$$x = \tan y, \quad -\frac{\pi}{2} < y < \frac{\pi}{2}.$$

For this substitution we use the reference triangle in Figure 3.1.5, from which we see that

$$dx = \sec^2 y\,dy,$$
$$\sqrt{1+x^2} = \sec y.$$

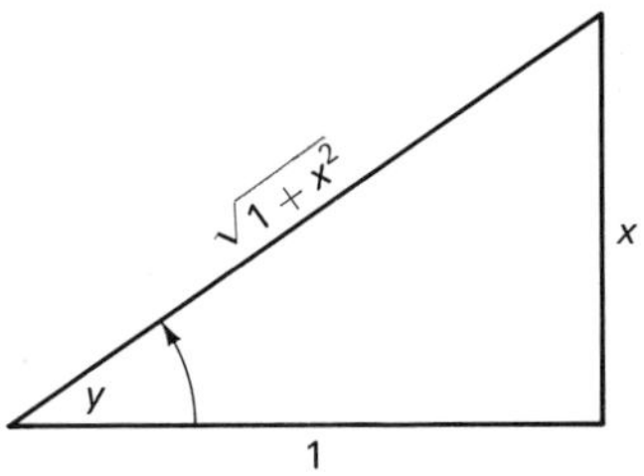

Figure 3.1.5

Since we are dealing with a definite integral, we must change the limits as well as the integrand in making the substitution. Since when $x = 0$, $y = \arctan 0 = 0$ and when $x = 1$, $y = \arctan(1) = \pi/4$, we have

(3.1.16) $$\int_0^1 \sqrt{1+x^2}\,dx = \int_0^{\pi/4} \sec^3 y\,dy.$$

Odd powers of secant or cosecant can be integrated by parts. Here the appropriate observation is that $\sec^2 y\,dy = d(\tan y)$. Thus

(3.1.17) $$\int_0^{\pi/4} \sec^3 y\,dy = \int_0^{\pi/4} \sec y\,d(\tan y)$$

$$= \sec y \tan y \Big|_0^{\pi/4} - \int_0^{\pi/4} \tan^2 y \sec y \, dy.$$

The difference $\sec y \tan y \mid_0^{\pi/4}$ is $\sqrt{2}$, and the integrand $\tan^2 y \sec y$ may be rewritten $\sec^3 y - \sec y$. Thus, (3.1.17) becomes

(3.1.18) $$\int_0^{\pi/4} \sec^3 y \, dy = \sqrt{2} - \int_0^{\pi/4} \sec^3 y \, dy + \int_0^{\pi/4} \sec y \, dy.$$

The rest is clear. Combining the like terms in (3.1.18) and evaluating the remaining integral by formula, we obtain

(3.1.19) $$\int_0^{\pi/4} \sec^3 y \, dy = \tfrac{1}{2}[\sqrt{2} + \log(\sqrt{2} + 1)].$$

By (3.1.15), (3.1.16), and (3.1.19), then, the required area is

$$A = \tfrac{1}{2}[\sqrt{2} + \log(\sqrt{2} + 1)].$$

Exercises

Evaluate the following integrals.

1 $\int \sqrt{9 - x^2} \, dx.$

2 $\int \sqrt{9 + x^2} \, dx.$

3 $\int \sqrt{x^2 - 16} \, dx.$

4 $\int_0^1 \frac{dx}{\sqrt{4 - x^2}}.$

5 $\int \frac{dx}{\sqrt{1 + (x + 2)^2}}.$

6 $\int \frac{dx}{\sqrt{x^2 - 2x}}.$

7 $\int_0^{\sqrt{5}} x^2 \sqrt{5 - x^2} \, dx.$

8 $\int x^3 \sqrt{5 + x^2} \, dx.$

9 $\int \frac{dx}{x\sqrt{x^2 - 9}}.$

10 $\int_{\sqrt{3}}^{3\sqrt{3}} \frac{dx}{x^2 \sqrt{x^2 + 9}}.$

11 $\int \frac{dx}{(x^2 - 4)^{3/2}}.$

12 $\int_0^a \frac{dx}{(a^2 + x^2)^{3/2}}.$

13 $\int \frac{\sqrt{x^2 - a^2}}{x} \, dx.$

14 $\int_1^3 \frac{dx}{x^4 \sqrt{x^2 + 3}}.$

15 $\int \frac{dx}{(4x^2 - 4x + 5)^{3/2}}.$

16 $\int \sqrt{4x - x^2} \, dx.$

3.2

Integration of Rational Functions

The techniques of the last section make it possible to integrate any rational function. To see this we must first borrow some facts from algebra which will enable us to reduce any rational function to a form in which these techniques apply.

A rational function is, you will recall, a function f such that $f(x)$ may be expressed as the ratio of two polynomials,

(3.2.1)
$$f(x) = \frac{a_0 + a_1x + \cdots + a_nx^n}{b_0 + b_1x + \cdots + b_mx_m}.$$

If n, the degree of the numerator, is greater than or equal to m, the degree of the denominator, then by a simple process of division we may express $f(x)$ as

$$f(x) = p(x) + \frac{r(x)}{q(x)},$$

where $p(x)$, $r(x)$, $q(x)$ are polynomials and the degree of $r(x)$ is *less than* the degree of $q(x)$. Since we may easily integrate the polynomial $p(x)$ by previous formulas, the problem of integrating $f(x)$ reduces to the problem of integrating a rational function g of the form

(3.2.2)
$$g(x) = \frac{r(x)}{q(x)}, \text{ degree of } r(x) < \text{degree of } q(x).$$

A special form of what is usually called the Fundamental Theorem of Algebra asserts that the polynomial $q(x)$ may always be factored into a product of terms of the following two forms:

(3.2.3)
$$(ax + b)^r,\ (ax^2 + bx + c)^s,$$

where r and s are positive integers and the coefficients a, b, c are real numbers. In other words, a polynomial can always be factored into real linear and quadratic factors. We assume that all occurrences of a given linear factor $ax + b$ or quadratic factor $ax^2 + bx + c$ are collected together in one term of the appropriate form in (3.2.3).

A rational function of the form (3.2.2) with its denominator factored into terms of the form (3.2.3) can always be expressed as a sum of *partial fractions*. The expression $r(x)/q(x)$ is a sum of a certain number of simpler fractions, which we shall call partial fractions, of the following types:

For *each* factor $(ax + b)^r$ there are r partial fractions forming a partial sum of the form

(3.2.4)
$$\frac{A_1}{ax + b} + \frac{A_2}{(ax + b)^2} + \frac{A_3}{(ax + b)^3} + \cdots + \frac{A_r}{(ax + b)^r},$$

where $A_1, A_2, \ldots, A_r$ are constants.

For *each* factor $(ax^2 + bx + c)^s$ there are s terms forming a partial sum of the form

(3.2.5) $$\frac{B_1x + C_1}{ax^2 + bx + c} + \frac{B_2x + C_2}{(ax^2 + bx + c)^2} + \cdots + \frac{B_sx + C_s}{(ax^2 + bx + c)^s},$$

where all of the B's and C's are constants.

Thus, a rational function of the form (3.2.2) is expressed as a sum of partial fractions when it is written as a sum of K terms, where K is the total number of linear and quadratic factors of $q(x)$, each factor being counted as many times as it occurs in the factorization of $q(x)$. The terms are of the form displayed in (3.2.4) and (3.2.5). The proof of the theorem from algebra which asserts this fact is not difficult, but it would be a greater digression than we can afford at this point.[1] We shall illustrate the calculation of this partial fraction representation of a rational function in several examples below.

[1] *The Fundamental Theorem of Algebra is discussed in many texts on abstract algebra. See, for example,* Topics in Algebra *by I. Herstein. The theorem on partial fractions follows from the fact that the constants A_1, A_2, etc. in (3.2.4) are solutions of a set of linear equations, where the number of equations is equal to or greater than the number of constants.*

But first let us examine where this partial fraction expression leads us. It certainly tells us that we can find the integral $\int f(x)\,dx$ of the rational function in (3.2.1) if we can find all integrals of the following two kinds:

(3.2.6) $$\int \frac{dx}{(ax + b)^r}, \quad \int \frac{(Bx + C)\,dx}{(ax^2 + bx + c)^s}.$$

The first of the integrals (3.2.6) presents no difficulty. It can readily be transformed into one of the form $(1/a)\int du/u^r$ by a substitution $u = ax + b$.

The second integral can be further broken down. It should be clear to you that by an appropriate choice of constants we may write the numerator $Bx + C$ as

$$Bx + C = D(2ax + b) + E,$$

where the factor $2ax + b$ is the derivative of the quadratic $ax^2 + bx + c$ in the denominator. Thus, we may write this second integral as a sum

(3.2.7) $$D\int \frac{(2ax + b)\,dx}{(ax^2 + bx + c)^s} + E\int \frac{dx}{(ax^2 + bx + c)^s}.$$

The first of the integrals in (3.2.7) can immediately be put into the form $\int dv/v^s$ and integrated. We thus see that *all* rational functions may be integrated provided that we can find any integral of the form

(3.2.8) $$\int \frac{dx}{(ax^2 + bx + c)^s},$$

where s is a positive integer. This is where the techniques of the last section come in.

We learned in Section 3.1 that an integral of the form

(3.2.9) $$\int \frac{du}{(1 \pm u^2)^s}$$

may be evaluated by a substitution of either $u = \tan y$ or $u = \sin y$ (see (3.1.3)). In one case we end up having to integrate an odd power of $\sec y$, in the other an even power of $\cos y$. Both may be handled by integration by parts (repeated several times, if necessary).

By completing the square and factoring out the appropriate constants, any integral (3.2.8) may be put in the form (3.2.9), where u is a linear expression in x. We hope that you are already somewhat familiar with the process of completing the square, and instead of discussing in detail this final reduction, we shall illustrate it in several specific cases below.

Before turning to some examples of the application of these techniques, we pause to comment on the significance of what we are doing. The reader will quickly recognize the limitations of these ideas as a practical approach to integrating rational functions. Knowing that any real polynomial *can* be factored into linear and quadratic factors is not much comfort to the person confronted with the integral

$$\int \frac{.53x + 1.7}{x^5 + 1.1x^3 + 2.8}\,dx.$$

He would certainly prefer to look in one of the many tables of standard integrals that are available. Or, if what he was really interested in was the definite integral

$$\int_1^2 \frac{.53x + 1.7}{x^5 + 1.1x^3 + 2.8}\,dx,$$

he would probably resort to a numerical approximation, perhaps of the sort discussed in Section 4.1.

But when one is faced by a rational function whose denominator can be factored, these techniques can be very helpful. Furthermore, it is of considerable intellectual interest (and is perhaps the main reason for discussing these ideas) to know that *any* rational function may be obtained as the derivative of some function built up from the classical elementary functions—polynomials, trigonometric and inverse trigonometric functions, and log—by algebraic combination and composition. For that is what we learn from the observed fact that any rational function may be integrated by a combination of *division* (to reduce the degree of the numerator below that of the denominator), *reduction to partial fractions*, *completing squares* in any quadratic factors in the denominator, and *algebraic or trigonometric substitutions*. Finally, as we learn from evaluating a reasonably complete list of trigonometric integrals, *all* of the above elementary functions are necessary for integration of rational functions. We now proceed to the promised examples.

Example 1 We wish to calculate

(3.2.10) $$\int \frac{3(x+1)}{(x-1)(x+2)^2}\,dx.$$

We may write the integrand as the sum of partial fractions of the form

(3.2.11) $$\frac{3(x+1)}{(x-1)(x+2)^2} = \frac{A}{x-1} + \frac{B}{(x+2)} + \frac{C}{(x+2)^2}.$$

Multiplying both sides by $(x-1)(x+2)^3$ we obtain a polynomial identity

(3.2.12) $$3(x+1) \equiv A(x+2)^2 + B(x-1)(x+2) + C(x-1).$$

Since (3.2.12) must hold for all $x \neq 1$, $x \neq -2$ and polynomials are continuous functions, the equation holds for *all* x. Thus, the coefficients A, B, C are determined. We may find their values either by collecting like powers of x on the right and equating the resulting expressions to the corresponding coefficients on the left (thereby providing simultaneous equations which determine A, B, C) or by substituting strategically chosen values for x directly into (3.2.12): In this case, a combination of these methods leads most easily to the desired values of A, B, C.

If we set $x = 1$ in (3.2.12), we find

$$6 = 9A, \text{ or } A = \tfrac{2}{3}.$$

If we set $x = -2$ in (3.2.12), we find

$$-3 = -3C, \text{ or } C = 1.$$

Then, noticing that the coefficient of x^2 on the right is $A + B$, we see that

$$A + B = 0.$$

Thus

$$B = -A = -\tfrac{2}{3}.$$

Hence

$$\frac{3(x+1)}{(x-1)(x+2)^2} = \frac{2}{3(x-1)} - \frac{2}{3(x+2)} + \frac{1}{(x+2)^2}.$$

Each of these terms is readily integrated:

$$\int \frac{3(x+1)}{(x-1)(x+2)^2}\,dx = \frac{2}{3}\log(x-1) - \frac{2}{3}\log(x+2) - \frac{1}{x+2} = \frac{2}{3}\log\left(\frac{x-1}{x+2}\right) - \frac{1}{x+2}.$$

Example 2 Suppose we wish to find

(3.2.13)
$$\int \frac{2x^5 - 2x^4 + 2x^3 + 3}{2x^4 - 2x^3 - x^2 + 1}\,dx.$$

We first notice that the degree of the numerator exceeds the degree of the denominator. Thus, we first divide, obtaining

(3.2.14)
$$\frac{2x^5 - 2x^4 + 2x^2 + 3}{2x^4 - 2x^3 - x^2 + 1} = x + \frac{x^3 + 2x^2 - x + 3}{2x^4 - 2x^3 - x^2 + 1}.$$

The next step is to factor the denominator into linear and quadratic factors, making sure that none of the quadratic terms can be further factored into linear polynomials. Without much difficulty we find that

$$2x^4 - 2x^3 - x^2 + 1 = (x - 1)^2(2x^2 + 2x + 1).$$

We must therefore integrate

(3.2.15)
$$\frac{x^3 + 2x^2 - x + 3}{2x^4 - 2x^3 - x^2 + 1} = \frac{x^3 + 2x^2 - x + 3}{(x - 1)^2(2x^2 + 2x + 1)}.$$

This function may be written as a sum of partial fractions:

(3.2.16)
$$\frac{A}{(x - 1)} + \frac{B}{(x - 1)^2} + \frac{Cx + D}{2x^2 + 2x + 1}.$$

If we multiply both (3.2.15) and (3.2.16) by

$$(x - 1)^2(2x^2 + 2x + 1),$$

we obtain

(3.2.17)
$$\begin{aligned} x^3 + 2x^2 - x + 3 = {} & A(x - 1)(2x^2 + 2x + 1) \\ & + B(2x^2 + 2x + 1) + (Cx + D)(x - 1)^2. \end{aligned}$$

Now collecting like terms on the right we can rewrite (3.2.17) as

(3.2.18)
$$\begin{aligned} x^3 + 2x^2 - x + 3 = {} & (2A + C)x^3 + (2B - 2C + D)x^2 \\ & + (-A + 2B + C - 2D)x \\ & + (-A + B + D). \end{aligned}$$

Equation (3.2.18) is true for all x, so coefficients of like terms on the left and right must be the same. Thus

(3.2.19)
$$\begin{aligned} 2A + C &= 1, \\ 2B - 2C + D &= 2, \\ -A + 2B + C - 2D &= -1, \\ -A + B + D &= 3. \end{aligned}$$

The simultaneous equations (3.2.19) have solution

$$A = 0,\ B = 1,\ C = 1,\ D = 2.$$

Thus, we conclude that

$$\frac{x^3 + 2x^2 - x + 3}{(x-1)^2(2x^2 + 2x + 1)} = \frac{1}{(x-1)^2} + \frac{x+2}{2x^2 + 2x + 1}. \tag{3.2.20}$$

The first of these partial fractions can be integrated immediately. The second requires a bit more attention. Since the denominator has derivative $4x + 2$, we write

$$\begin{aligned} \frac{x+2}{2x^2+2x+1} &= \frac{1}{4}\frac{(4x+2)+6}{2x^2+2x+1} \\ &= \frac{1}{4}\frac{(4x+2)}{2x^2+2x+1} + \frac{3}{2}\frac{1}{2x^2+2x+1}. \end{aligned} \tag{3.2.21}$$

Now all but the very last term are in form for integration by formula. We must ready this last term for a trigonometric substitution by completing the square in the denominator. We find that

$$2x^2 + 2x + 1 = \tfrac{1}{2}[(2x+1)^2 + 1],$$

so that

$$\frac{3}{2}\frac{1}{2x^2+2x+1} = \frac{3}{(2x+1)^2+1}. \tag{3.2.22}$$

Retracing our steps through (3.2.22), (3.2.21), (3.2.20), (3.2.15), (3.2.14), we see that the desired integral (3.2.13) is

$$\int x\,dx + \int \frac{dx}{(x-1)^2} + \frac{1}{4}\int \frac{(4x+2)\,dx}{2x^2+2x+1} + \frac{3}{2}\int \frac{2\,dx}{(2x+1)^2+1}.$$

Each of these is readily calculated. The third is of the form $\frac{1}{4}\int du/u$ and the last is $\frac{3}{2}\int du/(u^2+1)$. Applying the appropriate formulas we find that

$$\int \frac{2x^5 - 2x^4 + 2x^2 + 3}{2x^4 - 2x^3 - x^2 + 1}\,dx = \frac{x^2}{2} - \frac{1}{x-1} + \frac{1}{4}\log(2x^2 + 2x + 1) + \frac{3}{2}\arctan(2x+1).$$

Exercises

Evaluate the integrals.

1 $\displaystyle\int \frac{2x^2 - 4x + 1}{x-2}\,dx.$ **2** $\displaystyle\int \frac{2x^4 - x^3 + 3}{2x-1}\,dx.$

3 $\displaystyle\int \frac{3x-4}{(x-1)(x-2)}\,dx.$ **4** $\displaystyle\int \frac{3\,dx}{(2x+1)(x+2)}.$

5 $\displaystyle\int \frac{2x^3 + 3x^2 - 37x + 24}{x^2 - 4x + 3}\,dx.$ **6** $\displaystyle\int \frac{x^2 - 2x - 1}{x^2 - 4x + 4}\,dx.$

7 $\displaystyle\int \frac{3x - 2}{(x + 1)(x + 2)(x - 1)}\,dx.$ **8** $\displaystyle\int \frac{x - 3}{(x + 2)(x - 1)^2}\,dx.$

9 $\displaystyle\int \frac{x^2 + 4x + 6}{(x + 1)^3}\,dx.$ **10** $\displaystyle\int \frac{5x^2 - 2x + 5}{(x^2 - 1)^2}\,dx.$

11 $\displaystyle\int \frac{x^3 + 5x}{x^2 + 2}\,dx.$ **12** $\displaystyle\int \frac{x^3 + 2x + 3}{x^2 + 2}\,dx.$

13 $\displaystyle\int \frac{dx}{x^2 + 2x + 2}.$ **14** $\displaystyle\int \frac{dx}{(x^2 + 1)^2}.$

15 $\displaystyle\int \frac{x^3 + 3x^2 + 4x + 9}{(x^2 + 3)(x^2 + 4)}\,dx.$ **16** $\displaystyle\int \frac{x^3 + 3x^2 - 2x + 1}{x^4 + 5x^2 + 4}\,dx.$

17 $\displaystyle\int \frac{x^2 - 2x + 3}{(x - 1)^2(x^2 + 4)}\,dx.$ **18** $\displaystyle\int \frac{4x^2 + 2}{(4x^2 + 1)^2}\,dx.$

19 $\displaystyle\int \frac{3x^2 - 8x + 17}{x^3 - 3x^2 + 7x - 5}\,dx.$ **20** $\displaystyle\int \frac{x^5 + x^4 + 3x^3 + 2x + 1}{(x + 1)^2(x^2 + 1)^2}\,dx.$

3.3

The Definite Integral as a Limit

The chapter assumes a rather different flavor as we turn now to definite integrals. In this section we shall see an alternate approach to the definition of a definite integral that is necessary for certain applications and can also shed new light on the basic notion itself. The new approach also significantly extends the limit concept considerably beyond its present boundaries, for it represents the definite integral as a kind of limit.

Let us first recall briefly what a definite integral is. The full discussion can be found in Chapter 6 of *C 1*. Let

$$P = \{a = x_0, x_1, \ldots, x_n = b\}$$

be a partition of the closed interval $[a, b]$. A function f that is constant on each subinterval (x_{i-1}, x_i), $i = 1, 2, \ldots, n$, is called *simple* relative to P. If $f(x) = k_i$ for $x_{i-1} < x < x_i$, then the definite integral of f on $[a, b]$ is defined as

$$\int_a^b f(x)\,dx = \sum_{i=1}^{n} k_i(x_i - x_{i-1}). \tag{3.3.1}$$

Now let f be an arbitrary function bounded on the closed interval $[a, b]$. Then there can always be found simple functions $\underline{f}$ and $\overline{f}$ which "straddle" f in the sense that

$$\underline{f}(x) \le f(x) \le \overline{f}(x) \text{ for all } x \text{ in } [a, b].$$

Simple functions straddling a nonsimple function are illustrated in Figure 3.3.1.

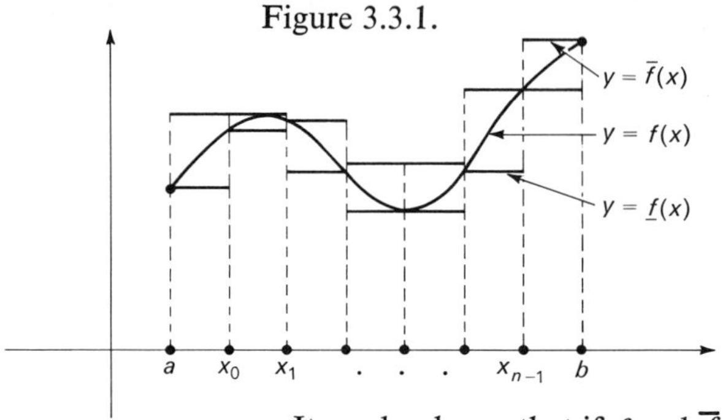

Figure 3.3.1

It can be shown that if $\underline{f}$ and $\bar{f}$ are any two simple functions which straddle f, then

$$\int_a^b \underline{f}(x)\,dx \le \int_a^b \bar{f}(x)\,dx.$$

From this fact and additional information about integrals of simple functions it can be shown that there must exist numbers I such that

$$\int_a^b \underline{f}(x)\,dx \le I \le \int_a^b \bar{f}(x)\,dx$$

for *all* simple functions $\underline{f}$ and $\bar{f}$ straddling f on $[a, b]$.

If it should happen that there is only *one* such number I, that is, if the simple functions lying above and below f determine a *unique* number I in this way, then f is said to be *integrable over* $[a, b]$. In this case we call I the *definite integral* of f and write

(3.3.2) $$I = \int_a^b f(x)\,dx.$$

In Chapter 6 of *C 1* we saw that every continuous function and every monotone function on $[a, b]$ is integrable. Thus, for large classes of functions definite integrals are known to exist. However, for a great many purposes—for most applications, for example—knowing that the integral exists is not of much use if one cannot actually calculate it.

Several methods are available for calculating definite integrals. One is given by the Fundamental Theorem, which asserts that if f is integrable and has an antiderivative F on $[a, b]$, then

$$\int_a^b f(x)\,dx = F(b) - F(a).$$

Thus, we can find the definite integral if we can find an indefinite integral. If we cannot find an antiderivative, then we are usually reduced to attempting to *approximate* $\int_a^b f(x)\,dx$ by one method or another. We have not said much about the possibilities of approximating $\int_a^b f(x)\,dx$ to a given degree of accuracy, except to note that by definition $\int_a^b f(x)\,dx$ may be approximated arbitrarily closely by integrals of simple functions, that is, by finite sums of the type (3.3.1). We shall discover our limit approach to the definite integral through a closer look at this method of approximation.

For the discussion which follows, it will be convenient to restrict our attention to functions which are continuous on $[a, b]$. In the first place, such functions are always integrable, so we shall always know that there *is* something to approximate. In the second place, continuous functions possess other desirable properties that will make our discussion easier. In the third place, the restriction is not serious, for almost all functions we shall encounter will be continuous. Nevertheless, we should be aware that the results that we shall obtain are valid (sometimes with minor rewording) for all integrable functions.

Let $P = \{a = x_0, x_1, \ldots, x_n = b\}$ be a partition of $[a, b]$. Then we may consider various simple functions relative to P that are below and above f. Clearly, the two such simple functions $\underline{f}$ and $\overline{f}$ whose integrals most closely approximate $\int_a^b f(x)\,dx$ are the functions

$$\begin{aligned} \underline{f}(x) &= m_i, \quad x \text{ in } [x_{i-1}, x_i), \ x = 1, 2, \ldots, n, \\ \overline{f}(x) &= M_i, \quad x \text{ in } [x_{i-1}, x_i), \ x = 1, 2, \ldots, n, \end{aligned}$$

where m_i is the minimum value of $f(x)$ on $[x_{i-1}, x_i]$ and M_i is the maximum value of $f(x)$ on $[x_{i-1}, x_i]$. The two integrals (finite sums) $\int_a^b \underline{f}(x)\,dx = \sum m_i(x_i - x_{i-1})$ and $\int_a^b \overline{f}(x)\,dx = \sum M_i(x_i - x_{i-1})$ are called the *lower sum* and *upper sum* respectively for f over $[a, b]$ relative to the partition P. We write

$$\begin{aligned} L(f, P) &= \sum_{i=1}^{n} m_i(x_i - x_{i-1}), \\ U(f, P) &= \sum_{i=1}^{n} M_i(x_i - x_{i-1}). \end{aligned}$$

Clearly, then,

(3.3.3) $$L(f, P) \leq \int_a^b f(x)\,dx \leq U(f, P),$$

for *any* partition P of $[a, b]$.

Note that both $L(f, P)$ and $U(f, P)$ are sums of the form

$$\sum_{i=1}^{n} f(x_i')(x_i - x_{i-1})$$

for some particular choice of number x_i' in the intervals $[x_{i-1}, x_i]$, for each $i = 1, 2, \ldots, n$. We shall be interested in *all* such sums, for *all* possible choices of points x_i' in the respective subintervals $[x_{i-1}, x_i]$.

Definition 3.3.1 If $P = \{a = x_0, x_1, \ldots, x_n = b\}$ is a partition of $[a, b]$ and x_i' is in $[x_{i-1}, x_i]$ for each $i = 1, 2, \ldots, n$, then the sum

$$\sum_{i=1}^{n} f(x_i')(x_i - x_{i-1})$$

is called a *Riemann sum for f relative to P*. We shall use the symbol $R(f, P)$ for any Riemann sum for f relative to P.

Note that $L(f, P)$ is a Riemann sum for f relative to P, the smallest possible Riemann sum for f relative to P. Similarly, $U(f, P)$ is the largest possible Riemann sum for f relative to P. That is,

(3.3.4) *If P is a partition of $[a, b]$ and $R(f, P)$ is any Riemann sum for f relative to P, then*

$$L(f, P) \leq R(f, P) \leq U(f, P).$$

By comparing (3.3.4) with (3.3.3) we can easily make one possible statement about approximating $\int_a^b f(x)\,dx$.

Lemma 3.3.1 *Let ϵ be a positive number. If P is a partition of $[a, b]$ such that*

$$|U(f, P) - L(f, P)| < \epsilon,$$

then any Riemann sum for f relative to P approximates $\int_a^b f(x)\,dx$ with an error of less than ϵ. That is,

$$\left| R(f, P) - \int_a^b f(x)\,dx \right| < \epsilon.$$

Proof By (3.3.4) and (3.3.3) we have

$$\left| R(f, P) - \int_a^b f(x)\,dx \right| \leq |U(f, P) - L(f, P)|.$$

The lemma then follows trivially.

As a method of approximating $\int_a^b f(x)\,dx$, Lemma 3.3.1 has one serious fault. Having chosen a partition P, one must calculate *two* Riemann sums L and U before knowing whether P will lead to a reasonable approximation. It would be much better if we had a criterion for determining when a given partition P will lead to a Riemann sum $R(f, \mathrm{P})$ which closely approximates $\int_a^b f(x)\,dx$ *before* going through the relatively laborious procedure of calculating both an upper sum and a lower sum.

We have. Moreover, the criterion is a simple and natural one: A very fine partition *guarantees* a good approximation, where a fine partition is one which divides $[a, b]$ into very short subintervals. In order to state this criterion and its conclusion precisely and to prove that our claim is justified we establish a quantitative measure of the fineness of a partition.

Definition 3.3.2 If $P = \{a = x_0, x_1, \ldots, x_n = b\}$ is a partition of $[a, b]$, the *norm* of P, written $|P|$ and sometimes called the *mesh-fineness* of P, is the length of the longest subinterval $[x_{i-1}, x_i]$, $i = 1, 2, \ldots, n$. That is,

$$|P| = \max |x_i - x_{i-1}|, i = 1, 2, \ldots, n.$$

Thus, a "fine" partition is one with a small norm $|P|$. The result stated above is that if $|P|$ is small, then *any* Riemann sum $R(f, P)$ constructed from this partition is close to $\int_a^b f(x)\,dx$. More precisely:

Theorem 3.3.2 *Let f be continuous on $[a, b]$. Then for every $\epsilon > 0$ there exists a $\delta > 0$ such that*

$$\left| R(f, P) - \int_a^b f(x)\,dx \right| < \epsilon \tag{3.3.5}$$

for any partition P with $|P| < \delta$ and any Riemann sum $R(f, P)$ relative to P.

A proof of this theorem under the stated hypotheses is elementary but quite messy. In order not to divert attention from the central issue by a lengthy and somewhat technical discourse, we shall give what amounts to a partial proof. We shall add to the hypothesis of continuity the slightly less innocent assumption that f is also differentiable with a bounded derivative on $[a, b]$. We are still left with the vast majority of functions which we are likely to encounter at this level, and in our first application of the theorem (in the next section) it will be necessary, for reasons unrelated to Theorem 3.3.2, to impose still stronger conditions.

Partial Proof Suppose there exists a constant B such that

$$|f'(x)| \leq B$$

for all x in $[a, b]$. Let $P = \{a = x_0, x_1, \ldots, x_n = b\}$. For any $i = 1, 2, \ldots, n$, let

$$m_i = f(u_i), \quad M_i = f(v_i),$$

where m_i and M_i are the minimum and maximum values of $f(x)$ on $[x_{i-1}, x_i]$. If we apply the Mean Value Theorem[1] to f on the interval

[1] *Theorem 4.1.3 in C 1: If f is differentiable on $[u, v]$, then there exists a number c in (u, v) such that $f(v) - f(u) = (v - u)f'(c)$.*

$[u_i, v_i]$ (or $[v_i, u_i]$ if $v_i < u_i$), we see that

$$M_i - m_i = |(u_i - v_i)f'(c_i)| \le (x_i - x_{i-1})B$$

for some c_i in (x_{i-1}, x_i). (Note that $|u_i - v_i| \le x_i - x_{i-1}$ and $|f'(c_i)| \le B$.) Thus, since $x_i - x_{i-1} \le |P|$ for all i,

(3.3.6)
$$M_i - m_i \le B|P|.$$

From (3.3.6) it follows easily that

(3.3.7)
$$U(f, P) - L(f, P) \le B|P|(b - a),$$

because $U(f, P) - L(f, P) = \sum(M_i - m_i)(x_i - x_{i-1})$, and

$$\sum(x_i - x_{i-1}) = b - a.$$

Now suppose that $\epsilon > 0$ is arbitrary. If $B|P|(b - a) < \epsilon$, then by (3.3.7) and Lemma 3.3.1 we see that (3.3.5) must hold. But $B|P|(b - a) < \epsilon$ is true if and only if

$$|P| < \frac{\epsilon}{B(b - a)}.$$

Thus, the number

$$\delta = \frac{\epsilon}{B(b - a)}$$

satisfies the requirements of the theorem and our partial proof is complete.

One cannot help noticing the similarity between the statement of Theorem 3.3.2 and the ϵ-δ definition of a limit given in Theorem 1.1.2. This leads to an interpretation of Theorem 3.3.2 as asserting that $\int_a^b f(x)\,dx$ is a *limit* of Riemann sums in a sense which is clarified in the following definition.

Definition 3.3.3 If for every $\epsilon > 0$ there exists a $\delta > 0$ such that

$$|R(f, P) - I| < \epsilon$$

for every partition P with $|P| < \delta$ and every Riemann sum $R(f, P)$ relative to P, then we say that

$$I = \lim_{|P| \to 0} R(f, P).$$

Thus, Theorem 3.3.2 asserts that

Theorem 3.3.3 *If f is continuous on $[a, b]$, then*

$$\int_a^b f(x)\,dx = \lim_{|P| \to 0} R(f, P).$$

This version of the theorem gives us the limit representation of a definite integral announced at the beginning of this section.

Note that Definition 3.3.3 is not just a repetition of the earlier ϵ-δ definition of a limit. The entire context has changed. We do not have here a function of $|P|$, $|P|$ being the independent variable which tends to zero. For any value of the norm there will be many partitions—infinitely many, in fact—all having precisely the fineness specified by the norm. And for any such partition P there are infinitely many possible Riemann sums, depending upon the choice of the numbers x_i' in the intervals $[x_{i-1}, x_i]$ for $i = 1, 2, \ldots, n$. Nevertheless the analogy between the statements in Theorems 3.3.2 and 1.1.2 is so nearly perfect that we are justified in describing Definition 3.3.3 as an extension of the limit concept to the present situation.

Exercises

1 Evaluate each of the following limits by considering Riemann sums for an appropriate function.

(a) $\lim\limits_{n\to\infty} \sum\limits_{i=1}^{n} \left(\frac{i}{n}\right)\frac{1}{n}$, (b) $\lim\limits_{n\to\infty} \frac{1}{n^3}\sum\limits_{i=1}^{n} i^2$,

(c) $\lim\limits_{n\to\infty} \frac{1}{n^6}\sum\limits_{i=0}^{n-1} i^5$, (d) $\lim\limits_{n\to\infty} \frac{\pi}{2n}\sum\limits_{i=1}^{n} \sin\left(\frac{i\pi}{2n}\right)$,

(e) $\lim\limits_{n\to\infty} \frac{1}{n}\sum\limits_{i=1}^{n} \left(\frac{i+n}{n}\right)^{1/2}$, (f) $\lim\limits_{n\to\infty} \sum\limits_{i=1}^{n} \frac{1}{i+n}$.

2 Express each of the following integrals as a limit of Riemann sums $R(f, P_n)$, where P_n is the regular partition of the interval of integration into n subintervals.

(a) $\int_0^1 x^3\,dx$, (b) $\int_0^1 e^x\,dx$,

(c) $\int_0^{\pi/2} x\cos x\,dx$, (d) $\int_0^2 e^{1/x}\,dx$,

(e) $\int_{-2}^{-1} \frac{1}{x^2}\,dx$, (f) $\int_0^1 \sqrt[3]{x}\,dx$.

3 Let f be continuous over $[a, b]$ and $\epsilon > 0$.

(a) Use the definition of integrable function to prove that there are two partitions P_1 and P_2 of $[a, b]$ which satisfy

$$U(f, P_1) - \int_a^b f(x)\,dx < \frac{\epsilon}{2}$$

and

$$\int_a^b f(x)\,dx - L(f, P_2) < \frac{\epsilon}{2}.$$

(b) Prove that there is a partition P of $[a, b]$ such that

$$U(f, P) - L(f, P) < \epsilon.$$

(*Hint:* Let P be a partition which is a refinement of both P_1 and P_2; that is, assume that every subdividing point in P_1 and every subdividing point in P_2 is a point in P.)

4 Use Theorem 3.3.3 to prove that if f and g are continuous on $[a, b]$, then

$$\int_a^b (f + g)(x)\, dx = \int_a^b f(x)\, dx + \int_a^b g(x)\, dx.$$

5 In our partial proof of Theorem 3.3.2 we have had to assume that $f'(x)$ is bounded on the interval $[a, b]$. This can be a bit bothersome because it excludes such common functions as $f(x) = \sqrt{1 - x^2}$ on the interval $[0, 1]$. But for that function, $f'(x)$ is bounded on any interval $[0, v]$, $0 < v < 1$ (it becomes unbounded at 1), and we can use this fact to arrive at the conclusion of Theorem 3.3.2. Inspecting the proof of the theorem in the text, make the necessary modifications to prove that *if f is differentiable on* $[a, b]$ *and $f'(x)$ is bounded on every interval* $[a, v]$, $a < v < b$ (even though the bound B_v for $|f'(x)|$ over $[a, v]$ depends on the choice of v), *then the conclusion of Theorem* 3.3.2 *holds.* (*Hint:* Let $V = M - m$, where M is the maximum value and m is the minimum value of $f(x)$ on $[a, b]$. Given $\epsilon > 0$, choose δ_1 so that $V\,\delta_1 < \epsilon/2$. Let B_1 be such that $|f'(x)| < B_1$ for all x in $[a, b - \delta_1]$. If $|P| < \delta_1$, write

$$U(f, P) - L(f, P) = \sum(M_i - m_i)(x_i - x_{i-1}).$$

Show that $|U(f, P) - L(f, P)| \leq B_1|P|(b - a) + V\,\delta_1$. Now find a δ which fits the requirements of the theorem.)

6 Use Exercise 5 to show that Theorems 3.3.2 and 3.3.3 apply to the function $f(x) = \sqrt{1 - x^2}$ on the interval $[-1, 1]$

3.4

Length of a Curve Revisited

Of the applications of definite integrals considered in Chapter 9 of *C 1* the application to finding the length of the graph of a function stood out as least satisfactory in one sense. We were forced to assert the integral construction of an arclength essentially by fiat, instead of showing, as with our other applications, that the formula followed necessarily from reasonable prior agreements. Theorem 3.3.3 will enable us to erase this blemish.

The discussion begins exactly as before. Let f be a function on $[a, b]$ which has a continuous derivative f' on $[a, b]$. If $P = \{a = x_0, x_1, \ldots, x_n = b\}$ is a partition of $[a, b]$, then we consider the polygonal path consisting of the line segments joining the pairs of consecutive points $(x_{i-1}, f(x_{i-1}))$, $(x_i, f(x_i))$ on the graph. This polygonal path has a length $L(P)$ such that

(3.4.1) $$L(P) = \sum_{i=1}^{n} \sqrt{(x_i - x_{i-1})^2 + (f(x_i) - f(x_{i-1}))^2}.$$

We may apply the Mean Value Theorem (see marginal note, Section 3.3) to the function f on *each* interval $[x_{i-1}, x_i]$, concluding that for each $i = 1, 2, \ldots, n$ there exists a number x_i' in (x_{i-1}, x_i) such that

(3.4.2) $$f(x_i) - f(x_{i-1}) = (x_i - x_{i-1})f'(x_i').$$

Substituting (3.4.2) into (3.4.1) and factoring, we have

(3.4.3) $$L(P) = \sum_{i=1}^{n} \sqrt{1 + [f'(x_i')]^2}\,(x_i - x_{i-1}).$$

Our difficulty in *C 1* arose in relating sums such as (3.4.3) to an integral. It was clear which integral was indicated, but the connections between this integral, (3.4.3), and the length of the graph were tenuous. But now we can build a stronger case. It is certainly clear that as we choose finer partitions the polygonal path shown in Figure 3.4.1 approximates the graph more and more closely. Thus, we are inclined to agree that if the polygonal lengths $L(P)$ should settle down around some number when $|P|$ is small, that number would be the outstanding candidate for the title "length of the graph."

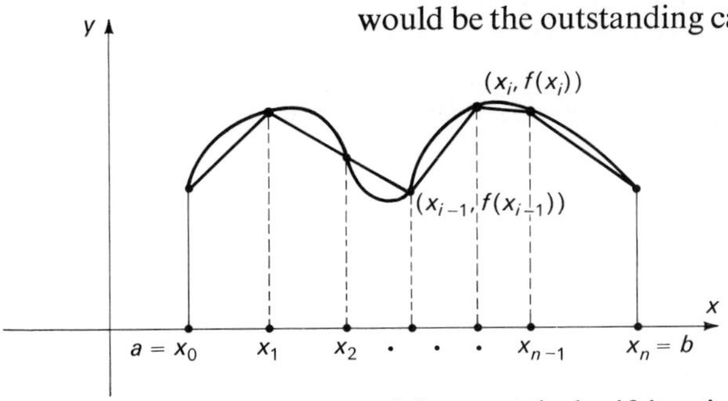

Figure 3.4.1

More precisely, if it exists, $\lim_{|P|\to 0} L(P)$ is clearly what we should call the length of the curve. But this limit *does* exist, for we recognize (3.4.3) as a Riemann sum. If

$$F(x) = \sqrt{1 + [f'(x)]^2},$$

then by (3.4.3) $L(P) = R(F, P)$, a Riemann sum for the function F relative to the partition P.

Since by assumption $f'(x)$ and hence $F(x)$ is continuous on $[a, b]$, we can apply Theorem 3.3.3, concluding that $\lim_{|P|\to 0} L(P)$ *does* exist and

(3.4.4) $$\lim_{|P|\to 0} L(P) = \lim_{|P|\to 0} R(F, P) = \int_a^b F(x)\,dx.$$

Thus, we define the length $L_a^b(f)$ of the graph of f between $(a, f(a))$ and $(b, f(b))$ as

(3.4.5) $$L_a^b(f) = \int_a^b \sqrt{1 + [f'(x)]^2}\, dx.$$

Formula (3.4.5) is, of course, the same formula which we observed in Chapter 9 of *C 1*. The addition is the limit statement (3.4.4), without which we could neither formulate satisfactorily the proper use of the sums (3.4.1) to define a length for the curve nor give more than a suggestion of why the above integral provides the most natural definition.

Exercises

Find the length of the specified arc of each of the following curves.

1 $f(x) = x^{3/2}$, $0 \leq x \leq 4/3$.

2 $f(x) = \frac{1}{2}x^2 - \frac{1}{4}\log x$, $1 \leq x \leq e$.

3 $f(x) = \log(1 - x^2)$, $\frac{1}{4} \leq x \leq \frac{3}{4}$.

4 $f(x) = \sqrt{1 - x^2}$, $0 \leq x \leq 1$.

5 $f(x) = \dfrac{a}{2}(e^{x/a} + e^{-x/a})$, $0 \leq x \leq a$.

6 $f(\theta) = 2 \sec \theta$, $0 \leq \theta \leq \pi/4$.

***7** $f(x) = e^x$, $0 \leq x \leq 1$.

8 Find an expression for the perimeter of the ellipse

$$\frac{x^2}{a^2} + \frac{y^2}{b^2} = 1.$$

(Do not try to evaluate it.)

9 Let f be a function with continuous derivative on $[a, b]$ and let s be the arclength function for f; i.e., $s(t)$ is defined as the arclength of $y = f(x)$ between $(a, f(a))$ and $(t, f(t))$.

(a) Show that s is an increasing function on $[a, b]$.

(b) What conditions on f would insure that the graph of s is concave upward at a point $(t_0, s(t_0))$?

(c) Can the graph of s have a point of inflection? If so, produce an example.

10 Not all continuous functions on a closed interval have graphs of finite length. Consider, for example,

$$f(x) = \begin{cases} x \sin \dfrac{\pi}{x}, & 0 < x \leq 1, \\ 0, & x = 0. \end{cases}$$

(a) Show that f is continuous on [0, 1].

(b) Show that the arclength of the curve $y = f(x)$ for x in

$$[1/n, 1/(n-1)]$$

must exceed $4/(2n-1)$. (*Hint:* Consider Figure 3.4.2.)

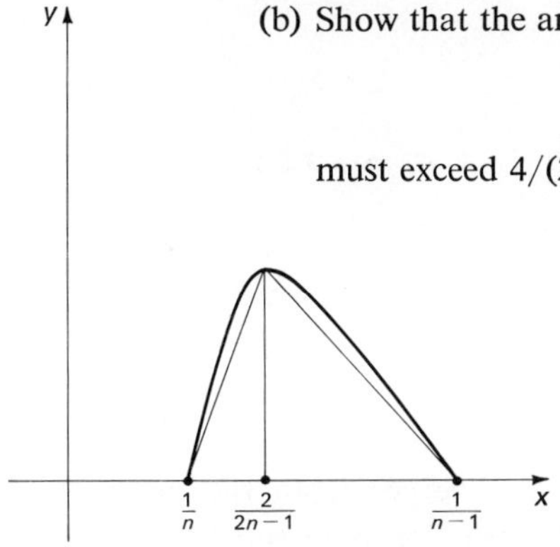

Figure 3.4.2

(c) Hence show that the arclength of $y = f(x)$ for x in the interval $[1/n, 1]$ exceeds $2(1 + \frac{1}{2} + \cdots + 1/n)$. Finally, show that the arclength of $y = f(x)$ for x in [0, 1] is not finite. (*Comment:* It will be shown in Chapter 5 that the sum $1 + \frac{1}{2} + \cdots + 1/n$ can be made arbitrarily large for a choice of n sufficiently large.)

(d) Does the last conclusion contradict the results of this section? Why?

3.5 *Volumes Revisited: Disks and Shells*

It will be instructive to reexamine the method of Chapter 9, *C 1*, for finding the volume of a solid of revolution from the point of view of a definite integral as a limit of Riemann sums. With this background we can then introduce an alternate method for finding such volumes which has some practical value. In the process we shall state one last theorem about limits of sums and the integral that is very useful in certain applications.

Example 1 Let R be the plane region in the first quadrant bounded by the graph of $y = \cos x$ and the two axes, between $x = 0$ and $x = \pi/2$. Let S be the solid formed by revolving R about the horizontal axis.

To find the volume V of S we begin exactly as before. For an arbitrary partition $P = \{0 = x_0, x_1, \ldots, x_n = \pi/2\}$ we can approximate the solid S both from within and from without by stacks of disks. See Figure 3.5.1. By calculating the volumes of these collections of cylinders, we obtain both an upper and a lower estimate for the volume V:

$$\sum_{i=1}^{n} \pi(\cos x_i)^2 \, \Delta x_i \le V \le \sum_{i=1}^{n} \pi(\cos x_{i-1})^2 \, \Delta x_i, \quad \Delta x_i = x_i - x_{i-1}. \tag{3.5.1}$$

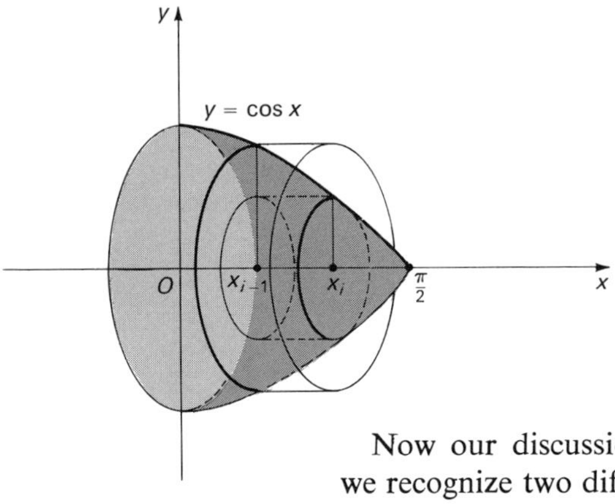

Figure 3.5.1

Now our discussion diverges from the previous one. In (3.5.1) we recognize two different Riemann sums for the function

(3.5.2) $$f(x) = \pi (\cos x)^2$$

relative to the partition P. One of these Riemann sums results from the choice $x_i' = x_i$ (the right-hand endpoint) in each interval $[x_{i-1}, x_i]$. Let us call this sum $R_1(f, P)$. The other Riemann sum, which we shall call $R_2(f, P)$, results from the choice $x_i' = x_{i-1}$ (the left endpoint) for each i.

Thus, (3.5.1) could be written

(3.5.3) $$R_1(f, P) \leq V \leq R_2(f, P)$$

for *certain* Riemann sums R_1 and R_2 relative to *any* partition P. By Theorem 3.3.3 we have

(3.5.4) $$\lim_{|P| \to 0} R_1(f, P) = \lim_{|P| \to 0} R_2(f, P) = \int_0^{\pi/2} \pi(\cos x)^2 \, dx.$$

Thus, by (3.5.3), (3.5.4), and the "Pinching Theorem," which applies to this situation as well as to ordinary limits of functions, we can conclude that

(3.5.5) $$V = \int_0^{\pi/2} \pi(\cos x)^2 \, dx.$$

Roughly speaking, since V is trapped between $R_1(f, P)$ and $R_2(f, P)$, V must equal their common limit.

We have found the integral (3.5.5) before (through integration by parts), with the result

$$V = \frac{\pi}{2} (\sin x \cos x + x)\Big|_0^{\pi/2} = \frac{\pi^2}{4}.$$

We shall call this method for finding volumes of solids of revolution the *disk method*, because we have approximated the solid by a stack of thin disks by partitioning the interval on the axis of revolution. (Recall that for certain other solids these disks have holes and might be more vividly described as *washers*. We mean to include this case when we refer to disks.)

Now let us consider another example in which things do not work out so well.

Example 2 Let S now be the solid formed by revolving the region R in the last example about the vertical axis (Figure 3.5.2). If we now form a partition P of the interval [0, 1] (which is the projection of R on the axis of revolution), we obtain the following estimates:

(3.5.6) $$\sum_{i=1}^{n} \pi(\arccos y_i)^2 \, \Delta y_i \le V \le \sum_{i=1}^{n} \pi(\arccos y_{i-1})^2 \, \Delta y_i,$$
$$\Delta y_i = y_i - y_{i-1}.$$

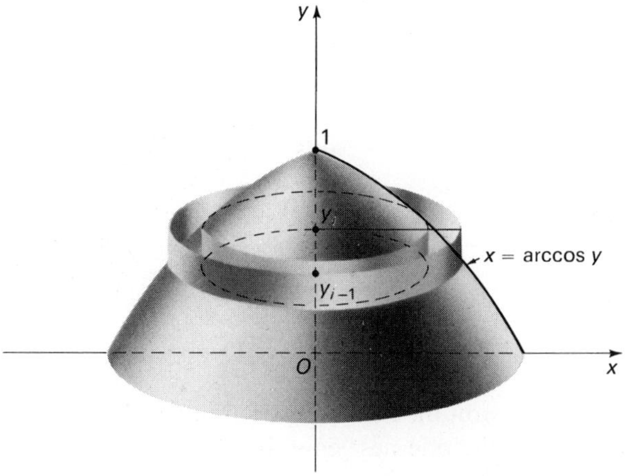

Figure 3.5.2

Recognizing the sums on the left and right in (3.5.6) as Riemann sums for the function

$$f(y) = \pi (\arccos y)^2$$

relative to the given partition P, we see as before that by taking the limit as $|P| \to 0$,

$$V = \int_0^1 \pi(\arccos y)^2 \, dy.$$

Thus as before we have succeeded in finding an integral expression for the desired volume. But now the integral is not so easy to find. We could work this out, using integration by parts, but we cannot be blamed for preferring not to!

Instead, let us see in our next example if we can find a different integral expression for this volume.

Example 3 R and S are as in Example 2. Recall from *C 1* that if for any reason we did not like an integral expression obtained for the area of a certain region we had the option of finding another integral expression by partitioning an interval on the other coordinate axis. We could find an integral *either in x or in y* for the given area. Could we perhaps do the same here?

Let us form a partition $P = \{0 = x_0, x_1, \ldots, x_n = \pi/2\}$ of the projection $[0, \pi/2]$ of R on the x-axis. On each subinterval $[x_{i-1}, x_i]$ form the two rectangles whose heights are the maximum and minimum values of $\cos x$ on that interval. These are the same rectangles which were used in Example 1. But now we rotate these rectangles the other way, about the y-axis, to form hollow cylindrical *shells* (*thin* shells if the lengths $x_i - x_{i-1} = \Delta x_i$ are small) which approximate a portion of S both from within and from without.

Figure 3.5.3

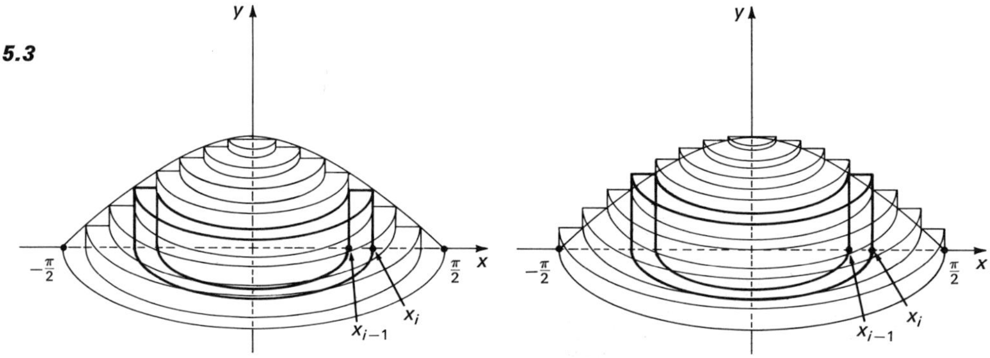

If we repeat this construction on each of the subintervals we shall obtain two *nests of thin shells;* one nest contains S and the other is contained in S (Figure 3.5.3). In order to obtain the estimates of the volume V of S to which this construction leads, we need a workable expression for the volume of a cylindrical shell. Actually, estimates of this volume will do. If such a shell (Figure 3.5.4) has interior radius r, exterior radius R, and height h, then its volume V_s is

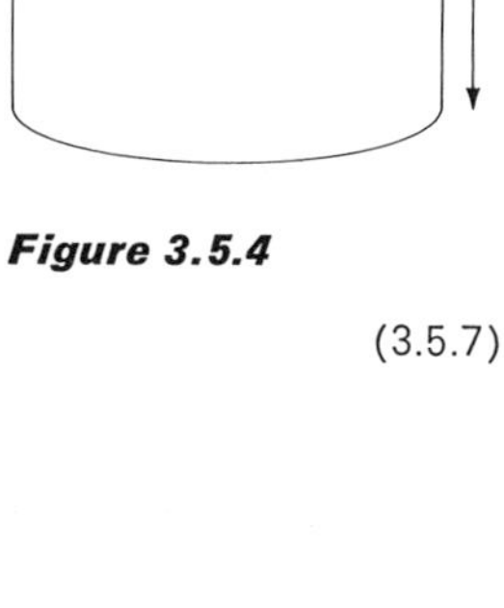

Figure 3.5.4

$$V_s = \pi(R^2 - r^2)h.$$

We may write $R^2 - r^2 = (R + r)(R - r)$. Then, if we let $R - r = \Delta r$ (the wall thickness of the shell) and observe that

$$2r < R + r < 2R,$$

we see that

(3.5.7) $$2\pi rh\,\Delta r < V_s < 2\pi Rh\,\Delta r.$$

Now let us return to the above nests of shells. In the larger nest, the typical shell has (see Figure 3.5.3) height $h = \cos x_{i-1}$, interior radius $r = x_{i-1}$, exterior radius $R = x_i$, and wall thickness $R - r = \Delta x_i$. If we call the volume of this shell V_i, then we have by (3.5.7)

$$V_i < 2\pi x_i \cos x_{i-1}\,\Delta x_i.$$

Thus, the volume V of S satisfies

(3.5.8) $$V \leq \sum_{i=1}^{n} V_i < \sum_{i=1}^{n} 2\pi x_i \cos x_{i-1}\,\Delta x_i.$$

By an analogous argument using the smaller shells in Figure 3.5.3 we see that

(3.5.9) $$\sum_{i=1}^{n} 2\pi x_{i-1} \cos x_i\,\Delta x_i < V.$$

Thus, combining (3.5.8) and (3.5.9) we have

(3.5.10) $$\sum_{i=1}^{n} 2\pi x_{i-1} \cos x_i\,\Delta x_i < V < \sum_{i=1}^{n} 2\pi x_i \cos x_{i-1}\,\Delta x_i.$$

It would appear that we are in a position to conclude our argument exactly as before, for the sums in (3.5.10) resemble Riemann sums for the function $F(x) = 2\pi x \cos x$. But a closer look reveals that they are *not* Riemann sums. If we write the function F as

$$F(x) = f(x)g(x), f(x) = 2\pi x,\ g(x) = \cos x,$$

then the sums in (3.5.10) are of the form

(3.5.11) $$\sum_{i=1}^{n} f(x_i')g(x_i'')\,\Delta x_i$$

for *different* choices of numbers x_i', x_i'' in $[x_{i-1}, x_i]$, whereas a Riemann sum for $F(x)$ has the form

$$\sum_{i=1}^{n} f(x_i')g(x_i')\,\Delta x_i.$$

However, all is not lost! Sums of the form (3.5.11) for a function which is a product of two other functions are called *generalized* Riemann sums. A theorem which we shall discuss in the next section asserts that any set of generalized Riemann sums converges, like Riemann sums, to the integral—in this case $\int_a^b f(x)g(x)\,dx$.

Thus we *may* conclude from (3.5.10) that V, being trapped between generalized Riemann sums for $2\pi x \cos x$ relative to any partition of $[0, \pi/2]$, must equal the corresponding integral. That is,

(3.5.12) $$V = \int_0^{\pi/2} 2\pi x \cos x\,dx.$$

This integral is easy to work out by an integration by parts. We obtain

$$V = 2\pi \int_0^{\pi/2} x\,d(\sin x) = 2\pi(x \sin x - \cos x)\Big|_0^{\pi/2} = 2\pi\left(\frac{\pi}{2} + 1\right).$$

The formula (3.5.12) which we have derived in this special case is general. So is the derivation. Suppose that f is any continuous nonnegative-valued function defined on the interval $[a, b]$ with $a > 0$. Let R be the region in the first quadrant bounded by the graphs of $y = f(x)$, $x = a$, $x = b$, $y = 0$. Let S be the solid formed by revolving R about the y-axis (Figure 3.5.5).

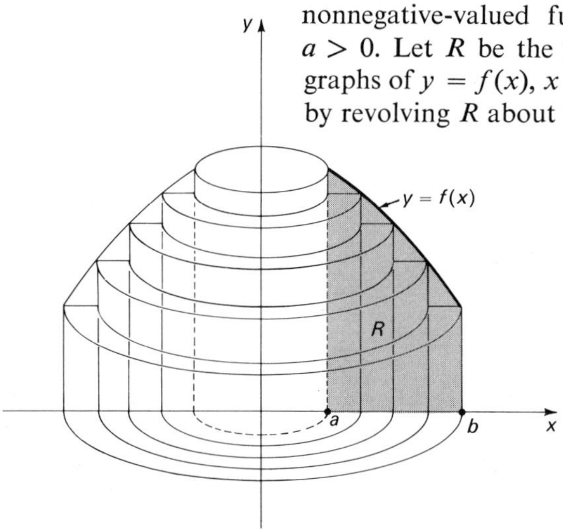

Figure 3.5.5

Relative to any partition P of $[a, b]$ we may approximate S by thin shells as suggested in Figure 3.5.5, with the resulting inequality for the volume V of S

$$\sum_{i=1}^{n} 2\pi x_{i-1} f(x_i') \, \Delta x_i \le V \le \sum_{i=1}^{n} 2\pi x_i f(x_i'') \, \Delta x_i,$$

where $f(x_i')$ and $f(x_i'')$ are the minimum and maximum values of f on $[x_{i-1}, x_i]$. These generalized Riemann sums converge to the same limit as $|P| \to 0$ (according to the theorem in the next section), and we see that

$$V = \int_a^b 2\pi x f(x) \, dx.$$

For obvious reasons, we call this method of finding certain volumes the *shell method*. As the above examples show, this method is sometimes more convenient for finding the volume of a solid of revolution than the disk method. In a given example it is seldom difficult to see which method is preferable.

Exercises

1 Find the volume of the solid generated by rotating each of the following regions about the x-axis.

(a) one arch of the curve $y = \sin x$,

(b) the region bounded by the curve $y = 1/(1 + x)$, the x-axis, and the lines $x = 1$ and $x = 5$,

(c) the region bounded by the curve $y = 1/(1 + x^2)$, the x-axis, and the lines $x = -1$ and $x = 1$.

2 Use the disk method to find the volume of the solid generated by rotating each of the following regions about the y-axis.

(a) the region bounded by the parabola $y^2 = x - 1$, the two coordinate axes, and the line $y = 1$,

(b) the region bounded by $y = x^3$, the y-axis, and the line $y = 2$,

(c) the region bounded by the curve $y = e^x$, the y-axis, and the line $y = 2$.

3 Use the shell method to find the volume of the solid generated by rotating each of the following regions about the y-axis.

(a) the region bounded by the curve $y = \sin x$ and the x-axis between $x = 0$ and $x = \pi$,

(b) the region bounded by $y = e^x$, the two coordinate axes, and the line $x = 2$,

(c) the region bounded by the curve $y = 1/(1 + x^2)$, the two coordinate axes, and the line $x = 1$.

4 Find the volume of the solid generated by rotating the region bounded by each of the following pairs of curves about the y-axis by both the disk and the shell method and compare the results.

(a) $y = \sqrt{x}$, and $y = x/2$, (b) $y = x^2$ and $y = x^{1/3}$,

(c) $y = x^3 + 1$ and $y = x + 1$.

5 Find the volume of the solid generated when the region bounded by the curve $y = x^{3/2}$, the x-axis, and the line $x = 4$ is rotated about

(a) the x-axis, (b) the y-axis,

(c) the line $x = 4$, (d) the line $y = 8$.

6 Find the volume of the solid generated when each of the following regions is rotated about the x-axis.

(a) the region bounded by the curves $y = x^2$ and $y^2 = x$,

(b) the region bounded by the circle $x^2 + y^2 = 1$ and the parabola $y^2 = 3x/2$,

(c) the region bounded by $y = 2x/\pi$ and $y = \sin x$,

(d) the region bounded by the curve $y = \frac{1}{2}(e^x + e^{-x})$, the two coordinate axes, and the line $x = 1$.

7 The region bounded by the ellipse

$$\frac{x^2}{a^2} + \frac{y^2}{b^2} = 1$$

is rotated about the y-axis. Find the volume of the solid generated.

8 A circle of radius r is centered at $(a, 0)$, $a > r$. Prove that the volume of the doughnut-shaped solid obtained by rotating the circle about the y-axis is equal to the area of the circle multiplied by the length of the circular path swept out by its center.

9 A hole of radius a is drilled through a sphere of radius r $(r > a)$. How much material is removed?

3.6

Generalized Riemann Sums

In the last section we introduced the notion of generalized Riemann sums for a function F that is a product of two other functions. The result which was asserted there can be made much more general. Suppose $F(u, v)$ is a function of two variables which is continuous in a region of the uv-plane. Suppose also that f and g are two continuous functions of a single variable x on the interval $[a, b]$. Suppose further that the point $(u, v) = (f(x), g(x))$ is in the region of continuity of F for each x in $[a, b]$. Then the composite function denoted $F(f, g)$ and defined by

$$F(f, g)(x) = F(f(x), g(x))$$

is continuous on $[a, b]$. Thus $F(f, g)$ is integrable on $[a, b]$, and its integral is the limit of Riemann sums of $F(f, g)$ as the norm of the defining partition tends to zero. But generalized Rieman sums can also be considered here.

Definition 3.6.1 Given the above condition any sum of the form

$$GR(F(f, g), P) = \sum_{i=1}^{n} F(f(x_i'), g(x_i''))\,\Delta x_i$$

for a partition P and any choices of x_i' and x_i'' in $[x_{i-1}, x_i]$ is called a *generalized Riemann sum for $F(f, g)$ relative to P.*

Now we can state the theorem predicted in the last section. Since we cannot prove it without initiating a more systematic discussion of continuous functions of several variables than would be wise here, we simply state the theorem, remarking only that continuity of F, f, g is enough to insure that when $|P|$ is small, generalized Riemann sums differ from the ordinary Riemann sums by insignificant amounts.

Theorem 3.6.1 *If $f(x)$ and $g(x)$ are continuous on $[a, b]$ and $F(u, v)$ is continuous at each point $(f(x), g(x))$, then*

$$\int_a^b F(f(x), g(x))\,dx = \lim_{|P|\to 0} GR(F(f, g), P),$$

where $GR(F(f, g), P)$ is any generalized Riemann sum for $F(f, g)$ relative to P.

3.7

Some Pumping Problems

An additional source of problems requiring the use of generalized Riemann sums concerns the pumping of liquids out of tanks. We illustrate such problems with two examples.

Example 1 A cylindrical oil tank with vertical axis of symmetry has radius 4 feet and height 10 feet. A pipe is inserted to the bottom of the tank with its outlet 5 feet above the top of the tank. How much work is required to pump the tank dry if the tank is initially full of oil of density 50 lb. per ft.^{3}?

The problem would be trivial if it were simply a matter of lifting a weight W of oil h feet. The work would be $W \cdot h$ ft.-lbs. But it isn't that simple, of course, because the oil in different parts of the tank must be lifted through different vertical distances. Note that by a fundamental fact of physics the oil will rise in the pipe *of its own accord* to the level of the surface of the oil in the tank. Thus, as the oil is pumped through the pipe with opening at the bottom of the tank it is as if the pipe were being lowered gradually so as just to touch the surface as it falls. Looked at in this way, we see that it is correct to picture the task of the pump as that of raising each bit of oil (each molecule, if we wish to think that finely) from its original position to a height of 5 feet above the top of the tank.

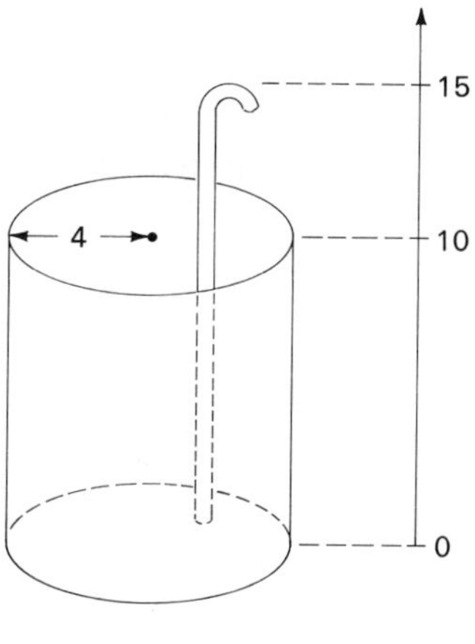

Figure 3.7.1

To measure these vertical distances, we establish a coordinate line as shown in Figure 3.7.1. Denote coordinates on the line by x. If we partition the interval [0, 10], which measures the initial tankful of oil, we can consider a corresponding partition of the oil itself into horizontal slices, or layers. Call the partition

$$P = \{0 = x_0, x_1, \ldots, x_n = 10\}$$

and observe the layer of oil between the levels x_{i-1} and x_i (Figure 3.7.2).

The volume of the oil in this layer is

$$V_i = 16\pi\,\Delta x_i = 16\pi(x_i - x_{i-1})\text{ ft.}^3,$$

so its weight w_i is

$$w_i = 800\pi\,\Delta x_i\text{ lbs.}$$

All this oil must be raised at *least* $15 - x_i$ ft. and not more than $15 - x_{i-1}$ ft. The total work W_i required to raise this layer to the mouth of the pipe must therefore satisfy the inequality

$$800\pi(15 - x_i)\,\Delta x_i \leq W_i \leq 800\pi(15 - x_{i-1})\,\Delta x_i.$$

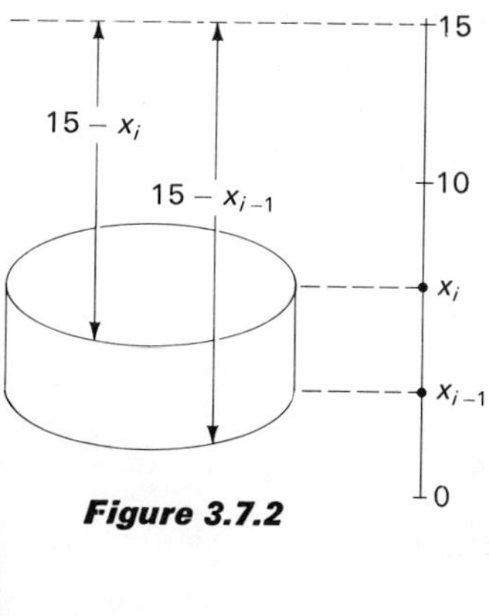

Figure 3.7.2

Since all the layers must be pumped to the mouth of the pipe, we obtain the following inequality for the total work W necessary:

$$\sum_{i=1}^{n} 800\pi(15 - x_i)\,\Delta x_i \leq W \leq \sum_{i=1}^{n} 800\pi(15 - x_{i-1})\,\Delta x_i.$$

We recognize Riemann sums immediately, of course, and since W is thus trapped between two of them for each partition P, we have

$$W = \lim_{|P|\to 0} \sum_{i=1}^{n} 800\pi(15 - x_i)\,\Delta x_i = \int_0^{10} 800\pi(15 - x)\,dx.$$

Thus

$$W = 800\pi\left(15x - \frac{x^2}{2}\right)\Big|_0^{10} = 80{,}000\pi \text{ ft.-lbs.}$$

We managed this without the use of generalized Riemann sums. But consider the following.

Example 2 How much work is required of your tongue and cheek muscles to draw through a straw the contents of a conical milk shake container?

We must make some reasonable assumptions here. Let us assume that the average milk shake container is a cone 6 in. deep with a top of radius 2 in. Let us further assume that the straw is 10 in. long and that the container is initially full to a depth of 5 inches of milk shake whose density is .5 oz./in.[3] Finally, assume that the straw is sufficiently wide or the liquid sufficiently "thin" so that the viscosity is not a factor.

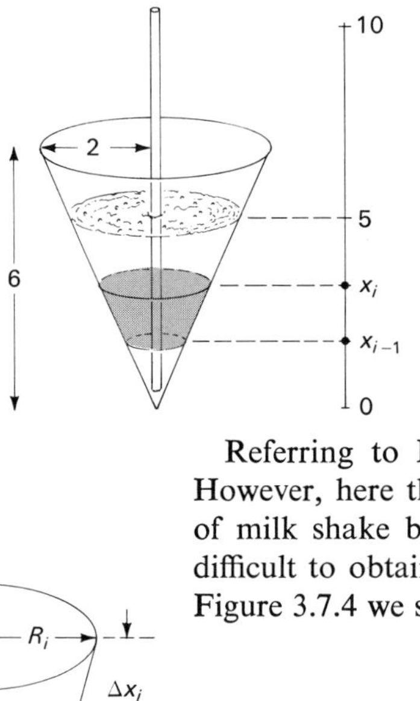

Figure 3.7.3

Referring to Figure 3.7.3 we proceed as in the last example. However, here there is an additional complication. Since the layer of milk shake between levels x_{i-1} and x_i is not cylindrical, it is difficult to obtain a reasonable expression for its weight. But from Figure 3.7.4 we see that the volume V_i of this layer satisfies

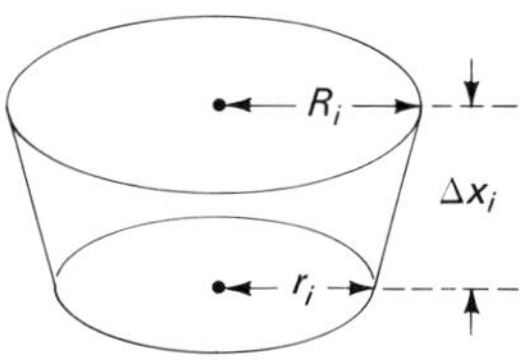

Figure 3.7.4

$$\pi r_i^2\,\Delta x_i \le V_i \le \pi R_i^2\,\Delta x_i.$$

By using similar triangles in Figure 3.7.3 we see that

$$R_i = \tfrac{1}{3}x_i,\; r_i = \tfrac{1}{3}x_{i-1};$$

so the weight $w_i = .5V_i$ of the layer satisfies

$$\frac{\pi}{18}\,x_{i-1}^2\,\Delta x_i \le w_i \le \frac{\pi}{18}\,x_i^2\,\Delta x_i.$$

Since all of the liquid in the layer must be raised at least $10 - x_i$ and no more than $10 - x_{i-1}$ inches, the work W_i necessary to pull that layer through the straw satisfies

$$\frac{\pi}{18} x_{i-1}^2(10 - x_i)\,\Delta x_i \leq W_i \leq \frac{\pi}{18} x_i^2(10 - x_{i-1})\,\Delta x_i,$$

and the total work then satisfies

$$\sum_{i=1}^{n} \frac{\pi}{18} x_{i-1}^2(10 - x_i)\,\Delta x_i \leq W \leq \sum_{i=1}^{n} \frac{\pi}{18} x_i^2(10 - x_{i-1})\,\Delta x_i. \tag{3.7.1}$$

The two sums in (3.7.1) are not Riemann sums. But they are generalized Riemann sums for the function $\left(\frac{\pi}{18} x^2\right)(10 - x)$. Thus, applying Theorem 3.6.1 and inequality (3.7.1) we see that

$$\begin{aligned} W &= \int_0^5 \frac{\pi}{18} x^2(10 - x)\,dx \\ &= \frac{\pi}{18}\left(\frac{10x^3}{3} - \frac{x^4}{4}\right)\Bigg|_0^5 \\ &= \frac{3125\pi}{216} = 45.8 \text{ in.-oz. (approximately).} \end{aligned}$$

Exercises

1 How much work is required to pump the tank in Example 1 dry if it is initially only half full?

2 The tank in Example 1 is filled halfway by pumping oil in through an opening in the bottom of the tank. How much work is done?

3 Imagine that the tank in Example 1 is filled with a large number of fluids of varying densities and allowed to stand. Suppose that the density of the fluid at a depth of x feet from the top of the tank is $(60 + x)$ lbs./ft.^{3}, the heavier fluids having settled to the bottom. Find the work done in pumping the tank dry. (For this exercise, imagine that the pipe merely sucks the liquid from the surface without mixing the remaining contents of the tank.)

4 How much work is done in pumping the contents of a hemispherical water tank of radius 6 ft. over the top?

5 Suppose the tank in Exercise 4 must be emptied with a pipe whose outlet is 2 ft. above the top of the tank. How much work is then required?

6 A tank is in the shape of a hemisphere of radius 6 ft. surmounted by a cylinder of the same radius and height 10 ft. The tank is filled with a fluid of density 75 lbs./ft.^{3} How much work is done in pumping the tank out over the top?

7 A tank is in the shape of the solid generated when a parabola is rotated about its axis. The tank is 8 feet deep and the radius of its circular top is

4 feet. If the tank is filled with water, how much work is done in pumping it out over the top? (*Hint:* Take the parabola to be rotated as $y = px^2$ and determine p.)

8 How much work is required to pump the tank in Exercise 7 full to a depth of 4 feet through an opening in the bottom?

3.8

Improper Integrals

We have learned that any continuous function on a closed interval $[a, b]$ is integrable. We have learned that some functions which are not continuous on $[a, b]$ are integrable. However, to be integrable in the sense of our definition a function must be bounded. Moreover, our definition permits consideration only of functions on bounded intervals, that is, intervals of finite length. Both of these conditions can be relaxed in some cases, and it is the purpose of this section to outline briefly how.

First, let us consider a function f which is not bounded on $[a, b]$, but is bounded and integrable on every interval $[u, b]$, $a < u < b$. We might say that $f(x)$ becomes unbounded at a. Clearly, f is not integrable on $[a, b]$ in the sense of our original definition (see the discussion preceding (3.3.2)). However, we can consider $I(u) = \int_u^b f(x)\,dx$ as a function of u, and it is perfectly possible that this function I has a limit as $u \to a+$ (that is, as u approaches a from the right). If it does, then we write

(3.8.1)
$$\int_a^b f(x)\,dx = \lim_{u \to a+} \int_u^b f(x)\,dx.$$

In this special sense, f has an integral from a to b. We call (3.8.1) a *convergent improper integral.*

Similarly, if f is integrable over any interval $[a, v]$, $a < v < b$, but is not integrable over $[a, b]$, then we can consider $\int_a^v f(x)\,dx$ as a function of v. If $\lim_{v \to b-} \int_a^v f(x)\,dx$ exists, we say that f has a convergent improper integral and write

(3.8.2)
$$\int_a^b f(x)\,dx = \lim_{v \to b-} \int_a^v f(x)\,dx.$$

Example 1 The function

$$f(x) = x^{-\frac{1}{2}}$$

is not integrable over $[0, 1]$. It is not defined at $x = 0$, and although it is bounded and continuous over every interval $[u, 1]$, $u > 0$, it is not bounded on $[0, 1]$. However,

$$\int_u^1 x^{-\frac{1}{2}}\, dx = 2x^{\frac{1}{2}}\Big|_u^1 = 2 - 2\sqrt{u},$$

so

$$\lim_{u\to 0+} \int_u^1 x^{-\frac{1}{2}}\, dx = 2.$$

Thus, f has a convergent improper integral over [0, 1], and we write

$$\int_0^1 x^{-\frac{1}{2}}\, dx = 2.$$

Example 2 The function

$$g(x) = \frac{1}{x}$$

also becomes unbounded at $x = 0$; hence is not integrable over [0, 1]. Here,

$$\int_u^1 g(x)\, dx = -\log u,$$

and thus

$$\lim_{u\to 0+} \int_u^1 g(x)\, dx = +\infty.$$

We say that the improper integral *diverges* and would not use the symbol $\int_0^1 g(x)\, dx$ at all.

Now consider a function defined on an unbounded interval $[a, +\infty)$ or $(-\infty, b]$. A definite integral is meaningless on such an infinite interval but an *improper* integral may exist. If f is integrable over any interval $[a, v]$, $v > a$, and the function $\int_a^v f(x)\, dx$ has a limit as $v \to +\infty$, then we say that f has a *convergent improper integral from a to* ∞ and write

$$\int_a^{+\infty} f(x)\, dx = \lim_{v\to+\infty} \int_a^v f(x)\, dx.$$

If f is integrable on $[u, b]$ for all $u < b$, then we write

$$\int_{-\infty}^b f(x)\, dx = \lim_{u\to-\infty} \int_u^b f(x)\, dx$$

if the limit exists.

The symbol $\int_{-\infty}^{+\infty} f(x)\, dx$ will be defined by

$$\int_{-\infty}^{+\infty} f(x)\, dx = \int_{-\infty}^{0} f(x)\, dx + \int_0^{+\infty} f(x)\, dx.$$

Hence, this "doubly improper" integral converges only when both integrals on the right converge.

Similarly, if f is integrable over any interval $[u, v]$ with $a < u < v < b$, but not over any interval of the form $[a, v]$ or $[u, b]$, then we shall define

(3.8.3) $$\int_a^b f(x)\,dx = \int_a^c f(x)\,dx + \int_c^b f(x)\,dx,$$

for any $a < c < b$, where the integrals on the right side of (3.8.3) are improper integrals of types (3.8.1) and (3.8.2).

Example 3 Consider three functions

$$f(x) = \frac{1}{x}, \quad g(x) = \frac{1}{\sqrt{x}}, \quad h(x) = \frac{1}{x^2}.$$

All of these are continuous and hence integrable on $[1, v]$ for any $v > 1$. We have

$$\int_1^v f(x)\,dx = \log v, \quad \int_1^v g(x)\,dx = 2\sqrt{u} - 2, \quad \int_1^v h(x)\,dx = 1 - \frac{1}{v}.$$

Thus, neither $\int_1^\infty 1/x\,dx$ nor $\int_1^\infty 1/x\,dx$ converges, but

$$\int_1^\infty \frac{1}{x^2}\,dx = 1.$$

Example 4 Consider the unbounded plane region R between the graph of $y = 1/x^2$ and the x-axis, extending from $x = 1$ infinitely far to the right (Figure 3.8.1). Although R is unbounded, it has finite area, in the sense that the area A_v of the part of the region from 1 to v approaches the limit 1 as $x \to +\infty$. This follows, of course, from Example 3.

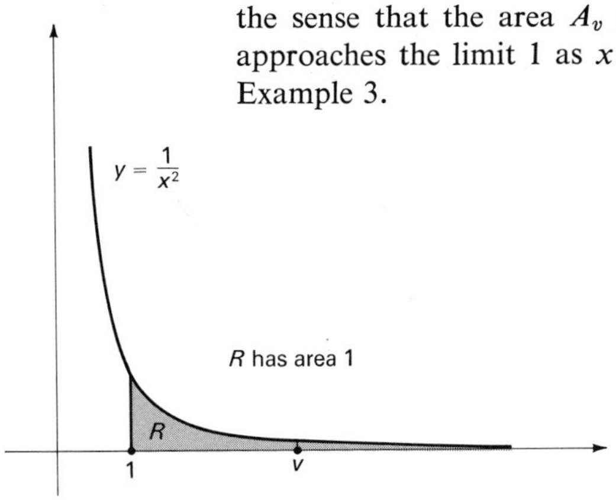

Figure 3.8.1

On the other hand, the region S beneath the graph of $y = 1/x$ from 1 to ∞ (Figure 3.8.2) has infinite area, because the area B_v of the part of S from 1 to v approaches $+\infty$ as $v \to +\infty$.

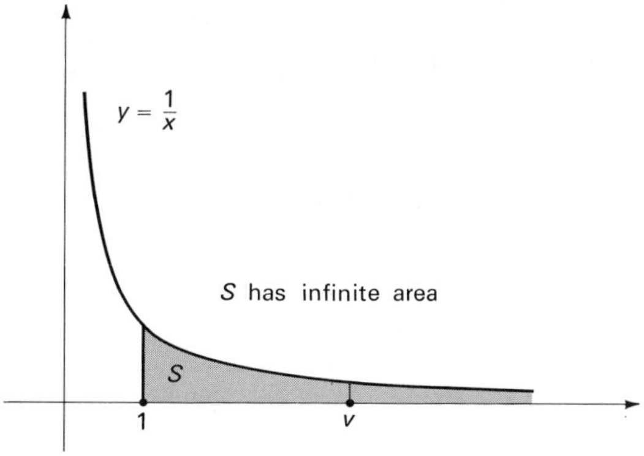

Figure 3.8.2

Example 5 The solid tube T swept out when the plane region S in Example 4 is rotated about the x-axis (Figure 3.8.3) is similarly unlimited in extent. However, since the integral

$$\int_1^\infty \frac{\pi}{x^2}\,dx$$

converges and has the value π, it is reasonable to say that this tube has finite volume π. A plane region of infinite area sweeps out a solid of finite volume! One can gain a great deal of insight into the meaning of the words *area* and *volume* by pondering this seemingly paradoxical statement.

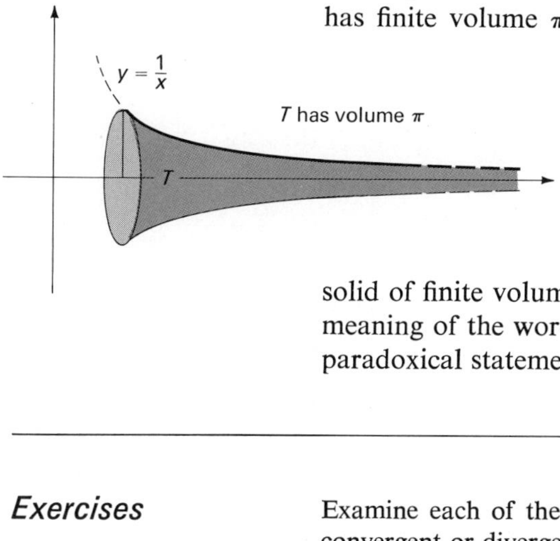

Figure 3.8.3

Exercises

Examine each of the following integrals to determine whether they are convergent or divergent. Evaluate those which are convergent.

1 $\displaystyle\int_0^1 \frac{dx}{\sqrt[3]{x}}$.

2 $\displaystyle\int_0^1 \frac{dx}{x\sqrt{x}}$.

3 $\displaystyle\int_2^4 \frac{dx}{x - 4}$.

4 $\displaystyle\int_2^4 \frac{dx}{\sqrt{x - 2}}$.

5 $\displaystyle\int_2^4 \frac{x\,dx}{\sqrt{x - 2}}$.

6 $\displaystyle\int_0^{\pi/4} \tan 2x\,dx$.

7 $\displaystyle\int_{\pi/4}^{\pi/2} \sec x\,dx.$

8 $\displaystyle\int_0^4 \frac{dx}{(x-2)^{1/3}}.$

9 $\displaystyle\int_1^2 (x^2-1)^{-1/2}\,dx.$

10 $\displaystyle\int_0^4 \frac{dx}{4-x^2}.$

11 $\displaystyle\int_0^1 \frac{\log x}{x}\,dx.$

12 $\displaystyle\int_{-a}^{a} \frac{dx}{\sqrt{a^2-x^2}}.$

13 $\displaystyle\int_0^2 \frac{dx}{\sqrt{4x-x^2}}.$

14 $\displaystyle\int_1^{+\infty} \frac{dx}{x^3}.$

15 $\displaystyle\int_1^{+\infty} \frac{dx}{\sqrt[3]{x}}.$

16 $\displaystyle\int_1^{+\infty} \frac{1}{1+x^2}\,dx.$

17 $\displaystyle\int_0^{+\infty} e^{-x}\,dx.$

18 $\displaystyle\int_{-\infty}^{0} \frac{dx}{x^2+4}.$

19 $\displaystyle\int_1^{+\infty} \frac{x\,dx}{(1+x^2)^2}.$

20 $\displaystyle\int_0^{+\infty} \sin x\,dx.$

21 $\displaystyle\int_{-\infty}^{0} \frac{dx}{x^2+2x+5}.$

22 $\displaystyle\int_{-\infty}^{+\infty} \frac{dx}{x^2+2x+2}.$

23 $\displaystyle\int_{-\infty}^{+\infty} \frac{dx}{x^2-2x+10}.$

24 $\displaystyle\int_2^{+\infty} \frac{dx}{x\sqrt{x^2-1}}.$

25 $\displaystyle\int_{-\infty}^{0} e^{x}\,dx.$

26 $\displaystyle\int_{-\infty}^{+\infty} \frac{dx}{(x+1)^2}.$

limits and approximations

chapter 4

4.0

Introduction Every limit statement is a statement about an approximation. The statement

$$\lim_{x \to c} f(x) = L$$

asserts that the number L may be approximated as closely as desired by values of the function for numbers x close to c. The statement

$$f \textit{ is differentiable at } c$$

asserts that the function may be closely approximated by a linear function or its graph approximated by a line in a neighborhood of c. The statement

$$f'(c) = m$$

asserts that the number m may be approximated arbitrarily closely by quotients $(f(c + h) - f(c))/h$ for small values of h. The statement

$$\int_a^b f(x)\, dx = \lim_{|P| \to 0} \sum_{i=1}^{n} f(x_i')\, \Delta x_i$$

asserts that the number $\int_a^b f(x)\, dx$ may be approximated by Riemann sums arbitrarily closely for sufficiently fine partitions of the interval.

These limit statements do two things: They assert the existence of certain numbers, the limits, and they also suggest that the values of these numbers can be determined approximately by calculable quantities. Throughout *C 1* and the first three chapters of *C 2* we have discussed such limit statements, their significance both in mathematics and in related areas, and methods for calculating the limits exactly.

We turn now to the other assertion implied in the limit statement. Often these numbers called limits cannot be calculated exactly. It is common in mathematics to prove conclusively the existence of a number which cannot be written down exactly, for example as a decimal. We know that

(4.0.1) $$e = \lim_{n \to 0} \left(1 + \frac{1}{n}\right)^n$$

exists, but what is the value of e? We know that

(4.0.2) $$4\int_0^1 \frac{dx}{1 + x^2} = \pi,$$

but how does this help us know the value of π?

It is often not enough for the mathematician or the user of mathematics to know that such a limit exists. He may need to know its value. Sometimes, however, he must be content to know an approximation of the value. Especially for the experimental scientist, knowing an approximation of the value is sometimes essentially as useful as knowing the exact value, provided that information about the accuracy of the approximation is available. The number $(1 + .01)^{100}$ may be so close to e that written as a decimal, it may be perfectly acceptable as a value for e. A calculated Riemann sum for the integral (4.0.6) with a partition of norm .02 may be so close to π that a scientist would be wasting his time looking for a better value.

To be useful, an approximation must be both *close* and *calculable.* Because of the limit statements (4.0.1) and (4.0.2), we feel that the above approximations of e and π are close, but it is important to know how close the approximations are to e and π. And although $(1.01)^{100}$ and the indicated Riemann sum could be calculated by entirely routine methods, would you want to attempt these calculations? Calculable approximations suggested by a limit statement are often of immediate interest to a person only as long as he can envision someone else doing the calculation. When he makes an honest attempt to estimate how long such a calculation might take his interest may disappear entirely.

In the two decades since World War II mankind has found a "someone else" who is both uncannily reliable and unbelievably fast. He is a machine, of course, and we need not pause here to chronicle the history of the development of the high speed electronic computer or to comment on its impact on our society. It suffices to say that its development has elevated the art of approximation to the ranks of the most important branches of mathematics. The science—some would object to the use of the word "art"—of developing approximating formulas, reducing them to algorithms, or sets of instructions for a machine, and estimating the errors inherent in the approximations is called *numerical analysis*. Its importance to science today can hardly be overstated.

Even though much of it has been developed so recently, numerical analysis is a very large subject, whose surface we cannot expect even to graze. But we can acquire a tiny insight into the art of approximation by examining two types which directly pertain to the calculus presented so far. These methods of approximation are not new, but have been known for several hundred years: a method of approximating a definite integral with much greater and more predictable accuracy than possible with ordinary Riemann sums, and a method of approximation of functions by polynomials.

This chapter, looking both backward and forward, is transitional. Section 4.1 might well have been part of Chapter 3, for it concerns the approximation of definite integrals. In Section 4.2 we present a theorem about differentiable functions, a variant of the Mean Value Theorem entirely suitable for Chapter 2. An immediate consequence of this theorem, appropriate to Chapter 1, presents a method for evaluating certain limits in situations not covered by the ordinary limit theorems. The theorem also has an application in the proof of Taylor's Theorem about polynomial approximations of functions, which explains its inclusion here. Taylor's Theorem, itself presented in Section 4.3, introduces the important notion of infinite series, the subject of Chapter 5.

4.1 Numerical Integration by Simpson's Method

As we have noted, a definite integral can be approximated by means of a Riemann sum. In principle, such an approximation can be made as accurate as we please. However, a Riemann sum $R(f, P)$, for a particular partition P of $[a, b]$ is really quite an inefficient approximation of the integral $\int_a^b f(x)\,dx$. In this section we consider methods of increasing efficiency here. As a first step, let us consider in detail how a Riemann sum approximation is produced. The procedure is to replace the function f by a constant function on each subinterval $[x_{i-1}, x_i]$ and then calculate the integral of the resulting *simple* (or step) function. The integral of such a simple function is, of course, just the kind of sum we call a Riemann sum. In geometric terms, we are approximating the graph of $y = f(x)$ by a separate horizontal line segment over each interval $[x_{i-1}, x_i]$, taking care only that each of the approximating segments crosses the graph over the corresponding subinterval. The single advantage of this kind of approximation of the graph or of the corresponding function is the ease of calculation of the resulting sums. The vindication of the method is the fact that it works, as expressed by Theorem 3.3.3, which asserts that if f is continuous then

$$\int_a^b f(x)\,dx = \lim_{|P|\to 0} R(f, P).$$

Still, we should be able to do better. We should be able on each subinterval to approximate the function by something other than a constant, and approximate the graph by a simple curve other than a horizontal line segment in such a way that the approximation is improved and at the same time the sums remain relatively easy to calculate. We can, and shall illustrate by constructing three possible approximations for $\int_a^b f(x)\,dx$, f continuous on $[a, b]$; the first involves Riemann sums, and the last leads to the so-called Simpson's method. We shall require that f be a continuous function on the closed interval $[a, b]$. Since we seek approximations for the integral $\int_a^b f(x)\,dx$ which can be readily calculated by a machine that cannot think or a man who would prefer not to, we must introduce the utmost regularity into our formulas consistent with reasonable accuracy. Accordingly, we shall use only regular partitions of $[a, b]$, agreeing to write for each integer n

(4.1.1)
$$x_i = a + \frac{i}{n}(b - a),$$
$$y_i = f(x_i),\ i = 0, 1, \ldots, n.$$

As a first attempt to approximate the integral we may select arbitrarily a Riemann sum relative to the regular partition in (4.1.1). Let us select the right-hand endpoint of each subinterval, so that the Riemann sum has the simple form

$$\sum y_i\,\Delta x_i = \frac{b - a}{n}(y_1 + y_2 + \cdots + y_n).$$

We thus have

(4.1.2)
$$\int_a^b f(x)\,dx \approx \frac{b - a}{n}(y_1 + y_2 + \cdots + y_n),$$

where we use the symbol "$\approx$" to denote "is approximately equal to."

The approximation (4.1.2), which results from the approximation of the graph of f by a series of horizontal line segments, will be very good if the integer n is large enough. However, the larger the value of n the more horrid the calculation, and we may prefer to improve the approximation by changing the method.

Generally speaking, the line segment joining the two endpoints of an arc of a curve is a closer approximation of that arc than the horizontal segment through one endpoint. The area under the arc is better approximated by the area of the trapezoid in Figure 4.1.1 than by that of the rectangle.

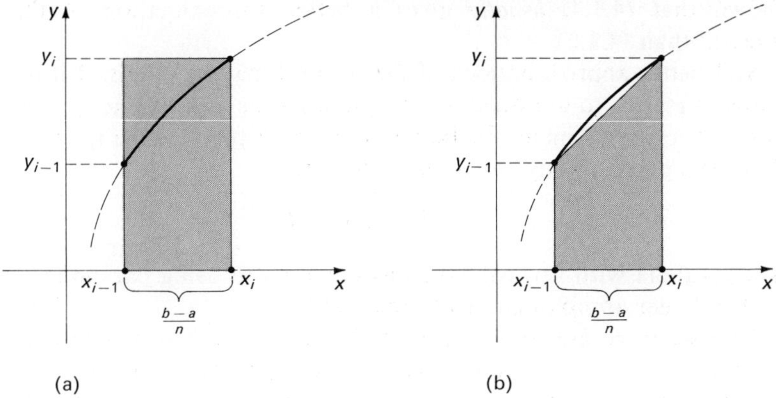

Figure 4.1.1

The area of the trapezoid is

$$\frac{b-a}{n}\left(\frac{y_{i-1}+y_i}{2}\right),$$

which we may use as an approximation of the area under the corresponding arc of the graph. Adding these numbers for $i = 1, 2, 3, \ldots, n$ we obtain another approximation of the integral (which is, in this geometric interpretation, the area under the curve between $x = a$ and $x = b$),

(4.1.3) $$\int_a^b f(x)\,dx \approx \frac{b-a}{n}\left(\frac{y_0+y_1}{2}+\frac{y_1+y_2}{2}+\cdots+\frac{y_{n-1}+y_n}{2}\right).$$

From another point of view, on the subinterval $[x_{i-1}, x_i]$ we are replacing the function f by a linear function so selected that its values at x_{i-1} and x_i agree with those of f. We then use the integral of this linear function as an approximation for the integral of f over $[x_{i-1}, x_i]$. This integral is, of course, the number

$$\frac{b-a}{n}\left(\frac{y_{i-1}+y_i}{2}\right),$$

the area of the trapezoid. Repeating for each i and adding, we obtain the integral of the resulting "piecewise linear" approximation of f, which provides the approximation (4.1.3).

Since each y_i, $i \neq 0, n$, occurs twice in (4.1.3), we can rewrite the approximating expression in a more compact form, yielding the approximating formula

(4.1.4) $$\int_a^b f(x)\,dx \approx \frac{b-a}{2n}(y_0+2y_1+2y_2+\cdots+2y_{n-1}+y_n).$$

The use of (4.1.4) to approximate the integral is often called the *trapezoid method,* because each small section of the region under the curve is replaced by a trapezoid as in Figure 4.1.1(b). It can be shown that (4.1.4) *usually* gives a better approximation of the integral than (4.1.2).

Still better approximations of the integral may be obtained if we allow ourselves to use quadratic polynomials as well as linear polynomials to approximate the function on small parts of the interval. Since the graph of any equation of the form

(4.1.5) $$y = Ax^2 + Bx + C$$

is a parabola with vertical axis, this amounts to using parabolic as well as linear approximations to the graph.

Because there are three constants in (4.1.5), each of which may be given any value, we can see easily that a parabola (or line) may be found which passes through any three given points Q_0, Q_1, Q_2 with different abscissas. Thus, returning to Figure 4.1.1, we see that a parabola (or line) may be fitted so as to pass through any three adjacent points Q_{i-1}, Q_i, Q_{i+1} on the graph. See Figure 4.1.2. It can be shown that *usually* the area of the shaded region in Figure 4.1.2 approximates the area under the curve better than the area of the trapezoids with tops $Q_{i-1}Q_i$ and Q_iQ_{i+1}.

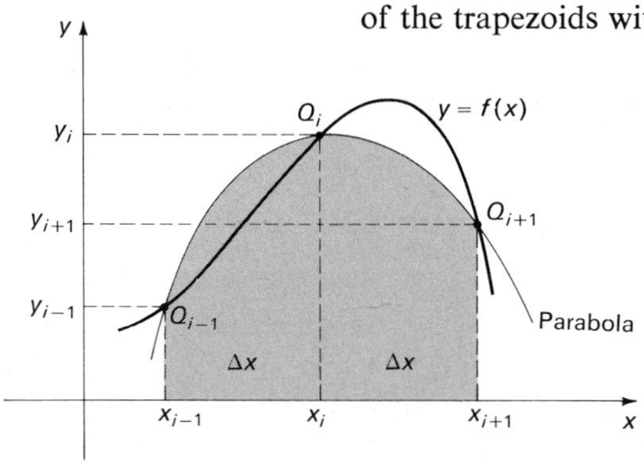

Figure 4.1.2

To approximate an integral using parabolic approximations with equations (4.1.5), we are naturally led to consider in adjacent pairs the subintervals created by the partition. We must therefore restrict our attention to partitions with an even number of subintervals.

Thus, consider $P_{2n} = \{a = x_0, x_1, \ldots, x_{2n} = b\}$ such that $\Delta x = x_i - x_{i-1} = (b - a)/2n$ for all i. Let $f(x_i) = y_i$ for $i = 1, 2, \ldots, 2n$. For each *odd* i consider Figure 4.1.2. In Exercise 4.1.13 you will see that if the graph of (4.1.5) passes through Q_{i-1}, Q_i, Q_{i+1}, then

(4.1.6) $$\int_{x_{i-1}}^{x_{i+1}} (Ax^2 + Bx + C)\,dx = \frac{\Delta x}{3}(y_{i-1} + 4y_i + y_{i+1}).$$

Using (4.1.6) as an approximation for $\int_{x_{i-1}}^{x_i} f(x)\,dx$ and adding these expressions over all odd i, we obtain

(4.1.7)
$$\int_a^b f(x)\,dx \approx \frac{\Delta x}{3}(y_0 + 4y_1 + y_2) + \frac{\Delta x}{3}(y_2 + 4y_3 + y_4) + \cdots + \frac{\Delta x}{3}(y_{2n-2} + 4y_{2n-1} + y_{2n}).$$

But $y_2, y_4, \ldots, y_{n-2}$ are all represented twice in (4.1.7). We may thus simplify the approximating expression, at the same time replacing Δx by $(b - a)/2n$, to obtain

(4.1.8)
$$\int_a^b f(x)\,dx \approx \frac{b-a}{6n}(y_0 + 4y_1 + 2y_2 + 4y_3 + \cdots + 4y_{2n-1} + y_{2n}).$$

The approximation formula (4.1.8), obtained by replacing $f(x)$ by a quadratic polynomial on each interval $[x_{i-1}, x_{i+1}]$, $i = 1, 3, 5, \ldots, 2n - 1$, is known as *Simpson's Rule.* A Simpson's Rule approximation of an integral is generally significantly better than either of the two approximations mentioned above. It can be shown that when $f(x)$ is a polynomial function of degree three or less the Simpson's Rule approximation is *exact.*

For an arbitrary function possessing sufficiently many derivatives an estimate may be found of the accuracy of a Simpson's Rule approximation. We shall describe this estimate without giving its derivation, which is too complex for inclusion here. Suppose that M is a positive number such that

$$|f^{(4)}(x)| \leq M \text{ for all } x \text{ in } [a, b].$$

($f^{(4)}$ is the fourth derivative of f.) Let us denote the approximating expression in (4.1.8) by S_{2n}. The *error* in a Simpson's Rule approximation is then

$$E_{2n} = \left|\int_a^b f(x)\,dx - S_{2n}\right|.$$

It can be shown that if $\Delta x = (b - a)/2n$ is the length of the subintervals created by the partition P_{2n}, then

(4.1.9)
$$E_{2n} \leq \frac{(b-a)M}{180}(\Delta x)^4.$$

Since the quantity $(b - a)M/180$ depends only on the function and the interval, (4.1.9) shows how the approximation $\int_a^b f(x)\,dx \approx S_{2n}$ improves as the norm $|P_{2n}|$ decreases.

Example 1 We illustrate the three approximations (4.1.2), (4.1.4), (4.1.8) (Simpson's Rule) for

(4.1.10)
$$\int_0^4 (x^3 - x)\,dx.$$

In each case we shall use the regular partition P_8. Notice that in using (4.1.2) and (4.1.4) $n = 8$, whereas in (4.1.8) $n = 4$. We have

$$x_i = \frac{i}{2}, \; y_i = \frac{i}{8}(i^2 - 4), \; i = 0, 1, 2, \ldots, 8.$$

Thus, we calculate:

i	0	1	2	3	4	5	6	7	8
y_i	0	$-\frac{3}{8}$	0	$\frac{15}{8}$	6	$\frac{105}{8}$	24	$\frac{315}{8}$	60

Using (4.1.2) with $n = 8$ we obtain

$$\int_0^4 (x^3 - x)\,dx \approx \tfrac{4}{8}(0 - \tfrac{3}{8} + 0 + \tfrac{15}{8} + 6 + \tfrac{105}{8} + 24 + \tfrac{315}{8}) = 72.$$

Using (4.1.4) with $n = 8$ we obtain

$$\int_0^4 (x^3 - x)\,dx \approx \tfrac{4}{16}(0 - \tfrac{3}{4} + 0 + \tfrac{15}{4} + 12 + \tfrac{105}{4} + 48 + \tfrac{315}{4} + 60) = 57.$$

Using (4.1.8) with $n = 4$ we obtain

$$\int_0^4 (x^3 - x)\,dx \approx \tfrac{4}{24}(0 - \tfrac{3}{2} + 0 + \tfrac{15}{2} + 12 + \tfrac{105}{2} + 48 + \tfrac{315}{2} + 60) = 56.$$

Compare these three approximations with the actual value of the integral:

$$\int_0^4 (x^3 - x)\,dx = \left.\frac{x^4}{4} - \frac{x^2}{2}\right|_0^4 = 56.$$

As we have mentioned before, when the integrand is a polynomial of degree three or less, a Simpson's Rule approximation is *exact*.

Example 2 Let us use Simpson's Rule to approximate the value of π by noticing, as suggested in the Introduction, that

$$4\int_0^1 \frac{dx}{1+x^2} = 4 \arctan x \Big|_0^1 = \pi$$

and using Simpson's Rule to approximate the integral.

Let $n = 5$. Before calculating, let us estimate how accurate the approximation will be. Here

$$f(x) = \frac{1}{1+x^2}, \; f^{(4)}(x) = \frac{24(1 - 10x^2 + 5x^4)}{(1+x^2)^5}.$$

It is an easy exercise to verify that if $0 \leq x \leq 1$, then

$$-4 \leq 1 - 10x^2 + 5x^4 \leq 1,$$

$$\frac{1}{32} \leq \frac{1}{(1+x^2)^5} \leq 1.$$

This is a straightforward review problem in graphing functions. Thus

(4.1.11) $$|f^{(4)}(x)| \leq 24 \cdot 4 = 96 \quad \text{for} \quad 0 \leq x \leq 1.$$

From (4.1.11) and (4.1.9) using $\Delta x = 0.1$, we have

(4.1.12) $$E_{10} \leq \tfrac{96}{180}(0.0001) = .000053 \ldots .$$

When we multiply the approximation for the integral by 4, the result differs from π by 4 times E_{10}. The error is thus no more than .00022, so we can have confidence in the first three and possibly even the fourth decimal place of our approximation.

We have for $i = 0, 1, 2, \ldots, 10$

$$x_i = \frac{i}{10}, \; y_i = (1 + x_i^2)^{-1},$$

and our approximation is

(4.1.13) $$\pi \approx \tfrac{4}{30}(y_0 + 4y_1 + 2y_2 + 4y_3 + \cdots + 4y_9 + y_{10}).$$

All that remains is the calculation of the y_i's. You can see that a desk calculator would be a great help here, since to realize the full accuracy of Simpson's Rule we should carry the calculations out to 4 or 5 decimal places. A chart shows the results of such calculations, including the final calculation of the approximation (4.1.13).

i	x_i	$y_i = (1 + x_i^2)^{-1}$	k	$k \cdot y_i$
0	0.0	1.000000	1	1.000000
1	0.1	0.990099	4	3.960396
2	0.2	0.961538	2	1.923076
3	0.3	0.917431	4	3.669724
4	0.4	0.862069	2	1.724138
5	0.5	0.800000	4	3.200000
6	0.6	0.735294	2	1.470588
7	0.7	0.671141	4	2.684564
8	0.8	0.609756	2	1.219512
9	0.9	0.552486	4	2.209944
10	1.0	0.500000	1	0.500000
		Σ		23.561946
		$\frac{4\Sigma}{30}$		3.141593

Our approximation is, then,

$$\pi \approx 3.141593.$$

The correct value of π to eight decimal places is

$$\pi = 3.14159265\ldots.$$

Thus, our approximation is actually correct to 6 decimal places, a bonus which is not entirely unexpected in view of the roughness of our estimate of E_{10} in (4.1.12).

Exercises

1 (a) Find a formula similar to (4.1.2) by using the left-hand endpoint of each subinterval to compute a Riemann sum.

(b) Draw a figure similar to those in Figure 4.1.1 for the approximation in (a).

(c) Show that the trapezoid method, formula (4.1.4), can be obtained as the average of the approximations in formula (4.1.2) and (a) above.

2 Compute each of the following integrals using the three methods given in (4.1.2), (4.1.4), and (4.1.8) and compare the results. In each case, use a partition with 8 subintervals.

(a) $\displaystyle\int_1^9 \frac{1}{x}\,dx,$ (b) $\displaystyle\int_1^5 \frac{1}{x^2}\,dx,$

(c) $\int_0^{\pi} \sqrt{\sin x}\, dx$, (d) $\int_0^{2\pi} \sin^2 x\, dx$.

3 (a) Use Simpson's Rule with 10 subintervals to approximate

$$\log 2 = \int_1^2 \frac{1}{x}\, dx.$$

(b) What is the maximum possible error in the approximation in (a)?

4 (a) Express the perimeter of the ellipse $x^2 + 4y^2 = 1$ in terms of an integral.

(b) Can Simpson's Rule be used to approximate the perimeter in (a)?

5 (a) Express the arclength of the curve $y = \cos x$ between $(0, 1)$ and $(\pi/2, 0)$ as an integral.

(b) Approximate the integral in (a) by using the trapezoid method and Simpson's Rule, each with 4 subintervals, and compare the results.

6 Approximate the integral $\int_0^{\pi} \sin x\, dx$ by Simpson's Rule with 6 subintervals and compare with the exact value found by integration.

7 Approximate the integral $\int_0^6 \sqrt{36 - x^2}\, dx$ by using the trapezoid method with 6 subintervals and compare with the exact value found by integration.

8 Use the error estimate (4.1.9) to show that Simpson's Rule is exact if the integrand is a polynomial of degree 3 or less.

9 The following values of a function $y = f(x)$ are given:

x	2.0	2.4	2.8	3.2	3.6	4.0	4.4
y	3.03	4.61	5.80	6.59	7.76	8.46	9.19

Use Simpson's Rule to approximate the area of the region bounded by the x-axis and the curve $y = f(x)$ between $x = 2$ and $x = 4.4$.

10 Use Simpson's Rule to approximate the volume of the solid generated when the region described in Exercise 9 is rotated about the x-axis.

11 Use Simpson's Rule and the shell method of Section 3.5 to approximate the volume of the solid generated when the region described in Exercise 9 is rotated about the y-axis.

12 Find a number n such that the error in approximating

$$\int_0^1 e^{-x^2}\, dx$$

by Simpson's method with n subintervals is less than .0005.

13 In order to establish formula (4.1.7) we shall use the following notation:

$$y = q(x) = Ax^2 + Bx + C,$$
$$\Delta x = x_1 - x_0 = x_2 - x_1,$$
$$y_0 = q(x_0),\ y_1 = q(x_1),\ y_2 = q(x_2),$$

we must prove that

(4.1.14) $$\int_{x_0}^{x_2} q(x)\,dx = \frac{\Delta x}{3}(y_0 + 4y_1 + y_2).$$

(a) Show that (4.1.14) is equivalent to (4.1.7).

(b) Prove (4.1.14) for the case $x_1 = 0$ by computing both sides of the equation directly.

(c) For the general case, move the parabola $y = q(x)$ horizontally so that the point $(x_1, 0)$ falls on the origin $(0, 0)$. This is a translation of the parabola through a distance $|x_1|$. The equation of the new parabola is $y = q(x + x_1) = \overline{q}(x)$. Prove that $y_0 = \overline{q}(-\Delta x)$, $y_1 = \overline{q}(0)$, $y_2 = \overline{q}(\Delta x)$.

(d) Prove that $\displaystyle\int_{-\Delta x}^{\Delta x} \overline{q}(x)\,dx = \int_{x_0}^{x_2} q(x)\,dx.$

(e) Now use (b) to prove that (4.1.14) holds.

4.2

The Cauchy Theorem and L'Hospital's Rule

In this section we shall consider a variant of the Mean Value Theorem, with one application to certain limit problems which have been extremely troublesome heretofore. Another application will follow in the next section.

The theorem concerns a pair of functions f and g defined for all t in a closed interval $[a, b]$. Soon it will be clear why we choose to call the independent variable t here, and it will also become clear that the theorem is a direct generalization of the Mean Value Theorem. It is usually referred to as the *Cauchy Theorem* or the Cauchy Mean Value Theorem.

Theorem 4.2.1 *If f and g are differentiable functions on $[a, b]$ and if $g(b) \neq g(a)$, then there exists at least one number $\bar{t}$ in (a, b) such that*

$$\frac{f(b) - f(a)}{g(b) - g(a)} = \frac{f'(\bar{t})}{g'(\bar{t})}.$$

Proof Consider the function G constructed as follows:

$$G(t) = [f(t) - f(a)][g(b) - g(a)] - [g(t) - g(a)][f(b) - f(a)].$$

Clearly, $G(a) = G(b) = 0$. Moreover, G is differentiable on $[a, b]$, because both f and g are. Thus, the hypotheses of Rolle's Theorem (Theorem 4.1.1 in *C 1*) are satisfied and we may conclude that, for some number $\bar{t}$ in (a, b), $G'(\bar{t}) = 0$. Since

$$G'(t) = f'(t)[g(b) - g(a)] - g'(t)[f(b) - f(a)],$$

the desired conclusion follows at once.

The proof of Theorem 4.2.1 is both concise and quite elementary. However, it obscures somewhat the strong geometric flavor of the theorem and its close connections with the original Mean Value Theorem. To place the theorem in its proper geometric context, set

(4.2.1) $$y = f(t), \qquad x = g(t)$$

and consider the set of all points

$$(x, y) = (g(t), f(t)), \qquad a \leq t \leq b.$$

This set of points forms a curve with endpoints (x_0, y_0), (x_1, y_1), where

$$(x_0, y_0) = (g(a), f(a)), \qquad (x_1, y_1) = (g(b), f(b)).$$

This specification of a curve by this method is called parametric representation, and the symbol t is called a parameter. Parametric representation will be discussed in some detail in Chapter 6. For the present, we point out that if the abscissa function g has an inverse g^{-1} on $[a, b]$ then the curve may be represented as the graph of a single function in the familiar way. For in this case we may solve the second of the equations in (4.2.1) for t in terms of x:

$$t = g^{-1}(x),$$

and express the relationship between x and y in (4.2.1) as

$$y = f(g^{-1}(x)) = F(x).$$

If f and g are differentiable, then by earlier theorems we know that the above function $F = f \circ g^{-1}$ is also differentiable. These theorems also tell us that

$$\frac{dF}{dx} = \frac{dy}{dt} \cdot \frac{dt}{dx} = \frac{dy}{dt} \Big/ \frac{dx}{dt} = \frac{f'(t)}{g'(t)}.$$

If we assume that $g'(t) > 0$ for all t in (a, b), so that $x_0 < x_1$, then F is defined on $[x_0, x_1]$ and we may apply the ordinary Mean Value Theorem (Theorem 4.1.3 in *C 1*) to F on $[x_0, x_1]$. We conclude that for some $\bar{x}$ in (x_0, x_1),

(4.2.2)
$$\frac{F(x_1) - F(x_0)}{x_1 - x_0} = F'(\bar{x}).$$

To see the connection with Theorem 4.2.1 we need only rewrite (4.2.2) using the above relationships. For, $x_1 = g(b)$, $x_0 = g(a)$, $F(x_1) = f(b)$, $F(x_0) = f(a)$, and if $\bar{x} = g(\bar{t})$, then

$$F'(\bar{x}) = \frac{f'(\bar{t})}{g'(\bar{t})}.$$

With these substitutions, (4.2.2) becomes the equation in Theorem 4.2.1. See Figure 4.2.1.

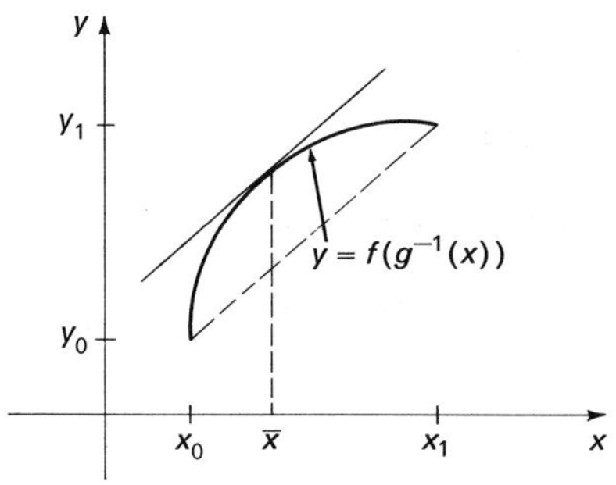

Figure 4.2.1

A very important application of the Cauchy Theorem arises in the following problem. Suppose we wish to find

$$\lim_{x \to 1} \frac{\log x}{x - 1}.$$

A quick examination reveals that none of our earlier theorems applies to this problem. There is no theorem on the limit of a quotient in which both numerator and denominator have limit zero, at least not yet! Moreover, there is no obvious way of squeezing this function between two other simpler functions with known limits.

We are asking for a limit of the form

$$\lim_{x \to 1} \frac{f(x)}{g(x)},$$

where

(4.2.3)
$$\lim_{x \to 1} f(x) = \lim_{x \to 1} g(x) = 0.$$

Moreover, since f and g are continuous for all $x \geq 0$, we conclude from (4.2.3) that

$$f(1) = g(1) = 0.$$

If we consider x and 1 as the (temporarily fixed) endpoints of an interval, we see that both f and g are differentiable on this closed interval, and the hypotheses of Theorem 4.2.1 are satisfied. Since

$$f'(x) = \frac{1}{x}, \quad g'(x) = 1,$$

we conclude that there is a number $\bar{t}$ between 1 and x such that

$$\frac{\log x - \log 1}{x - 1 - 0} = \frac{1}{\bar{t}}, \quad \text{or}$$

$$\frac{\log x}{x - 1} = \frac{1}{\bar{t}}.$$

We may write $\bar{t} = \bar{t}(x)$ and think of $\bar{t}$ as a function of the endpoint x. Since $\bar{t}$ is between x and 1, it is clear that

$$\lim_{x \to 1} \bar{t} = 1.$$

Thus

$$\lim_{x \to 1} \frac{\log x}{x - 1} = \frac{1}{\lim_{x \to 1} \bar{t}} = 1.$$

We formalize this application of the Cauchy Theorem in another theorem, commonly called l'Hospital's Rule.

Theorem 4.2.2 ***L'Hospital's Rule, first form*** *Let f and g be differentiable in a deleted neighborhood of c, and suppose that*

(i) $\lim_{x \to c} f(x) = \lim_{x \to c} g(x) = 0,$

(ii) $\lim_{x \to c} \dfrac{f'(x)}{g'(x)} = L.$

Then

$$\lim_{x \to c} \frac{f(x)}{g(x)} = L.$$

Proof We may assume that $f(c) = g(c) = 0$, i.e., that f and g are continuous at c, for if not we may replace them by new functions f_c and g_c such that $f_c(x) = f(x)$, $g_c(x) = g(x)$ for $x \neq c$, and $f_c(c) = g_c(c) = 0$. Then the theorem can be proved for f_c and g_c following the reasoning given below. The resulting limit conclusion for f_c and g_c will then obviously hold for f and g.

Both f and g are continuous on any interval $[x, c]$ or $[c, x]$ with x sufficiently close to c and differentiable on (x, c) or (c, x). Thus, the Cauchy Theorem applies, and we conclude that there exists a point $\bar{t} = \bar{t}(x)$ between x and c such that

(4.2.4)
$$\frac{f(x)}{g(x)} = \frac{f'(\bar{t})}{g'(\bar{t})}.$$

Because $c < \bar{t}(x) < x$ or $x < \bar{t}(x) < c$, we see by the Pinching Theorem[1] that

(4.2.5)
$$\lim_{x \to c} \bar{t}(x) = c;$$

but $\bar{t}(x) \neq c$ if $x \neq c$. Statements (4.2.5) and (ii) show that we may apply an earlier theorem on limits of composite functions, concluding that

(4.2.6)
$$\lim_{x \to c} \frac{f'(\bar{t}(x))}{g'(\bar{t}(x))} = L.$$

Statements (4.2.6) and (4.2.4) complete the proof.

Example 1 Let us find

$$\lim_{x \to 0} \frac{x}{e^x - 1}.$$

Here $\lim_{x \to 0} x = \lim_{x \to 0} (e^x - 1) = 0$, so l'Hospital's Rule applies. Thus

$$\lim_{x \to 0} \frac{x}{e^x - 1} = \lim_{x \to 0} \frac{1}{e^x} = \frac{1}{e}.$$

Example 2 What is

$$\lim_{x \to 0} \frac{e^x - 2 + e^{-x}}{1 - \cos 2x}?$$

Since $\lim_{x \to 0} (e^x - 2 + e^{-x}) = \lim_{x \to 0} (1 - \cos 2x) = 0$, we investigate the quotient of derivatives. What is

(4.2.7)
$$\lim_{x \to 0} \frac{e^x - e^{-x}}{2 \sin 2x}?$$

We cannot evaluate (4.2.7), because

$$\lim_{x \to 0} (e^x - e^{-x}) = \lim_{x \to 0} (2 \sin 2x) = 0.$$

This simply means that we must apply l'Hospital's Rule to find the limit (4.2.7). Using the rule, we have

(4.2.8)
$$\lim_{x \to 0} \frac{e^x - e^{-x}}{2 \sin 2x} = \lim_{x \to 0} \frac{e^x + e^{-x}}{4 \cos 2x} = \frac{2}{4} = \frac{1}{2}.$$

Thus, by (4.2.8) and another application of l'Hospital's Rule,

[1] *Exercise 1.7.17 of C 1: If $f(x) \le g(x) \le h(x)$ and $\lim_{x \to c} f(x) = \lim_{x \to c} h(x) = L$, then $\lim_{x \to c} g(x) = L$.*

$$\lim_{x\to 0} \frac{e^x - 2 + e^{-x}}{1 - \cos 2x} = \frac{1}{2}\cdot$$

Theorem 4.2.2 is but one form of l'Hospital's Rule, which occurs in several forms to fit several situations. We give a general statement of l'Hospital's Rule below in a simplified form sufficient for most applications. In this one statement we have combined up to 30 forms of the rule, depending on what one considers distinct forms. Many of these forms can be reduced by simple algebraic manipulations to Theorem 4.2.2. The others require separate proofs which are too complicated for inclusion here. Thus, except for the several proofs called for in the following exercises, we shall not attempt any further proof of l'Hospital's Rule.

The different forms of l'Hospital's Rule are obtained by selecting one symbol for each of the three place holders $\square$, $\triangle$, and $\bigcirc$.

The General l'Hospital's Rule

Let $\square$ be c, $c+$, $c-$, $+\infty$, or $-\infty$.
Let $\triangle$ be 0 or $+\infty$.
Let $\bigcirc$ be L, $+\infty$, or $-\infty$.

If $\displaystyle\lim_{x\to\square} |f(x)| = \lim_{x\to\square} |g(x)| = \triangle$

and $\displaystyle\lim_{x\to\square} \frac{f'(x)}{g'(x)} = \bigcirc,$

then $\displaystyle\lim_{x\to\square} \frac{f(x)}{g(x)} = \bigcirc.$

Example 3 Let us investigate the limit

$$\lim_{x\to 0+} x \log x.$$

Here, as is frequently the case, an algebraic manipulation is necessary before any form of l'Hospital's Rule applies. We may write

$$x \log x = \frac{\log x}{x^{-1}}\cdot$$

Now $\log x \to -\infty$ and $x^{-1} \to +\infty$ as $x \to 0+$, so the rule applies. We must look at

$$\frac{1/x}{-1/x^2} = -x,$$

which clearly has limit 0 as $x \to 0+$. Thus, we conclude that

$$\lim_{x\to 0+} x \log x = 0.$$

Example 4 Consider

$$\lim_{x\to+\infty} \frac{e^x}{x}.$$

This is already in the proper form for an application of l'Hospital's Rule, because $e^x \to +\infty$ and $x \to +\infty$ as $x \to +\infty$. Thus

$$\lim_{x\to+\infty} \frac{e^x}{x} = \lim_{x\to+\infty} \frac{e^x}{1} = +\infty.$$

Example 5 Consider

$$\lim_{x\to 0+} x^x.$$

This is not in a form to which the rule applies, but its logarithm can be put in proper form. For

$$\log x^x = x \log x,$$

and we need only refer to Example 3. We showed there that $\lim_{x\to 0+} x \log x = 0$, so

$$\lim_{x\to 0+} \log x^x = 0, \quad \lim_{x\to 0+} x^x = e^0 = 1,$$

the last statement following from an earlier theorem about limits of composites of continuous functions.[2]

[2] *If f is continuous and $\lim_{x\to c} g(x) = L$, then $\lim_{x\to c} f(g(x)) = f(\lim_{x\to c} g(x)) = f(L)$ (Theorem 1.7.5 in C 1).*

Example 6 The limit

$$\lim_{x\to+\infty} (x - 2\log x)$$

may be found by first considering $e^{x-2\log x} = e^x/x^2$, which is in a proper form for an application of l'Hospital's Rule; and we have

$$\lim_{x\to+\infty} \frac{e^x}{x^2} = \lim_{x\to+\infty} \frac{e^x}{2x} = \lim_{x\to+\infty} \frac{e^x}{2} = +\infty.$$

Notice that two applications of the rule were needed.

Thus,

$$\lim_{x\to+\infty} (x - 2\log x) = \lim_{x\to+\infty} \log\left(\frac{e^x}{x^2}\right) = +\infty,$$

where again we have moved the limit inside a continuous function.

Example 7 Let f be a function which has derivatives of all orders at c. That is, suppose that $f(c), f'(c), f''(c) = f^{(2)}(c), f'''(c) = f^{(3)}(c), f^{(4)}(c), \ldots$ all exist, and that as a result the functions $f^{(n)}(x)$, $n = 0, 1, 2, \ldots,$ are all continuous at $x = c$. (We use the notation $f^{(0)}(x)$ for $f(x)$ here.) If $f(c) = 0$ but $f'(c) \neq 0$, then

$$\lim_{x\to c} f(x) = 0, \quad \text{but}$$

$$\lim_{x \to c} \frac{f(x)}{(x-c)} = \lim_{x \to c} \frac{f'(x)}{1} = f'(c) \neq 0,$$

the latter equation following by an application of l'Hospital's Rule. In general, if $f(c) = f'(c) = f^{(2)}(c) = \cdots = f^{(n-1)}(c) = 0$ but $f^{(n)}(c) \neq 0$, then by a multiple application of l'Hospital's Rule we see that

(4.2.9) $$\lim_{x \to c} \frac{f(x)}{(x-c)^{n-1}} = \lim_{x \to c} \frac{f^{(n-1)}(x)}{(n-1)!} = \frac{f^{(n-1)}(c)}{(n-1)!} = 0;$$

and

(4.2.10) $$\lim_{x \to c} \frac{f(x)}{(x-c)^n} = \lim_{x \to c} \frac{f^{(n)}(x)}{n!} = \frac{f^{(n)}(c)}{n!} \neq 0,$$

(4.2.11) $$\lim_{x \to c} \left| \frac{f(x)}{(x-c)^{n+1}} \right| = \lim_{x \to c} \left| \frac{f^{(n)}(x)}{(n+1)!(x-c)} \right| = +\infty.$$

If (4.2.9) occurs, we say that $f(x)$ *tends to zero more rapidly than* $(x - c)^{n-1}$ *as* $x \to c$. If (4.2.10) occurs, we say that $f(x)$ and $(x - c)^n$ *tend to zero with the same order as* $x \to c$. If (4.2.11) occurs, we say that $f(x)$ *tends to zero less rapidly than* $(x - c)^{n+1}$ *as* $x \to c$.

If both (4.2.9) and (4.2.10) occur, we say that $f(x)$ *tends to zero with order n as* $x \to c$. Clearly, then, this order is a measure of the rapidity with which $f(x)$ tends to zero, for the function $(x - c)^n$ approaches zero, as $x \to c$, more rapidly for large values of n. It is easy to see that if $f(c) = f'(c) = \cdots = f^{(n)}(c) = 0$, then f tends to zero as $x \to c$ with order at least n. We leave the proof as an exercise.

We say that two functions f and g *fit each other with order n at* $x = c$ if the function $f(x) - g(x)$ tends to zero with order n as $x \to c$. From the above, it can easily be shown that f and g fit at $x = c$ with order n if and only if $f(c) = g(c)$, $f'(c) = g'(c), \ldots,$ $f^{(n)}(c) = g^{(n)}(c)$, but $f^{(n+1)}(c) \neq g^{(n+1)})(c)$. This proof is also left as an exercise. Thus, order of fit is a measure of how closely the graphs of two functions approximate each other, and l'Hospital's Rule makes it possible to develop a simple method of checking the order of fit of two functions.

Example 8 Let f be an arbitrary function having at least one derivative at $x = c$. If g is a linear function for which $g(c) = f(c)$, then g may be written

$$g(x) = m(x - c) + f(c).$$

As we see from Example 7, if $m \neq f'(c)$, then g fits f with order 0 ($g(c) = f(c)$, but $g'(c) \neq f'(c)$). On the other hand, if $m = f'(c)$, in which case the line (the graph of g) is tangent to the graph of f

at $x = c$, then g fits f with order at least 1. This makes more precise what we mean when we say that the tangent line to the graph of $y = f(x)$ is the best fitting line in a neighborhood of the point $(c, f(c))$.

Exercises

1 Show that if $f(c) = f'(c) = \cdots = f^{(n-1)}(c) = 0$, but $f^{(n)}(c) \neq 0$, then (4.2.9), (4.2.10), (4.2.11) hold.

2 Show that if $f(c) = f'(c) = \cdots = f^{(n-1)}(c) = 0$, then f tends to zero as $x \to c$ with order at least n.

3 Show that f and g fit at $x = c$ with order n if and only if $f(c) = g(c)$, $f'(c) = g'(c), \ldots, f^{(n-1)}(c) = g^{(n-1)}(c)$, but $f^{(n)}(c) \neq g^{(n)}(c)$.

4 Let f be a function with at least two continuous derivatives at $x = x_0$. If q is a quadratic polynomial function, then q may be expressed, after some algebraic manipulations, as

$$q(x) = A(x - x_0)^2 + B(x - x_0) + C.$$

Of all quadratic functions, there is precisely one which fits f at $x = x_0$ better than all the rest. Find the coefficients A, B, C of that quadratic function and describe the order of the fit. Its graph is the best fitting parabola for the graph of f in a neighborhood of $(x_0, f(x_0))$.

5 For each of the following pairs of functions, find a value of $\bar{t}$ which is guaranteed by Theorem 4.2.1.

(a) $f(x) = x,\ g(x) = x^2$ over $[1, 2]$,

(b) $f(x) = x^2,\ g(x) = x^3$ over $[a, b],\ a > 0$,

(c) $f(x) = \sin x,\ g(x) = \cos x$ over $[\pi/4, 3\pi/4]$.

6 In each case prove that a value of t exists which satisfies the stated inequality.

(a) $\dfrac{\sin x}{x} = \cos t, \quad 0 < t < x,$

(b) $\dfrac{\cos x - 1}{x} = \sin t, \quad 0 < t < x,$

(c) $\dfrac{\tan x}{x} = \sec^2 t, \quad 0 < t < x < \pi/2,$

(d) $\dfrac{\arctan x}{x} = \dfrac{1}{1 + t^2}, \quad 0 < t < x.$

7 Prove that $e^x > 1 + x$ for $x \neq 0$.

8 (a) Show that if $x > 0$, $\dfrac{\log(x + 1)}{x} = \dfrac{1}{t + 1}$ for some t, $0 < t < x$.

(b) Show that if $x > 0$, $\dfrac{x}{x + 1} < \log(x + 1) < x$.

9 Prove that $\lim_{k\to\infty} k^{1/k} = 1$.

In Exercises 10–34 evaluate the limits.

10 $\lim_{x\to 3} \dfrac{x^2 - 9}{x - 3}$.

11 $\lim_{x\to 0} \dfrac{\tan x}{x}$.

12 $\lim_{x\to 0} \dfrac{\sin x - x}{x}$.

13 $\lim_{x\to 0} \dfrac{\arctan x}{x}$.

14 $\lim_{x\to \pi/2} \dfrac{\cos x}{x - \pi/2}$.

15 $\lim_{x\to 1} \dfrac{\log x}{x - 1}$.

16 $\lim_{x\to 0} \dfrac{1 - \cos x}{x^2}$.

17 $\lim_{x\to \pi/2+} \dfrac{\log (x - \pi/2)}{\tan x}$.

18 $\lim_{x\to \pi/2} \dfrac{\log (\sin x)}{\cot x}$.

19 $\lim_{x\to +\infty} \dfrac{\log x}{x}$.

20 $\lim_{x\to +\infty} \dfrac{\log x}{\sqrt{x}}$.

21 $\lim_{x\to 0} \dfrac{\log (1 - x)}{e^x - 1}$.

22 $\lim_{x\to 0+} \dfrac{\log x}{\cot x}$.

23 $\lim_{x\to \pi/2-} \dfrac{\sec x + 1}{\tan x}$.

24 $\lim_{x\to +\infty} \dfrac{e^x}{x}$.

25 $\lim_{x\to +\infty} \dfrac{x + \log x}{x \log x}$.

26 $\lim_{x\to +\infty} \dfrac{e^x}{\log x}$.

27 $\lim_{x\to +\infty} xe^{-x}$.

28 $\lim_{x\to +\infty} x(e^{1/x} - 1)$.

29 $\lim_{x\to \pi/2} (\sec x - \tan x)$.

30 $\lim_{x\to +\infty} (1 + x^2)^{1/x}$.

31 $\lim_{x\to 0} (1 + x^2)^{1/x^2}$.

32 $\lim_{x\to 0+} x^x$.

33 $\lim_{x\to 0+} \left(\dfrac{1}{x}\right)^{\sin x}$.

34 $\lim_{x\to 0} (\cos x)^{1/x}$.

35 $\lim_{x\to \pi/2} (\sin x)^{\tan x}$.

36 L'Hospital's Rule is not all-powerful! Show that the limit

$$\lim_{x\to\infty} \frac{x - \sin x}{x + \sin x}$$

exists, but cannot be evaluated by l'Hospital's Rule.

4.3

Polynomial Approximations and Taylor's Theorem

In the discussion at the end of the last section we have a clear example of a limit statement being interpreted as a statement about an approximation. For the statement "g fits f with order n at $x = c$" is an approximation statement. It asserts the possibility of using the function g as an approximation of the function f in a neighborhood of $x = c$. The higher the order of fit the better the approximation. But the complete meaning of this approximation statement is contained in the limit statement

$$\lim_{x \to c} \frac{f(x) - g(x)}{(x - c)^n} = M \neq 0.$$

Exercise 4.2.3 introduces a systematic way of predicting and estimating the accuracy of an approximation of one function by another. We shall sharpen this estimate in this section.

As we remarked in the Introduction, the purpose of an approximation is to replace a difficult calculation with an easier one. An approximation, no matter how close, of one difficult function by another which is just as hard to deal with is of no use. One wishes to approximate general functions by special functions selected for ease of calculation. One such class of easily calculated functions is the class of polynomial functions. The object of this section is to develop from the discussion in the last section a simple method of selecting that polynomial of a given degree which best approximates a given function near a point and to construct an effective estimate of the accuracy of this approximation.

First, let us pose the problem in more precise terms. Suppose that f is a function which has derivatives of all orders in a neighborhood of $x = c$. Suppose further that at c these derivatives $f(c), f'(c), \ldots, f^{(n)}(c), \ldots$ are all known. The problem is to find for a given n the polynomial of degree n which best approximates $f(x)$ for x near c, and to estimate how close this approximation will be.

Example 1 Suppose we wish to find an approximate value for $\sqrt{25.5}$. It might be suggested that we look for easily calculated approximations of the function

(4.3.1) $$f(x) = \sqrt{x}$$

for values of x near 25. Polynomial approximations, if they can be found, should satisfy the calculability requirement, and since $f(25)$, $f'(25), f''(25), \ldots$ can all be found exactly, the situation is that described in the preceding paragraph.

Example 2 The same situation is encountered if we desire an approximation of $e^{-.01}$ of great accuracy. For the function

(4.3.2) $$f(x) = e^x$$

is easily differentiated, and we see that $f(0) = f'(0) = f''(0) = \cdots = 1$. A polynomial approximation of this function which is accurate for values of x near 0 should help us in finding the desired numerical approximation.

With the problem thus formulated the selection of the best-fitting polynomial becomes easy. For in Exercise 4.2.3 we found that in general the more derivatives two functions have in common at $x = c$, the better they fit in a neighborhood of $x = c$. Thus, if we wish to approximate $f(x)$ near c by means of a polynomial of degree n, then we should select one which has the maximum possible number of derivatives in common with f at c. This is easily done, as we see below.

First, observe that *any* polynomial of degree n may be written in the form

(4.3.3) $$a_0 + a_1(x - c) + a_2(x - c)^2 + \cdots + a_n(x - c)^n,$$

where c is any real number. The proof is left as an exercise (Exercise 4.3.17). If we denote the value of this polynomial for an arbitrary x by $P(x)$, then we see that

$$P(c) = a_0,\ P'(c) = a_1,\ P''(c) = 2a_2,\ P'''(c) = 6a_3, \ldots,$$

and in general

$$P^{(i)}(c) = (i!)a_i,\ i = 0, 1, \ldots, n.^1$$

This also is left as an exercise (Exercise 4.3.18).

[1] *Recall that $i!$ (read "i factorial") is the number $1 \cdot 2 \cdot 3 \cdot \ldots \cdot i$.*

Thus, by choosing the coefficients a_i appropriately, we may find a polynomial $P(x)$ whose derivatives at c of all orders up to and including n have any desired values. Then the polynomial is completely specified, for all of its coefficients are chosen. Clearly, the polynomial we are looking for, the one which best fits $f(x)$ near $x = c$, is found by setting $P^{(i)}(c) = i!a_i$ equal to $f^{(i)}(c)$. That is, we choose

$$a_i = \frac{f^{(i)}(c)}{i!},\ i = 0, 1, 2, \ldots, n.$$

(Keep in mind the usual conventions that $f^{(0)}(x) = f(x)$ and $0! = 1$.)

This polynomial is called the *Taylor polynomial of degree n* corresponding to f, and we shall denote it by T_n. Thus

(4.3.4) $$T_n(x) = f(c) + f'(c)(x - c) + \frac{f''(c)}{2!}(x - c)^2 + \cdots + \frac{f^{(n)}(c)}{n!}(x - c)^n.$$

By Exercise 4.2.3, T_n fits f at $x = c$ with order $n + 1$ at least. Any other polynomial of degree n fits f at $x = c$ with order n or less. Thus, we have

Theorem 4.3.1 *The Taylor polynomial T_n is the polynomial of degree n which best fits f in a neighborhood of $x = c$.*

Having found the best-fitting polynomial, we shall find it rather easy to estimate the error involved in the resulting approximation. We already know from Exercise 4.2.1 that

(4.3.5) $$\lim_{x\to c}\frac{f(x) - T_n(x)}{(x - c)^{n+1}} = \frac{f^{(n+1)}(c)}{(n + 1)!},$$

which says something about how fast $f(x) - T_n(x)$ tends to zero as $x \to c$, that is, how close $T_n(x)$ is to $f(x)$ for x near c. But we need more precise information, which we shall obtain by going back to the *source* of l'Hospital's Rule, the Cauchy Theorem.

The error whose magnitude we want to estimate is the difference

$$R_n(x) = f(x) - T_n(x).$$

We call R_n the *Taylor remainder* of order n. From the construction of T_n we know that

(4.3.6) $$R_n(c) = R_n'(c) = R_n''(c) = \cdots = R_n^{(n)}(c) = 0.$$

Statement (4.3.5) gauges the order of accuracy of the approximation by comparing $R_n(x)$ to the function

$$g(x) = (x - c)^{n+1}.$$

We wish now to make this comparison in a different and more explicit way. Notice that g also has property (4.3.6), that is,

$$g(c) = g'(c) = g''(c) = \cdots = g^{(n)}(c) = 0.$$

On the other hand,

(4.3.7) $$g(x) \neq 0,\ g'(x) \neq 0,\ \ldots,\ g^{(n)}(x) \neq 0, \text{ for } x \neq c,$$

so the stage is set for an n-fold application of the Cauchy Theorem. First, apply the theorem to the functions R_n and g on the interval between any x (we consider x as fixed, for the moment) and c. We conclude that

$$\frac{R_n(x)}{(x - c)^{n+1}} = \frac{R_n'(x_1)}{(n + 1)(x_1 - c)^n} = \frac{R_n'(x_1)}{g'(x_1)}$$

for some number x_1 between x and c. Since $x_1 \neq c$, we may by (4.3.7) apply the Cauchy Theorem again to the two functions R_n' and g' on the interval between x_1 and c, obtaining

$$\frac{R_n(x)}{(x - c)^{n+1}} = \frac{R_n'(x_1)}{(n + 1)(x_1 - c)^n} = \frac{R_n''(x_2)}{(n + 1)n(x_2 - c)^{n-1}}.$$

for some x_2 between x_1 and c and hence also between x and c. Continuing in this way, we eventually reach the conclusion that, for some number $x_{n+1} = \bar{x}$ between x and c,

(4.3.8) $$\frac{R_n(x)}{(x-c)^{n+1}} = \frac{R_n^{(n+1)}(\bar{x})}{(n+1)!}.$$

But since T_n is a polynomial of degree n, it is clear that $T^{(n+1)}(x)$ is identically zero, so we have

(4.3.9) $$R_n^{(n+1)}(x) = f^{(n+1)}(x) \quad \text{for all } x.$$

Statements (4.3.8) and (4.3.9) imply that

$$R_n(x) = \frac{f^{(n+1)}(\bar{x})}{(n+1)!}(x-c)^{n+1}$$

for some number $\bar{x}$ between x and c, giving us the estimate of $R_n(x)$ for which we have been searching. This expression for the Taylor remainder is given by the important theorem generally known as Taylor's Theorem.

Theorem 4.3.2 ***Taylor*** *If f has derivatives of all orders up to and including $n+1$ in an interval I containing $x = c$, then for every x in I there exists a number $\bar{x}$ between x and c such that*

(4.3.10) $$f(x) = T_n(x) + \frac{f^{(n+1)}(\bar{x})}{(n+1)!}(x-c)^{n+1},$$

where T_n is the Taylor polynomial (4.3.4).

The following immediate corollary is often more convenient than the theorem in the applications.

Corollary 4.3.3 *If f satisfies the hypotheses of Taylor's Theorem and if*

$$|f^{(n+1)}(x)| \leq M_n$$

for all x in I, then

(4.3.11) $$|R_n(x)| \leq \frac{M_n}{(n+1)!}|x-c|^{n+1}$$

for all x in I.

We have made significant and entirely unexpected progress in this theorem and corollary, results whose far reaching consequences we can only hint at here. The estimate of the error $|R_n(x)|$ in a Taylor polynomial approximation is a *global* statement. Heretofore, we have spoken of $T_n(x)$ approximating $f(x)$ *in a neighborhood of* c, and the presence of the term $|x-c|^{n+1}$ in (4.3.10) shows clearly how this approximation becomes very close as x gets near c. However, the estimate (4.3.10) holds for *any* x in the interval I on which f is defined and $(n+1)$-times differentiable. The presence of the term $(n+1)!$ in the denominator of the estimate suggests that if I is not too large and the numbers M_n do not get too large, then we may be able to find Taylor polynomials which approximate $f(x)$ closely *over the entire interval I*. It is impossible to overrate the importance of this first indication of the possibility of *global* approximations of a function by a polynomial.

We illustrate the uses of Taylor polynomials and Taylor's Theorem by continuing Examples 1 and 2.

Example 1 (continued) For the function (4.3.1) we have

$$f(x) = x^{1/2},\ f'(x) = \tfrac{1}{2}x^{-1/2},\ f''(x) = -\tfrac{1}{4}x^{-3/2},$$
$$f'''(x) = \tfrac{3}{8}x^{-5/2},\ f^{(4)}(x) = -\tfrac{15}{16}x^{-7/2}\ \ldots .$$

Thus,

$$f(25) = 5, f'(25) = \frac{1}{10}, f''(25) = -\frac{1}{500}, f'''(25) = \frac{3}{25{,}000}, \ldots,$$

so that we have, taking $n = 3$,

$$\sqrt{x} = T_3(x) + R_3(x),$$

where

$$T_3(x) = 5 + \frac{1}{10}(x-25) - \frac{1}{1000}(x-25)^2 + \frac{1}{50{,}000}(x-25)^3.$$

Our approximation is, then,

$$\sqrt{25.5} \approx T_3(25.5).$$

Before we calculate this approximation, let us estimate the accuracy. Since

$$|f^{(4)}(x)| = \tfrac{15}{16}x^{-7/2},$$

we see that for x between 25 and 25.5

$$|f^{(4)}(x)| \le \frac{15}{16}(25)^{-7/2} = \frac{3}{2^4 \cdot 5^6}.$$

Thus, by Corollary 4.3.3,

$$|R_3(25.5)| \le \frac{3}{(4!)2^4 5^6}(.5)^4 = \frac{1}{2^5 \cdot 10^6},$$

or

$$|R_3| \le .00000003 \ldots .$$

Therefore, the error can affect only the eighth decimal place, and our approximation will be good to *seven places*.

The rest is easy, for

$$\begin{aligned} T_3(25.5) &= 5 + \frac{.5}{10} - \frac{(.5)^2}{1000} + \frac{(.5)^3}{50{,}000} \\ &= 5.0497525, \end{aligned}$$

an approximation of $\sqrt{25.5}$ accurate to seven decimal places.

Example 2 (continued) For the function (4.3.2) we have

$$f(x) = f'(x) = f''(x) = \cdots = e^x.$$

Thus,

$$f(0) = f'(0) = f''(0) = \cdots = 1,$$

and we conclude that for any n

(4.3.12) $$e^x = 1 + x + \frac{x^2}{2!} + \frac{x^3}{3!} + \cdots + \frac{x^n}{n!} + R_n(x).$$

Suppose we wish to approximate $e^{-.01}$ to ten decimal place accuracy. Then we wish to use an n large enough so that $|R_n(x)| < 5 \cdot 10^{-11}$. In general, since

$$|f^{(n)}(x)| = e^x \leq e^0 = 1$$

for all $x \leq 0$, we have

$$|R_n| \leq \frac{(.01)^{n+1}}{(n+1)!} = \frac{10^{-2(n+1)}}{(n+1)!}.$$

It is easy to see that $n = 3$ is too small but that

$$|R_4| \leq \frac{10^{-10}}{120} < 10^{-12}.$$

Thus, $n = 4$ will give accuracy to at least 11 decimal places. We therefore have

$$\begin{aligned} e^{-.01} &\approx 1 - (.01) + \frac{(.01)^2}{2} - \frac{(.01)^3}{6} + \frac{(.01)^4}{24} \\ &= .990049833750. \end{aligned}$$

Such accuracy with so little effort was possible in each of these examples because the number x is quite close to the number c where the values of the function and its derivatives are known. As an example of the global power of Taylor's theorem consider the following.

Example 3 Suppose we use the Taylor polynomial in (4.3.12) to approximate e. We are then using $x = 1$, and the approximation (4.3.4) would become

$$e \approx 1 + 1 + \frac{1}{2!} + \frac{1}{3!} + \cdots + \frac{1}{n!}, \tag{4.3.13}$$

which is certainly interesting if for no reason other than its extreme simplicity. However, this value of x is much farther from 0 than was $x = -.01$, so we can expect to need a larger n to achieve as much accuracy.

In order to estimate the accuracy of an approximation based on (4.3.12) we must find an upper bound for $|f^{(n+1)}(x)| = e^x$ on [0, 1]. This upper bound is, of course, e itself—a bit embarrassing, since this is the number we are trying to approximate. We seem to have to know the number before we can say how accurately we are approximating it. However, recall that Corollary 4.3.3 does not require us to have the *best* upper bound for $|f^{(n+1)}(x)|$; any upper bound will give us some information. If we use a wasteful estimate of $|f^{(n+1)}(x)|$ and hence of $|R_n(x)|$, then the worst that can happen is that we achieve much more accuracy in our approximation than we could predict. Since we know from earlier considerations that $2 < e < 3$, we can thus be sure that

$$|f^{(n+1)}(x)| \leq 3 \quad \text{for } 0 \leq x \leq 1,$$

and therefore that

$$|R_n(1)| \leq \frac{3}{(n+1)!}.$$

Thus, we obtain at least three decimal place accuracy with $n = 6$. It is worth the effort to carry out this calculation and compare the result with a good decimal approximation of e.

Exercises

For each of the functions in Exercises 1–11 find the Taylor polynomial T_n of degree n and write out Taylor's formula (4.3.10).

1 $f(x) = \cos x$, $c = \pi/4$, $n = 4$.

2 $f(x) = \dfrac{1}{1+x}$, $c = 0$, $n = 4$.

3 $f(x) = \log x$, $c = 1$, $n = 4$.

4 $f(x) = \sec x$, $c = \pi/4$, $n = 3$.

5 $f(x) = e^{-x}$, $c = 0$, $n = 5$.

6 $f(x) = \dfrac{1}{\sqrt[3]{x}}$, $c = 8$, $n = 3$.

7 $f(x) = \tan x,\ c = \pi/4,\ n = 3.$

8 $f(x) = \sqrt{1 + x},\ c = 0,\ n = 5.$

9 $f(x) = \arcsin x,\ c = 0,\ n = 3.$

10 $f(x) = \dfrac{1}{\sqrt{1 - x}},\ c = 0,\ n = 4.$

11 $f(x) = \arctan x,\ c = 1,\ n = 3.$

12 Use Taylor's Theorem for $f(x) = \sqrt{1 + x}$ with $c = 0$ and $n = 4$ to approximate $\sqrt{1.2}$. Estimate the error in this approximation.

13 Use an appropriate Taylor polynomial of degree 4 to approximate sin 61°. Estimate the error in this approximation.

14 How many terms in Taylor's formula for $f(x) = e^x$, $c = 0$, are necessary to approximate $\sqrt[3]{e}$ with an error of less than .0005?

15 Show that if P_n is the Taylor polynomial of degree n for the function f at a point c, then P'_n is the Taylor polynomial of degree $n - 1$ for f' at c.

16 Use an appropriate Taylor polynomial to approximate $\sqrt[3]{28}$ with an error of less than 5×10^{-3}.

17 Prove that any polynomial p of degree n can be written as

$$p(x) = a_0 + a_1(x - c) + a_2(x - c)^2 + \cdots + a_n(x - c)^n,$$

where c is arbitrary. (*Hint:* Recall from algebra that $p(x) = (x - c)q(x) + p(c)$, where q is a polynomial of degree at most $n - 1$. Then use mathematical induction.)

18 Prove that if

$$p(x) = a_0 + a_1(x - c) + a_2(x - c)^2 + \cdots + a_n(x - c)^n,$$

then

$$p(c) = a_0,\ p'(c) = a_1,\ \ldots,\ p^{(i)}(c) = (i!)a_i,\ \ldots,\ p^{(n)}(c) = n!a_n.$$

19 On the same set of coordinate axes, graph the function $f(x) = e^x$ and its Taylor polynomials of degrees 0, 1, 2, 3 at the point $c = 1$.

20 Follow the instructions for Exercise 19 for the function $f(x) = \sqrt{x}$.

infinite series

chapter 5

5.0

Introduction In Taylor's Theorem at the close of the last chapter we considered certain polynomials $T_n(x)$ which approximate a given function well under certain conditions for all x in a specified interval. Let us look at this method of approximation in a slightly different way. Instead of looking at a specific integer n and considering the approximation $T_n(x) \approx f(x)$ for all x in an interval, let us look at a fixed x and consider the whole sequence of approximations $\{T_n(x)\}$, $n = 0, 1, 2, \ldots$, of the number $f(x)$. Since we believe, generally, in the basic principle that the Taylor polynomials become better approximations of the function as the degree n gets larger, it seems possible that the sequence $\{T_n(x)\}$ might approach $f(x)$ as a limit as $n \to \infty$. From Taylor's Theorem it is easy to see conditions under which this will be true; we shall investigate them more closely in Section 5.6.

For the moment, suppose that

(5.0.1) $$f(x) = \lim_{n\to\infty} T_n(x),$$

where $T_n(x)$ is the Taylor polynomial for f about $x = 0$. This is just another example of a sequence which has a limit. But it is a peculiar kind of sequence. Remember that

(5.0.2) $$T_n(x) = a_0 + a_1x + a_2x^2 + \cdots + a_nx^n,$$

where the coefficients a_i are related to f by the conditions

$$a_i = \frac{1}{i!}f^{(i)}(0).$$

So each term $T_n(x)$ of the sequence is obtained by adding a number a_nx^n to the previous term $T_{n-1}(x)$. Thus the terms $T_n(x)$ are ever-lengthening sums of the form

$$\begin{aligned} T_0(x) &= a_0, \\ T_1(x) &= a_0 + a_1x, \\ T_2(x) &= a_0 + a_1x + a_2x^2, \\ T_3(x) &= a_0 + a_1x + a_2x^2 + a_3x^3, \\ &\dots \end{aligned}$$

It is very tempting, then, to consider the limit of the sequence $\{T_n(x)\}$, namely $f(x)$, as a sort of infinite sum, that is, the sum of an infinite sequence of numbers $a_0, a_1x, a_2x^2, a_3x^3, \dots$. This sum is finite even though the number of numbers to be added together is infinite. It seems natural to write (5.0.1) and (5.0.2) together as

(5.0.3) $$f(x) = a_0 + a_1x + a_2x^2 + \cdots + a_nx^n + \cdots.$$

In a sense we have created no new ideas; (5.0.3) is merely a suggestive shorthand expression for the statements (5.0.1) and (5.0.2). But in another sense, we *have* created something new, a different way of thinking about an old problem. We have uncovered a new mathematical object, an infinite sum, or as we shall call it, an *infinite series*. Does this new mathematical object which is really just a different way of looking at an old mathematical problem represent a net gain in our insight?

It does. The very act of identifying a "new" object in a statement like (5.0.1) opens up the possibility of tremendous new penetrations. For, an exhaustive study of infinite series in their own right, an investigation of ways they can be manipulated and properties which they display, may reveal patterns and insights which shed new light on functions for which statements (5.0.1) or (5.0.3) are true.

This is indeed the case. Infinite series are among the most important tools of the mathematician in the study of functions. There are so many ways of manipulating and analyzing series that a statement of the form (5.0.3) becomes much more than just a convenient and suggestive shorthand. It actually carries with it a whole new set of insights concerning the function $f(x)$ which were not at all evident in the original statement of Taylor's Theorem.

In this chapter we shall first develop some of the basic ideas and theorems about series; then with this new point of view, in Section 5.6 we shall revisit Taylor's Theorem. Later, in Chapter 7 we shall see one situation (of many which are encountered in more advanced mathematics) in which certain functions are encountered *first* and perhaps *only* as infinite series.

5.1

Series

The idea of an infinite sum is patently absurd. The addition operation is a *finite* one; that is, it is defined only for *pairs* of numbers and is extended inductively to *finite* sets of numbers. A sum $a_1 + a_2 + \cdots + a_n$ makes sense if n is a positive integer, but $a_1 + a_2 + \cdots + a_n + \cdots$ does not, unless we add something extra to the purely algebraic notion of sum. That something extra is the notion of limit.

The ingredients of an infinite sum are, first of all, a sequence of numbers

$$a_1, a_2, a_3, \ldots .$$

These are the numbers to be added. Though it makes no sense to add them all, it does make sense to add any finite partial list. That is, we can consider the sums

$$a_1,\ a_1 + a_2,\ a_1 + a_2 + a_3,\ a_1 + a_2 + a_3 + a_4,\ \ldots .$$

These partial sums themselves form a second sequence of numbers. Now we may consider a limit, the limit of this second sequence, the sequence of partial sums. This limit, if it exists, will be defined to be the sum of the infinite set of numbers $a_1, a_2, a_3, \ldots$.

This sum is, of course, just a real number. But what is the particular object which is suggested by the words "infinite sum"? It must be more than just a real number, in much the same way that "$1 + 2 + 3 + 4$" is conceptually different from "10." School books often use the phrase "indicated sum" to reflect this distinction. We shall call the object which we are trying to describe an *infinite series*, and we shall write it thus:

$$a_1 + a_2 + a_3 + \cdots + a_n + \cdots .$$

As we have seen above, two sequences and a limit are involved in an adequate definition of the sum of such a series. So we shall define the series to be that pair of sequences.

Definition 5.1.1 An *infinite series* (or just *series*) is a pair of sequences of numbers

$$a_1, a_2, a_3, \ldots \tag{5.1.1}$$

and

$$S_1, S_2, S_3, \ldots , \tag{5.1.2}$$

which are related by the conditions

$$S_1 = a_1,\ S_2 = a_1 + a_2,\ S_3 = a_1 + a_2 + a_3,\ \ldots ,$$

$$S_n = a_1 + a_2 + \cdots + a_n,\ \ldots .$$

We use the shorthand notation

(5.1.3) $$a_1 + a_2 + a_3 + \cdots + a_n + \cdots \quad \text{or} \quad \sum_{k=1}^{\infty} a_k$$

for a series. The numbers in the sequence (5.1.1) are called the *terms* of the series. The numbers in the sequence (5.1.2) are called the *partial sums* of the series.

The sum of a series is, as we have suggested above, the limit of the sequence of partial sums.

Definition 5.1.2 If the sequence (5.1.2) of partial sums of the series (5.1.3) has a limit as $n \to \infty$, we say that the series *converges* or is *convergent*. Otherwise, we say that the series is *divergent*. If S is the limit of the sequence of partial sums, that is, if $S = \lim S_n$, then we say that the sequence *converges to the sum* S and we write

$$S = a_1 + a_2 + a_3 + \cdots + a_n + \cdots.$$

Obviously, we should be able to obtain some results about series by simply borrowing from earlier information about sequences. For example, consider the following special series.

Example 1 We investigate the *geometric* series

(5.1.4) $$1 + r + r^2 + \cdots + r^n + \cdots,$$

where r is any fixed real number. The terms of this series are 1, $r, r^2, \ldots$, and the nth partial sum is

(5.1.5) $$S_n = 1 + r + \cdots + r^{n-1}.$$

It is quite easy to see that if $r \neq 1$,

$$S_n = \frac{1 - r^n}{1 - r}.$$

For if we multiply out the expression $S_n(1 - r)$, we obtain a sum of $2n$ terms, all but two of which cancel in pairs. Or, as an alternative proof we may use mathematical induction to show that $S_n(1 - r) = 1 - r^n$ for all n. (See Exercise 5.1.9.) Thus, the convergence of the series (5.1.4) (for $r \neq 1$) depends upon whether the sequence (5.1.5) has a limit.

It is easy to show (see Exercise 5.1.10) that if $|r| < 1$ (that is, $-1 < r < 1$), then

$$r^n \to 0.$$

In this case,

$$S_n = \frac{1 - r^n}{1 - r} \to \frac{1}{1 - r} \text{ as } n \to \infty,$$

and we conclude that the series (5.1.4) converges to the sum $1/(1 - r)$. That is,

$$1 + r + r^2 + \cdots + r^n + \cdots = \frac{1}{1 - r} \text{ if } |r| < 1.$$

Several theorems about convergent and divergent series follow immediately from known results about sequences.

Theorem 5.1.1 *If the series $a_1 + a_2 + a_3 + \cdots$ converges, then $\lim a_n = 0$. That is, the sequence of terms of a convergent series has limit* 0.

Proof If S_n is the nth partial sum of the series, then $S_n \to S$, where S is the sum of the series. It follows that $S_{n+1} \to S$ also as $n \to \infty$. Thus, by one of the basic theorems on limits of algebraic combinations of sequences,

$$S_{n+1} - S_n \to S - S = 0 \quad \text{as} \quad n \to \infty.$$

To complete the proof we need only observe that by definition of S_{n+1} and S_n,

$$S_{n+1} - S_n = a_n.$$

Theorem 5.1.1 gives a *necessary* condition for a series to converge. It is therefore useful in showing that certain series diverge, for if the nth term a_n of a series does not tend to zero, then the theorem tells us that the series cannot converge.

Example 2 If $|r| \geq 1$, then an easy proof by induction shows that $|r^{n-1}| \geq 1$ for all positive integers n. Thus, obviously r^{n-1} cannot tend to zero as $n \to \infty$. Since r^{n-1} is the nth term of the series (5.1.5), we conclude that

$$1 + r + r^2 + \cdots + r^n + \cdots \text{ diverges for } |r| \geq 1.$$

But the necessary condition for convergence given in Theorem 5.1.1 is not a *sufficient* condition. Even if one verifies that $a_n \to 0$, one cannot be sure that $a_1 + a_2 + a_3 + \cdots$ converges, as the following example shows.

Example 3 We consider the so-called *harmonic series*

$$1 + \frac{1}{2} + \frac{1}{3} + \cdots + \frac{1}{n} + \cdots. \tag{5.1.6}$$

Here $a_n = 1/n$, so clearly $a_n \to 0$. However, we shall see that the series is divergent. We already know that in order to have a (finite) limit, a sequence must be bounded. That is, if $\lim S_n = S$, then there exists a constant M such that $S_n \leq M$ for $n = 1, 2, 3, \ldots$. We shall show that S_n is not bounded if

$$S_n = 1 + \frac{1}{2} + \frac{1}{3} + \cdots + \frac{1}{n}.$$

There are many ways of doing this, and the one we use here is neither the most elementary nor the most obvious. However, it is one of the shortest and most elegant, and it fits closely a forthcoming more general pattern.

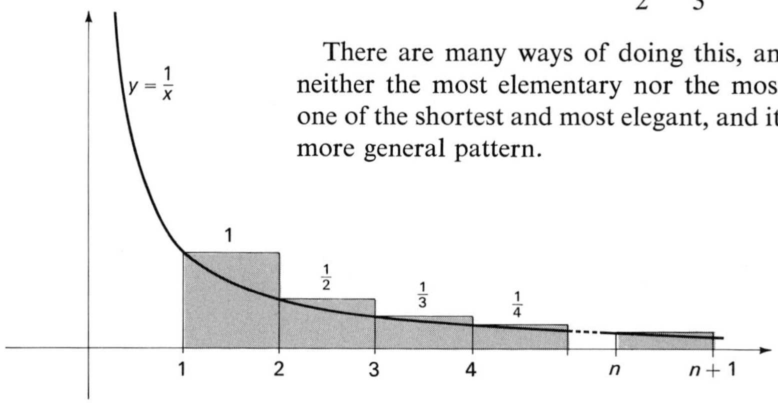

Figure 5.1.1

Consider the function f: $f(x) = 1/x$, for $x \geq 1$. The drawing in Figure 5.1.1 shows the region under the graph of f between $x = 1$ and $x = n + 1$ and also certain rectangles whose areas are $1, \frac{1}{2}, \frac{1}{3}, \ldots, \frac{1}{n}$ respectively. It is immediately evident from the figure that

$$\int_1^{n+1} \frac{dx}{x} \leq 1 + \frac{1}{2} + \cdots + \frac{1}{n} = S_n.$$

Evaluating the integral, we see that

$$\log(n + 1) \leq 1 + \frac{1}{2} + \cdots + \frac{1}{n} = S_n.$$

Since $\log(n + 1) \to \infty$ as $n \to \infty$, the sequence $\{S_n\}$ is not bounded and hence cannot converge. Thus, the series (5.1.6) is divergent even though its nth term tends to zero.

In the last example we showed that a series diverges by showing that its sequence of partial sums is not bounded. We recall a theorem which concerns the converse situation: Any increasing sequence which is bounded has a finite limit (Theorem 1.3.4). We can turn this immediately into a useful theorem about series.

Theorem 5.1.2 *Let $a_k \geq 0$ for $k = 1, 2, \ldots$. Then the series $a_1 + a_2 + a_3 + \cdots$ converges if and only if the sequence of partial sums $S_n = a_1 + a_2 + \cdots + a_n$ is bounded.*

Proof Since

$$S_{n+1} = S_n + a_{n+1},$$

we see by hypothesis that

$$S_{n+1} \geq S_n.$$

Thus the sequence $\{S_n\}$ is increasing. If this sequence is also bounded, Theorem 1.3.4 shows that it has a finite limit S. This means that $a_1 + a_2 + \cdots$ converges to the sum S. Conversely, if the series converges to S, then since $S_n \to S$ the sequence $\{S_n\}$ is bounded.

Example 4 Consider the series

(5.1.7)
$$1 + \frac{1}{2^2} + \frac{1}{3^2} + \cdots + \frac{1}{n^2} + \cdots.$$

Since this series is so similar to the series (5.1.6), it is not unnatural to analyze it in much the same way. If we consider the graph of $y = 1/x^2$ for $x \geq 1$, we can see from Figure 5.1.2 that

$$\frac{1}{2^2} + \frac{1}{3^2} + \cdots + \frac{1}{n^2} < \int_1^n \frac{dx}{x^2} = 1 - \frac{1}{n} < 1.$$

Thus

$$S_n = 1 + \frac{1}{2^2} + \frac{1}{3^2} + \cdots + \frac{1}{n^2} < 2$$

for all n, and we conclude by Theorem 5.1.2 that the series (5.1.7) is convergent.

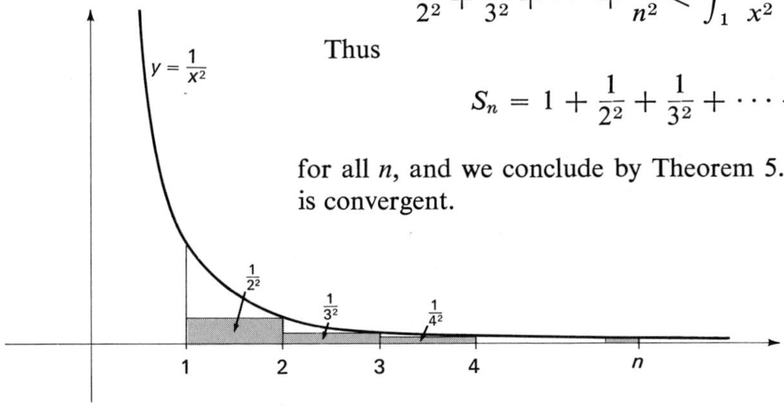

Figure 5.1.2

Note that this proof shows that the sum of the series (5.1.7) is not greater than 2, for the sum is the limit of a sequence of numbers each of which is less than 2. Clearly, also, the sum must be positive and larger than any partial sum. Thus, by actually computing S_n for large values of n we can find various estimates for the sum S such as

$$S_3 = \tfrac{49}{36} < S \leq 2.$$

Series may be added by merely adding together corresponding terms. Similarly, a series may be multiplied by a fixed number. The theorem which applies is exactly what you would expect, and it follows almost trivially from the basic theorem (Theorem 1.3.1) on algebraic combinations of sequences.

Theorem 5.1.3 *If $a_1 + a_2 + \cdots$ and $b_1 + b_2 + \cdots$ are convergent series with sums A and B respectively and if α, β are real numbers, then the series*

$$(\alpha a_1 + \beta b_1) + (\alpha a_2 + \beta b_2) + (\alpha a_3 + \beta b_3) + \cdots \tag{5.1.8}$$

converges to the sum $\alpha A + \beta B$.

The proof of this theorem is left as an exercise (Exercise 5.1.11). It depends on the fact that the partial sums of the series (5.1.8) are simply the obvious algebraic combinations of the corresponding sums for the two original series. Note that in the notation of Theorem 5.1.3, the series $a_1b_1 + a_2b_2 + a_3b_3 + \cdots$ does *not* have the number AB as its sum. Using the series in the theorem, can you write down a series which does converge to AB? Theorems about products of series are generally somewhat difficult to prove, and we shall not delve into this topic here.

However, you can show, using results we have already obtained, that if $a_n > 0$ and $b_n > 0$ for all n and if $a_1 + a_2 + \cdots$ and $b_1 + b_2 + \cdots$ both converge, so does the series $a_1b_1 + a_2b_2 + \cdots$. This will be left as an exercise.

Exercises

1 Decide whether each of the following series converges or diverges:

(a) $1 - 1 + 1 - \cdots + (-1)^{n+1} + \cdots$,
(b) $1 - 2 + 3 - \cdots + (-1)^{n+1}n + \cdots$,
(c) $1 + \frac{1}{3} + \frac{1}{9} + \cdots + \frac{1}{3^{n-1}} + \cdots$,
(d) $1 + (1.1) + (1.1)^2 + \cdots + (1.1)^{n-1} + \cdots$,
(e) $\frac{1}{2} + \frac{2}{3} + \frac{3}{4} + \cdots + \frac{n}{n+1} + \cdots$,
(f) $1 - \frac{3}{5} + \frac{9}{25} - \cdots + (-1)^{n-1}\frac{3^{n-1}}{5^{n-1}} + \cdots$.

2 Find the sum of each of the following (convergent) series:

(a) $1 + \frac{1}{5} + \frac{1}{25} + \cdots + \frac{1}{5^{n-1}} + \cdots$,
(b) $3 + \frac{3}{4} + \frac{3}{16} + \cdots + \frac{3}{4^{n-1}} + \cdots$,
(c) $1 - \frac{2}{3} + \frac{4}{9} - \cdots + (-1)^{n-1}(\frac{2}{3})^{n-1} + \cdots$,
(d) $1 + \frac{1}{\sqrt{3}} + \frac{1}{3} + \cdots + \frac{1}{\sqrt{3^{n-1}}} + \cdots$,
(e) $4 - \frac{4}{\sqrt[3]{3^2}} + \frac{4}{\sqrt[3]{3^4}} + \cdots + (-1)^{n-1}\frac{4}{\sqrt[3]{3^{2n-2}}} + \cdots$.

3 What rational number is represented by the decimal .141414 . . . ?

4 (a) Find the sum of the series $1 + r^2 + r^4 + \cdots + r^{2n} + \cdots$, where r is a real number, $|r| < 1$.

(b) Find the sum of the series $1 + 2r + 4r^2 + \cdots + 2^n r^n + \cdots$, where r is a real number, $|r| < \frac{1}{2}$.

5 (a) Show that if n is a positive integer,

$$\frac{1}{n(n+1)} = \frac{1}{n} - \frac{1}{n+1}.$$

(b) Find the sum of the series

$$\frac{1}{1 \cdot 2} + \frac{1}{2 \cdot 3} + \cdots + \frac{1}{n(n+1)} + \cdots.$$

(*Hint:* Use (a) to find a formula for S_n.)

6 Use the method of Exercise 5 to find the sum of the series

$$\frac{1}{1 \cdot 3} + \frac{1}{2 \cdot 4} + \cdots + \frac{1}{n(n+2)} + \cdots.$$

7 Use the method of Example 3 to show that the series

$$\frac{1}{\sqrt{1}} + \frac{1}{\sqrt{2}} + \frac{1}{\sqrt{3}} + \cdots + \frac{1}{\sqrt{n}} + \cdots$$

diverges.

8 Use the method of Example 4 to show that the series

$$1 + \frac{1}{2^3} + \frac{1}{3^3} + \cdots + \frac{1}{n^3} + \cdots$$

converges. Find an upper bound for its sum.

9 Use mathematical induction to prove that if n is a positive integer, r is a real number, and S_n is defined as in (5.1.5), then

$$S_n(1 - r) = 1 - r^n.$$

10 Prove that if $|r| < 1$, then

$$\lim_{n \to \infty} r^n = 0.$$

(*Hint:* If $r \neq 0$, the inequality $|r^n| < \epsilon$ can be solved for n by taking logarithms.)

11 Prove Theorem 5.1.3.

12 Decide whether the series

$$\frac{1}{2 \log 2} + \frac{1}{3 \log 3} + \cdots + \frac{1}{n \log n} + \cdots$$

converges or diverges. (*Hint:* Study Examples 3 and 4.)

13 Here is an alternate proof of the divergence of the harmonic series. Beginning with the term $\frac{1}{2}$, form successive blocks of $1, 2, 4, \ldots, 2^n, \ldots$ terms. Thus we have

$$\sum_{k=1}^{\infty} \frac{1}{k} = \left\{\frac{1}{2}\right\} + \left\{\frac{1}{3} + \frac{1}{4}\right\} + \left\{\frac{1}{5} + \frac{1}{6} + \frac{1}{7} + \frac{1}{8}\right\} + \cdots + \left\{\frac{1}{2^n + 1} + \cdots + \frac{1}{2^{n+1}}\right\} + \cdots.$$

Show that the sum of the terms within each block is greater than or equal to $\frac{1}{2}$, and consequently that by Theorem 5.1.2 the series diverges.

14 Let a_n be a sequence of nonzero numbers such that $\lim_{n\to\infty} a_n = +\infty$. Show that

(a) the series $\sum_{n=1}^{\infty} (a_{n+1} - a_n)$ diverges, and

(b) the series $\sum_{n=1}^{\infty} \dfrac{1}{a_n} - \dfrac{1}{a_{n+1}}$ converges.

15 Use Exercise 14 to show that the series $\sum_{n=1}^{\infty} \log(1 + 1/n)$ diverges.

16 Prove that if $a_n \geq 0$, $b_n \geq 0$ for all n and the series $\sum_{n=1}^{\infty} a_n$, $\sum_{n=1}^{\infty} b_n$ converge, then so does the series $\sum_{n=1}^{\infty} a_n b_n$. (*Hint:* Since $\sum_{n=1}^{\infty} a_n$ converges, its sequence S_n of partial sums is bounded. Since $\sum_{n=1}^{\infty} b_n$ converges, there is an M such that $b_n \leq M$ for all n. Show that the partial sums of $\sum_{n=1}^{\infty} a_n b_n$ are bounded and apply Theorem 5.1.2.)

5.2 Series of Positive Numbers

There are a number of important theorems on the convergence of series whose terms are positive numbers. We wish to consider four. The first is Theorem 5.1.2, which says that such a series converges if and only if its partial sums are bounded. The other three, which follow from Theorem 5.1.2 by very easy arguments, are the subject of this section.

One of the most convenient methods of testing a series for convergence or divergence is to compare it with other series known to converge or diverge. The following theorem outlines this method of comparison.

Theorem 5.2.1 ***Comparison Test*** *Let $a_k \geq 0$ and $b_k \geq 0$ for $k = 1, 2, \ldots,$ and suppose that $a_k \leq b_k$ for all $k \geq N$. Then*

(i) if $\sum_{k=1}^{\infty} b_k$ *converges,* $\sum_{k=1}^{\infty} a_k$ *also converges; whereas*

(ii) if $\sum_{k=1}^{\infty} a_k$ *diverges,* $\sum_{k=1}^{\infty} b_k$ *also diverges.*

Proof We denote the partial sums of the two series by A_n and B_n. That is,

$$A_n = a_1 + a_2 + \cdots + a_n,$$
$$B_n = b_1 + b_2 + \cdots + b_n.$$

If $n \geq N$, we have

$$\begin{aligned} A_n - A_N &= a_{N+1} + \cdots + a_n, \\ B_n - B_N &= b_{N+1} + \cdots + b_n; \end{aligned}$$

so, by hypothesis,

(5.2.1) $$0 \leq A_n - A_N \leq B_n - B_N.$$

From (5.2.1) we conclude by adding A_N throughout the inequality that

$$A_N \leq A_n \leq B_n - B_N + A_N$$

for all $n \geq N$. From this, it is an easy exercise (which we shall leave to you) to show that if the sequence $\{B_n\}$ is bounded, then the sequence $\{A_n\}$ is also bounded.

The proof of part (i) is easily concluded. If $\sum_{k=1}^{\infty} b_k$ converges, then by Theorem 5.1.2 the sequence $\{B_n\}$ is bounded. By the above remarks the sequence $\{A_n\}$ is also bounded, and another application of Theorem 5.1.2 shows that $\sum_{k=1}^{\infty} a_k$ converges.

Part (ii) is logically equivalent to part (i) in view of the fact that a sequence must be either convergent or divergent.

Example 1 Using Theorem 5.2.1 we can easily see that the series

(5.2.2) $$\frac{1}{1 \cdot 2} + \frac{1}{2 \cdot 3} + \frac{1}{3 \cdot 4} + \cdots + \frac{1}{n(n+1)} + \cdots$$

converges. For the general term of this series is

$$a_n = \frac{1}{n(n+1)} = \frac{1}{n^2 + n},$$

and since $n^2 + n > n^2$, we have

(5.2.3) $$a_n = \frac{1}{n^2 + n} < b_n = \frac{1}{n^2} \quad \text{for all } n \geq 1.$$

We showed in Example 4, Section 5.1 that the series $\sum_{k=1}^{\infty} b_k$ converges, so by (5.2.3) and Theorem 5.2.1, series (5.2.2) also converges.

Example 2 Consider the series

(5.2.4) $$\frac{1}{\log 2} + \frac{1}{\log 3} + \cdots.$$

The general term is

$$b_n = \frac{1}{\log n}, \; n \geq 2,$$

with $b_1 = 0$. Since

$$\log n < n \text{ for all } n \geq 1,$$

so that

$$b_n = \frac{1}{\log n} > a_n = \frac{1}{n} \text{ for } n \geq 2,$$

we can use the known divergence of the harmonic series and Theorem 5.2.1 to conclude that $\sum_{k=1}^{\infty} b_k$, that is, series (5.2.4), diverges.

A very simple corollary of Theorem 5.2.1 is often useful.

Corollary 5.2.2 *Let $a_k \geq 0$, $b_k \geq 0$ for all $k \geq 1$ and suppose that the sequence $\{b_k\}$ is bounded. Then if $\sum_{k=1}^{\infty} a_k$ converges, the sequence $\sum_{k=1}^{\infty} a_k b_k$ also converges.*

Proof If $0 \leq b_k < M$ for all k, then the series $\sum a_k b_k$ may be compared to the series $\sum MA_k$, using Theorem 5.2.1. The series $\sum Ma_k$ converges, according to Theorem 5.1.3.

Example 3 The general term of the series

$$101 + \frac{201}{8} + \frac{301}{27} + \cdots + \frac{100n + 1}{n^3} + \cdots \tag{5.2.5}$$

may be written

$$a_n = \frac{1}{n^2} \cdot \frac{100n + 1}{n}.$$

Since it is easily seen that

$$0 < \frac{100n + 1}{n} \leq 101 \text{ for } n \geq 1$$

and we know that $\sum_{k=1}^{\infty} 1/k^2$ converges, we conclude that (5.2.5) is also convergent.

Corollary 5.2.2 merely presents a disguised method of comparing one series with another. The next theorem presents another comparison in disguise.

Theorem 5.2.3 *Let $a_k > 0$ for all $k \geq 1$. If there exists a number $r < 1$ and an integer N such that*

$$\frac{a_{k+1}}{a_k} \leq r \tag{5.2.6}$$

for all $k \geq N$, then the series $\sum_{k=1}^{\infty} a_k$ converges.

If there exists a number $s > 1$ and an integer M such that

$$\frac{a_{k+1}}{a_k} \geq s$$

for all $k \geq M$, then the series $\sum_{k=1}^{\infty} a_k$ diverges.

Proof If we have (5.2.6) for $k \geq N$, then

$$a_{k+1} \leq ra_k.$$

Moreover,

$$a_k \leq ra_{k-1},\ a_{k-1} \leq ra_{k-2},\ \ldots .$$

Combining these inequalities, we obtain

$$a_{k+1} \leq ra_k \leq r^2 a_{k-1} \leq \cdots \leq r^{k+1-N} a_N.$$

Since

$$a_{k+1} < r^{k+1-N} a_N \quad \text{for all } k \geq N$$

and $r^{k+1-N} a_N$ is the $(k - N)$th term of a convergent series (Example 1, Section 5.1, and Theorem 5.1.3), Theorem 5.2.1 shows that $\sum a_k$ converges.

We leave the proof of the second part as an exercise.

The following corollary of Theorem 5.2.3 is usually known as the *ratio test* for convergence, although the theorem itself could also perfectly well go by that name.

Corollary 5.2.4 ***Ratio Test*** *Suppose the ratio a_{k+1}/a_k has a limit L as $k \to \infty$.* If $0 \leq L < 1$, *then the series $\sum a_k$ converges. If $L > 1$, then the series $\sum a_k$ diverges.*

Proof By the definition of the limit of a sequence, if $0 < L < 1$ then there exists an N such that

$$0 < \frac{a_{k+1}}{a_k} < \frac{L+1}{2} < 1$$

for all $k \geq N$. For if $L < 1$, then it can easily be seen that $(L + 1)/2 > L$ and a_{k+1}/a_k must eventually get arbitrarily close to L. Apply the theorem with $r = (L + 1)/2$. A similar observation reduces the corollary to the theorem in case $L > 1$.

Example 4 Consider the series

$$100 + \frac{100^2}{2!} + \frac{100^3}{3!} + \cdots + \frac{100^n}{n!} + \cdots .$$

For any $k \geq 1$ we have

$$\frac{a_{k+1}}{a_k} = \frac{100^{k+1}}{(k+1)!} \cdot \frac{k!}{100^k} = \frac{100}{k+1}.$$

Thus

$$\lim_{k\to\infty} \frac{a_{k+1}}{a_k} = 0 < 1,$$

and we conclude from Corollary 5.2.4 that the series converges. In Section 5.6 we shall see what its sum is.

Finally, we have another frequently useful test, whose proof, depending upon Theorem 5.1.2, is also easy. We shall not give the details, because the basic ideas of the proof have already been used. Specifically, a direct and obvious adaptation of the arguments used in Examples 3 and 4, Section 5.1, shows that

Theorem 5.2.5 ***Integral Test*** *If f is a decreasing function with positive values, then the series*

(5.2.7)
$$\sum_{k=1}^{\infty} f(k)$$

converges if the improper integral

(5.2.8)
$$\int_1^{\infty} f(x)\,dx$$

converges. If the integral (5.2.8) *diverges, then so does the series* (5.2.7).

Example 5 Let p be a positive real number. We have already considered the series

$$1 + \frac{1}{2^p} + \frac{1}{3^p} + \cdots + \frac{1}{n^p} + \cdots$$

for the values $p = 1, 2$. This series can be written in the form (5.2.7) for the function

$$f(x) = x^{-p}.$$

This function is decreasing and positive-valued for $x \geq 1$, so we may apply Theorem 5.2.5. Since

$$\int_1^n x^{-p}\,dx = \begin{cases} \dfrac{1}{1-p}\,(n^{(1-p)} - 1) & \text{if } p \neq 1, \\ \log n & \text{if } p = 1, \end{cases}$$

we see that the integral $\int_1^{\infty} (1/x^p)\,dx$ converges to $-1/(1-p)$ if $p > 1$ and diverges if $p \leq 1$.

Thus, the series $1 + 1/2^p + 1/3^p + \cdots + 1/n^p + \cdots$ converges if $p > 1$ and diverges if $p \leq 1$.

Exercises

Test the series in Exercises 1–20 for convergence.

1 $\displaystyle\sum_{n=1}^{\infty} \frac{1}{n^2 + 3}$.

2 $\displaystyle\sum_{n=1}^{\infty} \frac{n}{n^3 + n + 1}$.

3 $\displaystyle\sum_{n=5}^{\infty} \frac{1}{n - 4}$.

4 $\displaystyle\sum_{n=1}^{\infty} \frac{\sqrt{n}}{2n - 3}$.

5 $\sum_{n=3}^{\infty} \frac{1}{\log(\log n)}$.

6 $\sum_{k=1}^{\infty} \frac{1}{2^k + 3k}$.

7 $\sum_{k=1}^{\infty} \frac{k}{k^2 + 2}$.

8 $\sum_{k=1}^{\infty} \frac{k^2}{k^4 - 2k + 1}$.

9 $\sum_{k=1}^{\infty} \frac{k}{2^k}$.

10 $\sum_{n=1}^{\infty} \frac{n}{e^n}$.

11 $\sum_{n=1}^{\infty} \frac{1}{\sqrt{3n} - 2}$.

12 $\sum_{n=1}^{\infty} \frac{\log n}{n}$.

13 $\sum_{n=1}^{\infty} \frac{1}{(n+2)(n+3)}$.

14 $\sum_{n=1}^{\infty} \frac{n+2}{(n+3)3^n}$.

15 $\sum_{n=1}^{\infty} \frac{1}{\sqrt{n^3 + 7}}$.

16 $\sum_{k=1}^{\infty} \frac{1}{(2k)^\pi}$.

17 $\sum_{n=2}^{\infty} \frac{1}{\log n^2}$.

18 $\sum_{n=1}^{\infty} \frac{n!}{n^n}$.

19 $\frac{1}{1 \cdot 3} + \frac{2}{3 \cdot 5} + \frac{3}{5 \cdot 7} + \cdots$.

20 $\frac{1}{2} + \frac{1 \cdot 3}{2 \cdot 5} + \frac{1 \cdot 3 \cdot 5}{2 \cdot 5 \cdot 8} + \frac{1 \cdot 3 \cdot 5 \cdot 7}{2 \cdot 5 \cdot 8 \cdot 11} + \cdots$.

21 Show that if $\{A_n\}$ and $\{B_n\}$ are the sequences defined in Theorem 5.2.1 and if $\{B_n\}$ is bounded, then $\{A_n\}$ is bounded.

22 Criticize the following "proofs" that the harmonic series $\sum_{n=1}^{\infty} 1/n$ converges.

(a) For each n, $a_{n+1}/a_n = \frac{n}{n+1} < 1$, so by Theorem 5.2.3 the series converges.

(b) The limit of the nth term $1/n$ as $n \to \infty$ is 0. Thus by Theorem 5.1.1 the series converges.

(c) For $n \geq 2$, $1/n < 1/\log n$, so by Theorem 5.2.1 the series converges.

23 Prove the second part of Theorem 5.2.3.

24 Under the hypotheses of Theorem 5.2.1 (i), show that

$$\sum_{k=N}^{\infty} a_k \leq \sum_{k=N}^{\infty} b_k.$$

25 (a) Let $a_k > 0$ for $k \geq 1$. Prove that if there exists a number $r < 1$ and an integer N such that for $k \geq N$

$$\sqrt[k]{a_k} \leq r,$$

then the series $\sum_{k=1}^{\infty} a_k$ is convergent. On the other hand, if there exists a number $s > 1$ and an integer N such that if $k \geq N$, $\sqrt[k]{a_k} \geq s$, then the series is divergent. (*Hint:* Show that for $k \geq N$ the series can be compared with a geometric series.)

(b) Suppose that $a_k > 0$ for $k \geq 1$ and $\lim_{k\to\infty} \sqrt[k]{a_k} = L$. Then if $0 \leq L < 1$, the series $\sum_{k=1}^{\infty} a_k$ converges; if $L > 1$, the series diverges.

26 (a) Prove Theorem 5.2.5.

(b) Show that if $\int_1^{\infty} f(x)\,dx$ converges, then the *remainder* of the series after *n* terms are summed,

$$R_n = \sum_{k=n+1}^{\infty} f(k),$$

satisfies

$$R_n \leq \int_n^{\infty} f(x)\,dx.$$

27 (a) Use part (b) of Exercise 26 to estimate the error committed in approximating $\sum_{n=1}^{\infty} 1/n^2$ by the sum of its first 100 terms.

5.3 Absolute and Conditional Convergence

We turn now to a consideration of series whose terms are arbitrary real numbers, not necessarily positive. If there is some regularity in the signs of the terms, then we can frequently determine whether the series converges. In the simplest case, in which the sign alternates, the test is often particularly simple. We call such a series an alternating series, which may be written

$$\pm(a_1 - a_2 + a_3 - a_4 + \cdots), \tag{5.3.1}$$

where $a_k > 0$ for all k. We have the following simple test.

Theorem 5.3.1 *If the sequence $\{a_k\}$ is a positive decreasing sequence with limit zero, then the alternating series* (5.3.1) *converges. If S is the sum of this series and S_n is its nth partial sum, then*

$$|S - S_n| < a_{n+1}.$$

Proof We consider only the + sign in (5.3.1). Clearly both these series converge or both diverge. We examine the even- and odd-numbered partial sums separately. If $n = 2m$, then

$$S_n = S_{2m} = (a_1 - a_2) + (a_3 - a_4) + \cdots + (a_{2m-1} - a_{2m}); \tag{5.3.2}$$

whereas if $n = 2m + 1$, then

$$S_n = S_{2m+1} = a_1 - (a_2 - a_3) - (a_4 - a_5) - \cdots - (a_{2m} - a_{2m+1}). \tag{5.3.3}$$

All numbers enclosed in parentheses in (5.3.2) and (5.3.3) are positive if the sequence $\{a_k\}$ is decreasing. Since

(5.3.4)
$$\begin{aligned} S_{2m+2} &= S_{2m} + (a_{2m+1} - a_{2m+2}) > S_{2m}, \\ S_{2m+3} &= S_{2m+1} - (a_{2m+2} - a_{2m+3}) < S_{2m+1}, \end{aligned}$$

we see that the sequence $S_2, S_4, S_6, \ldots$ is *increasing* and the sequence $S_1, S_3, S_5, \ldots$ is decreasing.

Moreover, for each m

(5.3.5)
$$S_{2m+1} = S_{2m} + a_{2m+1} > S_{2m}.$$

By (5.3.4) and (5.3.5) we see that each interval $[S_{2m}, S_{2m+1}]$ contains the next interval $[S_{2m+2}, S_{2m+3}]$ and hence all subsequent intervals of the form $[S_{2p}, S_{2p+1}]$ for $p > m$. That is, the numbers S_n display themselves as in Figure 5.3.1.

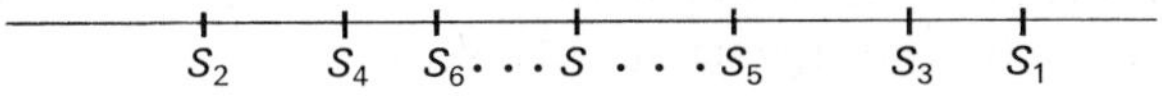

Figure 5.3.1

Thus the sequence $S_2, S_4, S_6, \ldots$ is an increasing sequence bounded above by all of the numbers S_{2m+1}. The sequence $S_1, S_3, S_5, \ldots$ is a decreasing sequence bounded below by all of the numbers S_{2m}.

We conclude from Theorem 1.3.4 that both sequences S_{2m} and S_{2m+1} have limits, and from the fact that

$$S_{2m+1} - S_{2m} = a_{2m+1} \to 0,$$

that these limits must be the same. If we call this common limit S, so that

$$\lim_{m\to\infty} S_{2m} = S = \lim_{m\to\infty} S_{2m+1},$$

it is an easy exercise to show that S is also the limit of the original (complete) sequence S_n:

(5.3.6)
$$\lim_{n\to\infty} S_n = S.$$

Thus the series (5.3.1) converges to a sum S, as suggested by Figure 5.3.1.

Notice that the distance from any S_n to S is less than the distance from S_n to S_{n+1}. Since $S_{n+1} - S_n = a_{n+1}$, this proves the last part of the theorem.

Example 1 The harmonic series (5.1.6) is divergent, as we have seen. However, the *alternating* harmonic series

(5.3.7)
$$1 - \tfrac{1}{2} + \tfrac{1}{3} - \tfrac{1}{4} + \cdots$$

fits the conditions of Theorem 5.3.1, because the sequence $\{1/n\}$ is decreasing and has limit zero. Therefore (5.3.7) converges to a sum S. We shall see later on what this number S is. Right now we can say from Theorem 5.3.1 that

$$\left|S - \left(1 - \frac{1}{2} + \frac{1}{3} - \cdots + (-1)^{n+1}\frac{1}{n}\right)\right| < \frac{1}{n+1}.$$

For series whose terms are not all positive or all negative and which are not alternating series we cannot usually say much about convergence in the ordinary sense. However, we can sometimes establish a stronger kind of convergence which we shall now define.

Definition 5.3.1 The series $\sum_{k=1}^{\infty} a_k$ is said to be *absolutely convergent* if the series of positive numbers $\sum_{k=1}^{\infty} |a_k|$ is convergent.

Example 2 As we have seen, the alternating harmonic series (5.3.7) is convergent but not absolutely convergent; and the series

(5.3.8) $$1 - r + r^2 - r^3 + r^4 + \cdots$$

is absolutely convergent if $0 < r < 1$. It is also convergent, for if $0 < r < 1$, the sequence r^n decreases to the limit 0. It is easy to see that (5.3.8) has the sum $(1 + r)^{-1}$, whereas the series of absolute values of the terms has the sum $(1 - r)^{-1}$. (See Example 1, Section 5.1.)

Theorem 5.3.2 *If $\sum_{k=1}^{\infty} a_k$ is absolutely convergent, it is also convergent. In this case, $|\sum a_k| \leq \sum |a_k|$.*

Proof The trick here is to write the series $\sum a_k$ as the difference between two series of nonnegative numbers. For each $k = 1, 2, \ldots,$ define b_k and c_k as follows:

$$b_k = \begin{cases} a_k & \text{if } a_k \geq 0, \\ 0 & \text{if } a_k < 0, \end{cases}$$

$$c_k = \begin{cases} 0 & \text{if } a_k \geq 0, \\ -a_k & \text{if } a_k < 0. \end{cases}$$

Obviously, we have

(5.3.9) $$b_k \geq 0, \; c_k \geq 0, \; a_k = b_k - c_k$$

for all k. Also, for each k,

(5.3.10) $$\begin{aligned} b_k &= \text{either } |a_k| \text{ or } 0, \text{ and} \\ c_k &= \text{either } |a_k| \text{ or } 0. \end{aligned}$$

Since $\sum |a_k|$ converges, we know that the partial sums of this series are bounded by, let us say, the positive number M. From (5.3.10) we see that the partial sums of both series $\sum b_k$ and $\sum c_k$

are also bounded by M. Using (5.3.9) and Theorem 5.1.2, we conclude that

$$\sum_{k=1}^{\infty} b_k \quad \text{and} \quad \sum_{k=1}^{\infty} c_k \text{ are convergent.}$$

Thus, by Theorem 5.1.3 the series

$$\sum_{k=1}^{\infty} a_k = (b_1 - c_1) + (b_2 - c_2) + (b_3 - c_3) + \cdots$$

also converges.

The second conclusion follows immediately from the fact that $\sum_1^n (b_k - c_k) \le \sum_1^n (b_k + c_k) = \sum_1^n |a_k|$ for all n.

Example 1 above shows that the converse of Theorem 5.3.2 is not true; a convergent series is not necessarily absolutely convergent. A series like the alternating harmonic series which is convergent but not absolutely convergent is often called *conditionally convergent.*

The following test for absolute convergence follows directly from the ratio test given in Corollary 5.2.4.

Theorem 5.3.3 *If $a_k \neq 0$ for all k and*

$$\lim_{k\to\infty} \frac{|a_{k+1}|}{|a_k|} = L,$$

then the series $\sum_{k=1}^{\infty} a_k$ is absolutely convergent if $L < 1$ and divergent if $L > 1$.

Proof The first part of the theorem is merely a statement of Corollary 5.2.4 for the series $\sum|a_k|$. As for the second part, Corollary 5.2.4 only states that $\sum|a_k|$ diverges, not necessarily that $\sum a_k$ diverges. But an inspection of the proof of this corollary and its parent theorem reveals that if $L > 1$ then the sequence $\{|a_k|\}$ has limit ∞. Thus, the sequence $\{a_k\}$ cannot have limit 0, so $\sum a_k$ diverges by Theorem 5.1.1.

If the limit L in this theorem is 1, then anything can happen. The series may diverge, it may converge conditionally, or it may actually converge absolutely.

Example 3 The series

$$\sum_{k=1}^{\infty} \frac{x^k}{k!}$$

is absolutely convergent. Here

$$a_k = \frac{x^k}{k!}, \text{ so}$$

$$\frac{|a_{k+1}|}{|a_k|} = \frac{|x|}{k+1} \to 0 \text{ as } k \to \infty$$

for any value of x.

Example 4 Consider the series

(5.3.11)
$$\sum_{k=1}^{\infty} \frac{x^k}{k}.$$

Here

$$\frac{|a_{k+1}|}{|a_k|} = |x| \frac{k}{k+1} \to |x| \text{ as } k \to \infty.$$

Thus, by Theorem 5.3.3 the series (5.3.11) converges absolutely if $|x| < 1$ and diverges if $|x| > 1$. If $x = 1$, we see that (5.3.11) is the harmonic series; hence it diverges. If $x = -1$, the series is the alternating harmonic series, so it is conditionally convergent.

Exercises

Determine whether each of the series in Exercises 1–13 is absolutely convergent, conditionally convergent, or divergent.

1 $\sum_{k=1}^{\infty} (-1)^k \frac{1}{2^k}$.

2 $\sum_{k=1}^{\infty} (-1)^k \frac{k+1}{2k+1}$.

3 $\sum_{n=1}^{\infty} \frac{(-1)^n}{n^2}$.

4 $\sum_{n=1}^{\infty} \frac{(-1)^n}{\sqrt{n}}$.

5 $\sum_{n=1}^{\infty} (-1)^n \frac{3^{n-2}}{2^n}$.

6 $\sum_{m=1}^{\infty} \frac{\sin m}{2^m}$.

7 $\sum_{n=1}^{\infty} \frac{\cos n}{n^2}$.

8 $\sum_{n=2}^{\infty} \frac{(-1)^n}{n \log n}$.

9 $\sum_{n=1}^{\infty} \frac{(-1)^n}{2n-1}$.

10 $\sum_{n=2}^{\infty} (-1)^n \frac{\log n}{\log n^3}$.

11 $\sum_{k=1}^{\infty} \frac{(-1)^k k^2}{2^k}$.

12 $\sum_{k=1}^{\infty} \frac{(-1)^k (k+1)}{\sqrt{k}\,(k+2)}$.

13 $1 - \frac{1}{4} - \frac{1}{9} + \frac{1}{16} - \frac{1}{25} - \frac{1}{36} + \frac{1}{49} - \cdots$ (a plus sign alternating with two minus signs).

14 For which values of x does the series $\sum_{k=0}^{\infty} (-1)^k x^k$ converge absolutely?

15 How many terms of the alternating harmonic series should be summed in order to approximate the sum of the whole series to within 1/1000?

16 For which values of x does the series $\sum_{n=1}^{\infty} nx^n$ converge?

17 Find a divergent alternating series $a_1 - a_2 + a_3 - \cdots$ such that $a_n > 0$ and $\lim_{n\to\infty} a_n = 0$.

18 Rearrange the terms of the alternating harmonic series

$$1 - \frac{1}{2} + \frac{1}{3} - \cdots + \frac{(-1)^{n+1}}{n} + \cdots$$

so that the resulting series diverges.

19 Complete the proof of (5.3.6).

20 Show that under the hypotheses of Theorem 5.3.1, the remainder

$$R_n = S - S_n$$

has the same sign as the $(n + 1)$th term. Thus the error made in approximating such a series by its first n terms has the same sign as the first omitted term. (*Hint:* $R_n = \sum_{k=n+1}^{\infty} (-1)^k a_k$ is itself an alternating series to which Theorem 5.3.1 may be applied. Consider

$$|R_n - (-1)^{n+1}a_{n+1}| < a_{n+2} < a_{n+1}.)$$

5.4 Power Series

Up to this point, we have been discussing series whose terms are real numbers. It is possible to extend the discussion to series whose terms are functions. If for each $k = 1, 2, \ldots$, we are given a function f_k and if we assume that all of the functions f_k are defined on a common interval $[a, b]$, then we may consider the series

$$f_1(x) + f_2(x) + \cdots + f_n(x) + \cdots$$

for any x in $[a, b]$. If the series converges for certain values of x, the sum will define a new function

$$f(x) = f_1(x) + f_2(x) + \cdots. \tag{5.4.1}$$

To determine the values of x for which such a series converges, we may apply all of the tests we have developed. In Examples 3 and 4, Section 5.3, we followed this procedure. In addition, there are several new questions we could ask. Under what conditions, for example, will a function f defined by a series (5.4.1) be differentiable or integrable? What will be the relationship between the derivative $f'(x)$ (when it exists) and the derivatives of the terms? How can the integral of f be found from the integrals of the f_k?

We shall not discuss these questions in the general case but shall restrict our attention to a special case for which answers are easily obtained. We shall consider only the case of series of "power functions," that is, functions which have the form $a_k(x - c)^k$ for some fixed real number c.

A *power series* is, then, a series of functions

(5.4.2) $$a_0 + a_1(x - c) + a_2(x - c)^2 + \cdots + a_n(x - c)^n + \cdots,$$

which we can also write as

(5.4.3) $$\sum_{k=0}^{\infty} a_k(x - c)^k.$$

Note that the series (5.4.3) begins with a term with index 0, instead of the usual 1. As we shall see later, there are very good reasons for a "zeroth term," but it may cause some confusion if we are not careful. The term $a_n(x - c)^n$ is not now the nth term, but the $(n + 1)$th term. The partial sum

$$S_n(x) = a_0 + a_1(x - c) + \cdots + a_n(x - c)^n$$

is not the nth partial sum; it is the $(n + 1)$th in the list of sums $S_0(x), S_1(x), \ldots, S_n(x)$. So we shall refer only to "the term $a_n(x - c)^n$" or "the partial sum $S_n(x)$."

The series (5.4.2) will be called the *power series with center c*, for reasons which will become clear soon. Frequently we can restrict our discussion to power series with center 0, for a simple change of variable $x' = x - c$ reduces (5.4.2) to the 0-centered series

$$a_0 + a_1x' + a_2(x')^2 + \cdots + a_n(x')^n + \cdots.$$

In this section we shall discuss convergence of a power series; in the next two sections, the differentiation and integration of functions defined as sums of power series and the contribution of power series to the study of a wide class of functions.

Consider first the power series

(5.4.4) $$\sum_{k=0}^{\infty} a_kx^k$$

and the set S of numbers x for which (5.4.4) converges. Let x_0 be in this set, so that

$$\sum_{k=0}^{\infty} a_kx_0^k$$

converges. Then by Theorem 5.1.1 $a_kx_0{}^k \to 0$ as $k \to \infty$. A sequence which has a limit is bounded; thus there exists a number M such that

$$|a_kx_0^k| \leq M$$

for all k.

Now suppose that $|x| < |x_0|$ and let

$$r = \frac{|x|}{|x_0|} < 1.$$

Then

(5.4.5) $$|a_k x^k| = |a_k x_0^k| \left|\frac{x}{x_0}\right|^k \leq Mr^k$$

for all k.

As we saw in Example 1, Section 5.1, Mr^k is the general term of a convergent series with positive terms. Thus, by (5.4.5) and the comparison test (Theorem 5.2.1), the series

$$\sum_{k=0}^{\infty} |a_k x^k|$$

is convergent. Thus (5.4.4) is absolutely convergent and hence convergent for all numbers x such that $|x| < |x_0|$.

This can be stated another way. If x_0 is in S, then the entire open interval $(-|x_0|, |x_0|)$ is contained in S. It can be shown that a set S having this property must be a set of one of the following forms:

$$(-\infty, \infty), \ [-r, r], \ [-r, r), \ (-r, r], \ (-r, r).$$

That is, S is an interval, finite or infinite, open or closed at either end, and symmetric about the origin. A rigorous proof would divert us significantly from our present discussion, so we omit it. The result is certainly plausible, however.

We have therefore partially proved

Theorem 5.4.1 *Given any power series* (5.4.4) *with center* 0 *there exists a nonnegative "number"* r *(which may be* ∞*) such that the series converges absolutely for* $|x| < r$ *and diverges for* $|x| > r$*. The series may converge or diverge for* $x = r$*.*

The simple change of variable $x' = x - c$ reduces a power series (5.4.3) to one with center 0. We thus have an immediate corollary of this theorem:

Corollary 5.4.2 *Given any power series* $\sum a_k(x - c)^k$*, there exists an* r*, which may be either a nonnegative real number or* $+\infty$*, such that the series converges absolutely for all* x *in the open interval* $(c - r, c + r)$ *and diverges for all* x *outside the closed interval* $[c - r, c + r]$*. At the endpoints of the interval the series may converge or diverge.*

The interval $[c - r, c + r]$ is called the *interval of convergence* of the power series, and the number r is called the *radius of convergence.*

Example 1 We saw in Example 3, Section 5.3 that the series $\sum(1/k!)x^k$ has an infinite radius of convergence. Its interval of convergence is $(-\infty, \infty)$.

Example 2 We saw in Example 4, Section 5.3 that the series $\sum(1/k)x^k$ has radius of convergence 1. The series converges at one endpoint of the interval of convergence ($x = -1$) and diverges at the other ($x = 1$).

The ratio test for absolute convergence gives us one simple method for finding this radius of convergence in many cases.

Theorem 5.4.3 *Let $a_k \neq 0$ for all k. If*

$$r = \lim_{k\to\infty} \frac{|a_k|}{|a_{k+1}|},$$

then r is the radius of convergence of the series $\sum_{k=0}^{\infty} a_k(x - c)^k$.

Proof If we apply the ratio test to the series, we conclude that it converges for all x for which

$$\lim_{k\to\infty} \left|\frac{a_{k+1}(x - c)^{k+1}}{a_k(x - c)^k}\right| = |x - c| \lim_{k\to\infty} \left|\frac{a_{k+1}}{a_k}\right| = \frac{|x - c|}{r}$$

is less than 1. It diverges when this limit $|x - c|/r$ is greater than 1.

Thus, the series $\sum a_k(x - c)^k$ converges (absolutely) when $|x - c| < r$ and diverges when $|x - c| > r$. But this is exactly the condition which makes r the radius of convergence of the series.

Example 3 The series

$$\sum_{k=0}^{\infty} k^2x^k$$

has radius of convergence r, where

$$r = \lim_{k\to\infty} \frac{k^2}{(k + 1)^2} = \lim_{k\to\infty} \left(\frac{k}{k + 1}\right)^2 = 1.$$

Since for $x = \pm 1$ the term k^2x^k does not tend to zero, the series diverges at both endpoints of its interval of convergence $[-1, 1]$.

Example 4 Consider the series

$$\sum_{k=0}^{\infty} k^kx^k.$$

The radius of convergence is

$$r = \lim_{k\to\infty} \frac{k^k}{(k + 1)^{k+1}} = \lim_{k\to\infty} \left(\frac{k}{k + 1}\right)^k \cdot \left(\frac{1}{k + 1}\right) = \frac{1}{e} \cdot 0 = 0.$$

(For the calculation of $\lim\,(k/(k + 1))^k$ see Example 6, Section 1.3.)

Here the interval of convergence reduces to a point 0. Clearly the series converges for $x = 0$, but by Theorem 5.4.1 it diverges for $x \neq 0$.

Exercises

Determine the interval of convergence for each of the following power series.

1 $\displaystyle\sum_{k=0}^{\infty} \frac{x^k}{2^k}$.

2 $\displaystyle\sum_{k=0}^{\infty} \frac{x^k}{(k+1)3^k}$.

3 $\displaystyle\sum_{k=0}^{\infty} \frac{(x-2)^k}{(k+1)(k+2)}$.

4 $\displaystyle\sum_{k=0}^{\infty} \frac{(-1)^k x^k}{k^2+1}$.

5 $\displaystyle\sum_{n=1}^{\infty} \frac{(-1)^n (x+3)^n}{\sqrt{n}}$.

6 $\displaystyle\sum_{n=0}^{\infty} \frac{x^{2n}}{(n+1)}$.

7 $\displaystyle\sum_{k=0}^{\infty} (-1)^k k x^k$.

8 $\displaystyle\sum_{n=0}^{\infty} \frac{x^{2n+1}}{(2n+1)!}$.

9 $\displaystyle\sum_{n=0}^{\infty} \frac{(n+1)x^n}{(n+2)^2}$.

10 $\displaystyle\sum_{k=0}^{\infty} a^{k^2} x^k$, where $0 < a < 1$.

11 $\displaystyle\frac{3}{1^2 \cdot 4} x + \frac{3^2}{3^2 \cdot 4^2} x^2 + \frac{3^3}{5^2 \cdot 4^3} x^3 + \frac{3^4}{7^2 \cdot 4^4} x^4 + \cdots$.

12 $\displaystyle\frac{x+2}{2 \cdot 4} - \frac{(x+2)^2}{4 \cdot 6} + \frac{(x+2)^3}{6 \cdot 8} - \frac{(x+2)^4}{8 \cdot 10} + \cdots$.

13 $\displaystyle\sum_{n=0}^{\infty} \frac{(-1)^2}{(n+1)^3} x^n$.

14 Determine the interval of convergence for the power series

$$1 + \frac{1}{2}x + \frac{1}{2 \cdot 3}x^2 + \frac{1}{2^2 \cdot 3}x^3 + \frac{1}{2^2 \cdot 3^2}x^4 + \frac{1}{2^3 \cdot 3^2}x^5 + \cdots.$$

Find the sum of the series for each x within its interval of convergence. (*Hint:* The series can be written as the sum of two geometric series.)

5.5 Functions Defined by Power Series

We have seen that a function f defined as the sum of a power series is always defined on some open interval which is symmetric about the center of the series. The length of the interval is twice the radius of convergence of the power series. In this section we shall show that such a function possesses derivatives of all orders, and that these derivatives are themselves sums of power series obtained in a natural way from the original series.

Given the power series

(5.5.1) $$\sum_{k=0}^{\infty} a_k(x - c)^k,$$

we shall call the series

(5.5.2) $$\sum_{k=1}^{\infty} ka_k(x - c)^{k-1}$$

the *derived series* for (5.5.1). Its terms are the derivatives of the terms of (5.5.1).

If series (5.5.1) has radius of convergence r and if

(5.5.3) $$f(x) = \sum_{k=0}^{\infty} a_k(x - c)^k$$

for x in $(-r, r)$, then we shall show that series (5.5.2) also has radius of convergence r and that the sum of this series is the derivative $f'(x)$. Repeating this reasoning we shall conclude that $f'(x)$ also has a derivative $f''(x)$, and so on. Thus f has derivatives of all orders.

This observation, perhaps the most important to be made about series at least in their relation to functions, is the focus of this entire chapter. It shows that a function that can be defined by a power series is of necessity one of our "best behaved" functions. In the next section we shall turn this question around, asking whether all infinitely differentiable functions may be represented as sums of power series.

The theorem we are seeking is not a trivial one. For a proof, we require some preliminary information. First, we must establish that the two series (5.5.1) and (5.5.2) do have the same radius of convergence.

Lemma 1 *The power series $\sum_0^\infty a_k(x - c)^k$ and its derived series $\sum_1^\infty ka_k(x - c)^k$ have the same radius of convergence.*

Proof As we have observed earlier, it is sufficient to prove the lemma for the case $c = 0$, to which other cases can be reduced by simple changes of variable. Thus, let r be the radius of convergence of

$$\sum_{k=0}^{\infty} a_k x^k$$

and R be the radius of convergence of

(5.5.4) $$\sum_{k=1}^{\infty} ka_k x^{k-1}.$$

Let $|x| < r$ and pick a number $d > 1$ such that

$$0 < d|x| = x_0 < r.$$

In Exercise 4.2.9 we established that

$$\lim_{k \to \infty} k^{1/k} = 1,$$

so there exists an N such that

(5.5.5) $$0 < k^{1/k} < d \quad \text{for } k \geq N.$$

From (5.5.5) it follows that

$$0 < k < d^k \quad \text{for } k \geq N$$

and therefore that

$$|ka_k x^k| < |a_k|(d|x|)^k = |a_k|\,|x_0|^k.$$

Since the series $\sum a_k {x_0}^k$ is absolutely convergent, it follows by comparison that

$$\sum_{k=1}^{\infty} |ka_k x^k| = |x| \sum_{k=1}^{\infty} k|a_k|\,|x|^{k-1}$$

converges. Thus (5.5.4) converges for this x and we conclude that (5.5.4) converges for all $|x| < r$.

Thus $r \leq R$. The same proof, using the fact that $k^{1/k} \to 1$ as $k \to \infty$, shows that $R \leq r$. Thus, the lemma is proved.

Also, for the proof of our major theorem we must know that a function defined by (5.5.3) is continuous on its domain.

Lemma 2 *If $f(x) = \sum_{k=0}^{\infty} a_k(x - c)^k$ on $(c - r, c + r)$, then $f(x)$ is continuous on the interval.*

Proof Again, it is sufficient to consider $\sum_1^{\infty} a_k x^k$ on $(-r, r)$. Let $-r < x_0 < r$ and pick $d > 0$ so that

(5.5.6) $$|x_0| < d < r.$$

If x is any number such that

(5.5.7) $$|x| < d$$

and k is a positive integer, then

$$|x^k - x_0^k| = |x - x_0|\,|x^{k-1} + x^{k-2}x_0 + \cdots + x_0^{k-1}|.$$

The sum on the right has k terms, so we see by (5.5.6), (5.5.7) that

$$|x^k - x_0^k| < |x - x_0|(kd^{k-1}).$$

Thus, since

$$|f(x) - f(x_0)| = \left|\sum_{k=0}^{\infty} a_k(x^k - x_0^k)\right| \le \sum_{k=0}^{\infty} |a_k|\,|x^k - x_0^k|,$$

we have

$$|f(x) - f(x_0)| < |x - x_0| \sum_{k=0}^{\infty} k|a_k|d^{k-1},$$

provided the latter series converges. By Lemma 1 it does converge, so we may take the limit on both sides as $x \to x_0$ and conclude that

$$f(x) \to f(x_0) \quad \text{as } x \to x_0.$$

Thus f is continuous at x_0 and the lemma is proved.

Theorem 5.5.1 *Let $f(x) = \sum_0^\infty a_k x^k$ for x in the interval of convergence $(-r, r)$. Then f is differentiable and*

$$\text{(i)}\quad f'(x) = \sum_{k=1}^{\infty} k a_k x^{k-1},$$

$$\text{(ii)}\quad \int_0^x f(t)\,dt = \sum_{k=0}^{\infty} \frac{a_k}{k+1} x^{k+1}$$

for all x in $(-r, r)$.

Proof It is more convenient to establish the integral formula (ii) first. We can then use (ii) and the Fundamental Theorem to prove (i). Since by Lemma 2 f is continuous, we know that it is integrable and that the Fundamental Theorem applies.

Let $0 < x < r$. Since $\sum a_k x^k$ is absolutely convergent, we know that

$$R_n = \sum_{k=n+1}^{\infty} |a_k| x^k$$

tends to zero as $n \to \infty$. For any $0 \le t \le x$, we have

$$\left|f(t) - \sum_{k=0}^{n} a_k t^k\right| = \left|\sum_{k=n+1}^{\infty} a_k t^k\right| \le \sum_{k=n+1}^{\infty} |a_k| x^k = R_n,$$

or

$$-R_n < f(t) - \sum_{k=0}^{n} a_k t^k < R_n.$$

Thus, we have by a well-known theorem about integrals (Theorem 6.6.5 in *C 1*)

(5.5.8) $$-R_n x < \int_0^x f(t)\,dt - \sum_{k=0}^{n} \frac{a_k}{k+1} x^{k+1} < R_n x.$$

Since $-R_n x$ and $R_n x$ tend to zero as $n \to \infty$, we see by (5.5.8) that

$$\int_0^x f(t)\,dt = \lim_{n\to\infty} \sum_{k=0}^{n} \frac{a_k}{k+1} x^{k+1} = \sum_{k=0}^{\infty} \frac{a_k}{k+1} x^{k+1},$$

proving part (ii) of the theorem.

To prove part (i) let us define

(5.5.9) $$g(x) = \sum_{k=1}^{\infty} k a_k x^{k-1};$$

then $g(x)$ is defined for all x in $(-r, r)$ by Lemma 1.

If we apply part (ii) to the series and function in (5.5.9), we conclude that

$$\int_0^x g(t)\,dt = \sum_{k=1}^{\infty} a_k x^k,$$

or

$$\int_0^x g(t)\,dt = f(x) - a_0.$$

Thus, by one of the forms of the Fundamental Theorem,

(5.5.10) $$f'(x) = \frac{d}{dx}\int_0^x g(t)\,dt = g(x).$$

Comparing (5.5.10) with (5.5.9) we see that part (i) is proved.

Corollary 5.5.2 *If f is defined by*

(5.5.11) $$f(x) = \sum_{k=0}^{\infty} a_k (x-c)^k,$$

then f is differentiable everywhere in the interval of convergence and

$$f'(x) = \sum_{k=1}^{\infty} k a_k (x-c)^{k-1}.$$

The proof of this corollary from Theorem 5.5.1 and the chain rule for differentiation of composite functions is left as an exercise.

Corollary 5.5.3 *The function f defined by* (5.5.11) *has derivatives of all orders. For any n and any x in the interval of convergence, $f^{(n)}(x)$ is the sum of the power series obtained by differentiating the terms of* (5.5.11) *n times.*

The proof of this corollary is trivial; we shall not even call it an exercise.

Exercises

1 Prove Corollary 5.5.2.

2 State and prove a theorem which generalizes Theorem 5.5.1(ii) to the case in which the center of the interval of convergence is not 0.

3 (a) Use Theorem 5.5.1 to find the sum of the series

$$x + \frac{1}{2}x^2 + \frac{1}{3}x^3 + \cdots + \frac{x^n}{n} + \cdots.$$

(We already know its radius of convergence from Example 4 in Section 5.3.)

(b) Can Theorem 5.5.1 be extended to include the endpoints of the interval of convergence? (*Hint:* Compare Examples 2 and 4, Section 5.3.)

4 Use the identities

$$\frac{1}{2-x} = \frac{1}{1-(x-1)},$$

$$\frac{1}{2-x} = \frac{1}{2}\left(\frac{1}{1-\frac{1}{2}x}\right)$$

to find two power series expansions for $f(x) = 1/(2-x)$, one about $c = 1$ and one about $c = 0$.

5 Find power series about $c = 0$ whose sums are the following functions.

(a) $f(x) = \dfrac{2}{1+x}$, (b) $f(x) = \dfrac{1}{1-3x}$,

(c) $f(x) = \dfrac{1}{3-x}$, (d) $f(x) = \dfrac{1}{1+x^2}$,

(e) $f(x) = \dfrac{1}{1+x} + \dfrac{1}{1-x}$, (f) $f(x) = \dfrac{1}{2+3x^2}$.

6 Find the intervals of convergence for the power series in Exercise 5.

7 Find a power series about $c = 0$ whose sum is $1/(1-x)^2$. What is its interval of convergence? (*Hint:* $1/(1-x)^2$ is the derivative of $1/(1-x)$.)

8 Find a power series about $c = 0$ whose sum is arctan x. What is its radius of convergence? (*Hint:* $\arctan x = \int_0^x (1/1+t^2)\,dt$.)

9 Find a power series about $c = 0$ whose sum is $\log(1+x)$. What is its radius of convergence? (*Hint:* Can you express $\log(1+x)$ as a derivative or an integral?)

10 Use the power series derived in Exercise 9 and elementary properties of logarithms to find power series about $c = 0$ whose sums are

(a) $\log(1-x)$, (b) $\log(1-x^2)$,

(c) $\log\left(\dfrac{1}{1+x}\right)$, (d) $\log\left(\dfrac{1+x}{1-x}\right)$.

11 Approximate

$$\int_0^{1/2} \frac{x}{1 + x^3}\,dx$$

to 4 decimal places.

12 Use Exercise 9 above to find a power series about $c = 0$ whose sum is

$$\int_0^x \frac{\log (1 + t)}{t}\,dt.$$

5.6

Taylor and Maclaurin Series

We have shown that a function which can be expressed as the sum of a power series is necessarily infinitely differentiable. Moreover, we have seen how the derivatives of such a function may be obtained from the series representation of the function.

Now let us turn this problem around. Suppose a function has derivatives of all orders for all x in some interval I, and c is in the interior of I. We can ask whether it is always possible to find a power series with center c such that $f(x)$ is the sum of this series for all x in some interval $(c - r, c + r)$ contained in I.

The answer is, Not always. But we can recognize criteria for the existence of such a series.

We may also ask a related question: If such a series exists, how may it be found in terms of the given function? The answer is easy to see from results in the last section. For if

(5.6.1) $$f(x) = \sum_{k=0}^{\infty} a_k(x - c)^k,$$

then

$$f(c) = a_0.$$

Moreover, by Corollary 5.5.3 we have

$$\begin{aligned} f'(x) &= a_1 + 2a_2(x - c) + 3a_3(x - c)^2 + \cdots, \\ f''(x) &= 2a_2 + 2 \cdot 3a_3(x - c) + 3 \cdot 4a_4(x - c)^2 + \cdots, \\ f'''(x) &= 2 \cdot 3a_3 + 2 \cdot 3 \cdot 4a_4(x - c) + 3 \cdot 4 \cdot 5a_5(x - c)^2 + \cdots, \\ &\cdots \\ f^{(n)}(x) &= n!a_n + (n + 1)!a_{n+1}(x - c) + \cdots. \end{aligned}$$

Thus,

$$\begin{aligned} f'(c) &= a_1, \\ f''(c) &= 2!a_2, \\ &\cdots \\ f^{(n)}(c) &= n!a_n \\ &\cdots; \end{aligned}$$

or

$$\begin{aligned} a_0 &= f(c), \\ a_1 &= \frac{f'(c)}{1!}, \\ a_2 &= \frac{f''(c)}{2!}, \\ &\cdots \\ a_n &= \frac{f^{(n)}(c)}{n!}, \\ &\cdots . \end{aligned}$$

Thus, all the coefficients of the series (5.6.1) have been found in terms of f and its derivatives. We have proved

Theorem 5.6.1 *If the function f can be represented as the sum of a power series with center c in some interval $(c - r, c + r)$, then that series must be*

(5.6.2)
$$f(x) = \sum_{k=0}^{\infty} \frac{f^{(k)}(c)}{k!}(x - c)^k.$$

Theorem 5.6.1 says that if a function has a representation as the sum of a power series, that representation is *unique*. The series is called the *Taylor series* for f about c, or the *Taylor series expansion* of f about c. In the special case $c = 0$, it is usually called the *Maclaurin series* expansion; so the Maclaurin series for the function f is

$$\sum_{k=0}^{\infty} \frac{f^{(k)}(0)}{k!} x^k.$$

The foregoing should look familiar. The partial sums of the Taylor series for f with center c are just the *Taylor polynomials*

$$T_n(x) = \sum_{k=0}^{n} \frac{f^{(k)}(c)}{k!}(x - c)^k$$

studied in Section 4.3. The Taylor series (5.6.2) for f converges to $f(x)$ for all x in its interval of convergence if and only if

$$f(x) = \lim_{n\to\infty} T_n(x)$$

for all x in this interval. This in turn is true if and only if

$$\lim_{n\to\infty} R_n(x) = 0$$

for all x in the interval, where $R_n(x)$ is the Taylor remainder discussed in Section 4.3.

Thus, the Taylor theorem, which gives an estimate for $R_n(x)$, provides one criterion for the convergence to $f(x)$ of the Taylor series for f.

Theorem 5.6.2 *Let f have derivatives of all orders on the interval $(c - r, c + r)$ and let x be a number in that interval. Let $M_n(x)$ be the maximum value of $|f^{(n)}(t)|$ for t between c and x. If*

(5.6.3)
$$\frac{M_n(x)}{n!}(x - c)^n \to 0 \text{ as } n \to \infty,$$

then

$$f(x) = \sum_{k=0}^{\infty} \frac{f^{(k)}(c)}{k!}(x - c)^k.$$

Proof By the corollary to the Taylor Theorem, Theorem 4.3.2,

$$|R_n(x)| \leq \frac{M_{n+1}(x)}{(n+1)!}(x - c)^{n+1}.$$

Thus, (5.6.3) implies that $R_n(x) \to 0$.

Example 1 Let

$$f(x) = \alpha_0 + \alpha_1 x + \alpha_2 x^2 + \cdots + \alpha_n x^n$$

be a polynomial function. By direct calculation it is easy to see that the polynomial coincides with the Taylor series for f about 0. This fact is also easily deduced from the above. For, all derivatives of f of order $n \geq m + 1$ vanish *for all* x. Thus, $M_n(x) = 0$ for all $n \geq m + 1$ ($M_n(x)$ as in Theorem 5.6.2); we conclude that for any x, $f(x)$ is the sum of its Taylor series about 0. But the polynomial is itself a power series with center 0; therefore by Theorem 5.6.1 it must be the Taylor series.

Example 2 The function

$$f(x) = e^x$$

has derivatives of all orders. We have

$$\begin{cases} f^{(k)}(x) = e^x, \\ f^{(k)}(0) = 1, \end{cases} \quad k = 0, 1, 2, \ldots.$$

Thus, the Maclaurin series for e^x is

(5.6.4)
$$\sum_{k=0}^{\infty} \frac{1}{k!} x^k = 1 + x + \frac{x^2}{2!} + \frac{x^3}{3!} + \cdots.$$

Clearly, the maximum value of e^t for t between x and 0 is either e^x or 1, whichever is largest. Thus

$$\frac{M_n(x)}{n!}x^n = \begin{cases} e^x \dfrac{x^n}{n!} & \text{or} \\ \dfrac{x^n}{n!} \end{cases}$$

For any fixed x, $x^n/n! \to 0$ as $n \to \infty$; thus e^x is the sum of its Maclaurin series for any x;

$$e^x = 1 + x + \frac{x^2}{2!} + \frac{x^3}{3!} + \cdots + \frac{x^n}{n!} + \cdots.$$

Hence the series (5.6.4) is usually called the exponential series.

Example 3 The function

$$f(x) = \sin x$$

is also infinitely differentiable.

We find

$$f(x) = \sin x,\ f'(x) = \cos x,\ f''(x) = -\sin x,$$
$$f'''(x) = -\cos x,\ f^{(4)}(x) = \sin x,\ \ldots.$$

The derivatives repeat cyclically in groups of four. Thus

$$f(0) = 0,\ f'(0) = 1,\ f''(0) = 0,\ f'''(0) = -1,$$
$$f^{(4)}(0) = 0,\ f^{(5)}(0) = 1,\ \ldots;$$

and the Maclaurin series for $\sin x$ is

$$0 + x + 0 - \frac{x^3}{3!} + 0 + \frac{x^5}{5!} + \cdots.$$

Since $|f^{(n)}(x)|$ is either $|\sin x|$ or $|\cos x|$, we see that $|M_n(x)| \le 1$ for any x. Thus,

$$\left|\frac{M_n(x)}{n!}x^n\right| \le \left|\frac{x^n}{n!}\right| \to 0;$$

and $\sin x$ is the sum of its Maclaurin series for all x:

(5.6.5) $$\sin x = x - \frac{x^3}{3!} + \frac{x^5}{5!} - \frac{x^7}{7!} + \cdots + \frac{(-1)^n x^{2n+1}}{(2n+1)!} + \cdots.$$

Example 4 Since the Maclaurin series for $\sin x$ converges for all x, we can easily find the Taylor series expansion of $\cos x$ about 0. By Theorem 5.5.1

(5.6.6) $$\cos x = \frac{d}{dx}\sin x = 1 - \frac{x^2}{2!} + \frac{x^4}{4!} - \frac{x^6}{6!} + \cdots + \frac{(-1)^n x^{2n}}{(2n)!} + \cdots,$$

and this expansion is valid also for all x. The series (5.6.5) and (5.6.6) are often called the sine and cosine series, for obvious reasons.

Exercises

1 Use the method of Example 3 to determine the Maclaurin series for $f(x) = \cos x$.

2 Find the Maclaurin series for

(a) $f(x) = \sin 2x$, (b) $f(x) = e^{x^2}$,

(c) $f(x) = \dfrac{1}{e^x}$, (d) $f(x) = x \cos x$,

(e) $f(x) = \dfrac{\sin x}{x}$, (f) $f(x) = \cos \sqrt{x}$.

3 Use the identity $\cos^2 x = \frac{1}{2}(1 + \cos 2x)$ to find the Maclaurin series for $f(x) = \cos^2 x$.

4 Find the first four nonzero terms of the Maclaurin series for $f(x) = e^x \sin x$ by (a) formula (5.6.2), and (b) by formally multiplying the Maclaurin series for e^x and $\sin x$. (In (b) one does not simply multiply coefficients of like powers of x, but rather treats the power series for e^x and $\sin x$ as infinite polynomials:

$$(a_0 + a_1x + a_2x^2 + \cdots)(b_0 + b_1x + b_2x^2 + \cdots)$$
$$= a_0b_0 + (a_0b_1 + a_1b_0)x + (a_0b_2 + a_1b_1 + a_2b_0)x^2 + \cdots).$$

5 Can you find a Maclaurin series for $f(x) = |x|$? Why?

6 Find the Taylor series for
(a) $f(x) = 1/x$ about $c = 1$,
(b) $f(x) = 2/(x + 3)$ about $c = -1$,
(c) $f(x) = \sin x$ about $c = \pi/2$,
(d) $f(x) = e^x$ about $x = 3$.

7 Approximate the value of $\sin 0.4$ to 5 decimal places.

8 Approximate each of the following to 4 decimal places:
(a) $\sin 62°$, (b) $\cos 5°$,
(c) $e^{0.1}$, (d) $\cos 48°$.

9 Approximate $\int_0^{0.4} \sin x^2\, dx$ to 6 decimal places.

10 Find the Taylor series for $\cos x$ in powers of $(x - \pi/4)$.

11 Find the Maclaurin series for $f(x) = (e^x - 1)/x$. What is its interval of convergence?

12 Find the Maclaurin series for $f(x) = (\cos x - 1)/x$.

5.7 Logarithm and Binomial Series

In spite of our many assurances about the importance of infinite series, you may well have wondered just what we have accomplished by the journey from Taylor polynomials to Taylor series. Except for the observation that a power series expansion of a function is unique, we made little use in Section 5.6 of the information about series carefully assembled in the earlier sections. You are justified in wondering whether acquiring the ability to differentiate and integrate a power series, the culmination of a rather lengthy discussion, is worth the effort if this is its only application.

Of course, a deep understanding of the reasons for a careful study of series requires repeated exposure to their uses. One such use will be seen briefly in Chapter 7 in connection with solutions of a differential equation. But the following two examples illustrate how Taylor's Theorem sometimes lets us down and how a knowledge of the properties of power series comes to our rescue. In these examples, Taylor polynomials cannot do the job alone.

Example 1 We have found Taylor series for most of the elementary functions, e.g., exp, sin, cos, ... However, the function log is missing from our list. Since such series expansions have many obvious uses, not the least of which is in finding approximations of the values of these functions, it would be good to have one for log. Obviously, it is hopeless to look for a Taylor series about 0 for $\log x$, which is not even defined for $x = 0$. However, $\log(1 + x)$ is defined and infinitely differentiable for $-1 < x < 1$, and we can consider the Taylor series about 0 for this function.

Differentiating, we find that if

$$f(x) = \log(1 + x),$$

then

$$f'(x) = \frac{1}{1+x}, \; f''(x) = \frac{-1}{(1+x)^2}, \; f'''(x) = \frac{2}{(1+x)^3}, \; \ldots,$$

and in general

$$f^{(n)}(x) = \frac{(-1)^{n+1}(n-1)!}{(1+x)^n}.$$

Thus, the Taylor series for $\log(1 + x)$ is

$$x - \frac{x^2}{2} + \frac{x^3}{3} - \frac{x^4}{4} + \cdots + \frac{(-1)^n x^{n+1}}{n+1} + \cdots.$$

By the ratio test we easily see that this series converges for $|x| < 1$. However, does it converge to $\log(1 + x)$? If we use Theorem 5.6.2, we obtain the following upper bounds for the absolute value of the Taylor remainders

(5.7.1)
$$\frac{M_n(x)}{n!} x^n = \frac{x^n}{n} \quad \text{for } x \geq 0,$$
$$\frac{M_n(x)}{n!} x^n = \frac{1}{n}\left(\frac{x}{1+x}\right)^n \quad \text{for } x < 0.$$

(See Exercise 5.7.5.) These tend to 0 only if $x \geq -\frac{1}{2}$. Thus, we cannot be sure that the Maclaurin series converges to its function throughout the entire interval of convergence.

However, another approach, producing the series and the desired information about its sum, is possible. We already know from Example 1, Section 5.1, that

$$\frac{1}{1+t} = 1 - t + t^2 - t^3 + \cdots \quad \text{for } |t| < 1.$$

Thus, using Theorem 5.5.1 we have

(5.7.2) $$\log(1+x) = \int_0^x \frac{dt}{1+t} = x - \frac{x^2}{2} + \frac{x^3}{3} - \frac{x^4}{4} + \cdots$$

for all $|x| < 1$. What could be more simple! Here, the ability to integrate a convergent power series term by term is of paramount importance.

The series expansion (5.7.2) allows us to find series representing the logarithms of numbers between 0 and 2. Should we wish to approximate other logarithms, the following ingenious procedure is available.

Replacing x by $-x$ in (5.7.2), we see that

(5.7.3) $$\log(1-x) = -x - \frac{x^2}{1} - \frac{x^3}{3} - \frac{x^4}{4} - \cdots.$$

Since both series (5.7.2) and (5.7.3) converge for $-1 < x < 1$, by Theorem 5.1.3 they may be subtracted, yielding

$$\log\left(\frac{1+x}{1-x}\right) = (x + x) + \left(\frac{x^3}{3} + \frac{x^3}{3}\right) + \cdots,$$

or

(5.7.4) $$\log\left(\frac{1+x}{1-x}\right) = 2\left(x + \frac{x^3}{3} + \frac{x^5}{5} + \cdots\right) \quad \text{for all } |x| < 1.$$

As x ranges from -1 to 1 the quantity $(1 + x)/(1 - x)$ ranges through all positive real numbers. Hence, for any positive number (5.7.4) may be used to find a series which converges to its logarithm. For example,

$$\log 3 = 2\left(\frac{1}{2} + \frac{1}{3 \cdot 8} + \frac{1}{5 \cdot 32} + \frac{1}{7 \cdot 128} + \cdots\right)$$

because $(1 + x)/(1 - x) = 3$ when $x = \frac{1}{2}$.

Example 2 You have undoubtedly encountered the Binomial Theorem in your past mathematical experience, one form of which says that if m is a positive integer, then

(5.7.5)
$$(1 + x)^m = 1 + mx + \frac{m(m-1)}{2}x^2 + \frac{m(m-1)(m-2)}{3!}x^3 + \cdots + x^m.$$

The coefficients of this polynomial are the so-called binomial coefficients

$$\binom{k}{m} = \frac{m(m-1)\cdots(m-k+1)}{k!}.$$

You can easily see that (5.7.5) is actually the Maclaurin series expansion of the function $(1 + x)^m$, all terms in x^k for $k \geq m + 1$ vanishing, because the corresponding derivatives vanish. This observation suggests an obvious generalization of the Binomial Theorem.

Suppose α is any real number. Successive differentiation of the function $(1 + x)^\alpha$ reveals that the Maclaurin series looks exactly like (5.7.5), except that it does not terminate. For,

$$\frac{d^k}{dx^k}(1 + x)^\alpha = \alpha(\alpha - 1)\cdots(\alpha - k + 1)(1 + x)^{\alpha - k};$$

so, if $f(x) = (1 + x)^\alpha$,

$$\frac{f^{(k)}(0)}{k!} = \frac{\alpha(\alpha - 1)\cdots(\alpha - k + 1)}{k!},$$

which vanishes for large k if and only if α is a positive integer.

Thus, the Maclaurin series for $(1 + x)^\alpha$ is

(5.7.6)
$$1 + \alpha x + \frac{\alpha(\alpha - 1)}{2}x^2 + \cdots + \frac{\alpha(\alpha - 1)\cdots(\alpha - k + 1)}{k!}x^k + \cdots.$$

Looking at the coefficients in this series we see that, since $a_k = \alpha(\alpha - 1)\cdots(\alpha - k + 1)/k!$,

$$\left|\frac{a_k}{a_{k+1}}\right| = \left|\frac{k + 1}{\alpha - k}\right| \to 1 \quad \text{as } k \to +\infty,$$

so (5.7.6) has a radius of convergence $r = 1$.

But does this Maclaurin series converge to its function? We shall spare you the details, but an attempt to use Taylor's Theorem to estimate the Taylor remainders does not yield the conclusion that they tend to zero for all $|x| < 1$. The trouble is much like the trouble which we met in the last example; in fact, it is exactly the same when $\alpha = -1$. So once again we have found the Maclaurin series easily, only to encounter doubts about its usefulness.

But again there is a way out. This time there is no other function with known series expansion from which we can obtain the desired one by integration or differentiation. So we must proceed differently. We do know that the convergent power series (5.7.6) defines a function, call it g:

(5.7.7) $$g(x) = 1 + \alpha x + \frac{\alpha(\alpha - 1)}{2!}x^2 + \frac{\alpha(\alpha - 1)(\alpha - 2)}{3!}x^3 + \cdots.$$

Let us try to find how the derivative $g'(x)$ is related to $g(x)$. Differentiating (5.7.7) we have

(5.7.8) $$g'(x) = \alpha + \frac{\alpha(\alpha - 1)}{1}x + \frac{\alpha(\alpha - 1)(\alpha - 2)}{2!}x^2 + \frac{\alpha(\alpha - 1)(\alpha - 2)(\alpha - 3)}{3!}x^3 + \cdots.$$

Thus,

(5.7.9) $$xg'(x) = \alpha x + \frac{\alpha(\alpha - 1)}{1}x^2 + \frac{\alpha(\alpha - 1)(\alpha - 2)}{2!}x^3 + \cdots.$$

Now add (5.7.8) and (5.7.9) to obtain

(5.7.10) $$(1 + x)g'(x) = \alpha\left[1 + \alpha x + \frac{\alpha(\alpha - 1)}{2!}x^2 + \frac{\alpha(\alpha - 1)(\alpha - 2)}{3!}x^3 + \cdots\right].$$

Comparing (5.7.10) with (5.7.7) we observe that

(5.7.11) $$(1 + x)g'(x) = \alpha g(x)$$

for all $|x| < 1$.

Condition (5.7.11) is a differential equation of a type which we shall study in Chapter 7. Fortunately, it is a very simple one which can be solved by a number of methods. The simplest is to show that (5.7.11) implies that the function

$$F(x) = g(x)(1 + x)^{-\alpha}$$

has a derivative which vanishes everywhere on $(-1, 1)$. For

$$\begin{aligned} F'(x) &= g'(x)(1 + x)^{-\alpha} - \alpha g(x)(1 + x)^{-\alpha-1} \\ &= [(1 + x)g'(x) - \alpha g(x)](1 + x)^{-\alpha-1} \\ &= 0. \end{aligned}$$

Thus, F is a constant function, and since $g(0) = 1$, we have $F(0) = 1$. Therefore, $F(x) = 1$ for all x, and we have

$$g(x) = (1 + x)^{\alpha}.$$

Therefore, by (5.7.7) we have

(5.7.12) $$(1 + x)^{\alpha} = 1 + \alpha x + \frac{\alpha(\alpha - 1)}{2}x^2 + \frac{\alpha(\alpha - 1)(\alpha - 2)}{3!}x^3 + \cdots.$$

This series is called for obvious reasons the *binomial series*.

Exercises

1 Find a power series for

(a) $f(x) = \sqrt{1 + x}$, (b) $f(x) = \sqrt{1 - 2x}$,

(c) $f(x) = \sqrt[3]{2 + x}$, (d) $f(x) = \sqrt{1 - x^2}$,

(e) $f(x) = \dfrac{1}{(1 + 3x)^2}$, (f) $f(x) = \dfrac{1}{\sqrt{1 - x}}$,

(g) $f(x) = \dfrac{1}{\sqrt[3]{2 - 3x^2}}$, (h) $f(x) = \dfrac{x}{\sqrt{1 - x^3}}$.

2 Use an appropriate power series to approximate to five decimal places

(a) $\sqrt{1.02}$, (b) $\log (1.05)$,

(c) $\log 0.98$,

(d) $\sqrt{8.82} = \sqrt{9 - 0.18} = 3\sqrt{1 - 0.02}$.

3 Approximate the value of $\int_0^{1/2} \sqrt{1 - x^3}\, dx$ to 4 decimal places.

4 Approximate the value of $\int_0^{0.2} \dfrac{dx}{\sqrt{1 - x^4}}$ to 4 decimal places.

5 Derive the formulas in (5.7.1).

6 Approximate the value of $\int_0^{0.1} \dfrac{\log (1 + x)}{x}\, dx$ to 4 decimal places.

7 Show that

$$\arcsin x = x + \frac{1}{2}\frac{x^3}{3} + \frac{1 \cdot 3}{2 \cdot 4}\frac{x^5}{5} + \frac{1 \cdot 3 \cdot 5}{2 \cdot 4 \cdot 6}\frac{x^7}{7} + \cdots$$

by using

$$\arcsin x = \int_0^x \frac{dt}{\sqrt{1 - t^2}}.$$

8 Find a power series for $f(x) = 1/(1 + x)^2$ by

(a) differentiating term-by-term a known series for $1/(1 + x)$,

(b) using the binomial series formula (5.7.12).

Are the resulting series the same?

9 Use multiplication of power series as in Exercise 5.6.4 to find the first four nonzero terms of the Maclaurin series for

$$f(x) = e^x \log (1 + x).$$

10 Show that (5.7.5) is the Maclaurin series for $(1 + x)^m$.

11 Use (5.7.3) and (5.7.4) to obtain two infinite series whose sums are log 2. Which series seems better suited for computation? Why?

smooth curves; elementary differential geometry

chapter 6

6.0

Introduction

In Chapter 1 of *C 1* we used the intuitive notion of *continuous curve* to introduce and motivate the definition of continuous function. Since then functions have been our primary interest, although curves in the form of graphs of functions have been used frequently to illustrate properties of functions and to facilitate the presentation of examples. Now we shall reverse our point of view to consider briefly what we shall call *smooth curves*, using our information about functions to formulate and elucidate a few geometrical properties of curves. Time and space will severely limit both the depth and breadth of our inquiry—we shall only scratch the surface of a branch of mathematics called *differential geometry*. You may think as you read this chapter that the switch in point of view is hardly perceptible. The material will seem to flow naturally from the preceding in *C 1* and in *C 2*. Indeed, differential geometry does straddle the fence between analysis, that is, the calculus, and geometry; it is a study of curves and surfaces in the plane, space, and higher dimensional spaces, using the tools of calculus. At times—now, for example—it is difficult if not impossible to distinguish between calculus and differential geometry, the difference being more one of attitude than of substance.

In Section 6.1 we shall review briefly curves as graphs of functions, developing conditions which will insure the degree of smoothness we seek. But in the succeeding sections we shall quickly move on to other methods of describing curves, methods which lift some of the restrictions imposed on the ordinary graph of a function. The resulting extended descriptions of smooth curves will allow us to fully exploit differential geometry. At the end of the chapter it will also be possible for us to investigate briefly the physics of motions along smooth curves.

6.1

Graphs of Functions; Smooth Curves

We have never actually defined a curve. In Section 6.2 we shall come fairly close to a workable definition of a curve in the plane, but up to now the notion has been intuitive. We have made a tacit agreement that the graph of any function is a curve, but this has created some strange bedfellows, as we have seen. In fact, each of the geometrical figures in Figure 6.1.1 is the graph of a function (assuming vertical and horizontal coordinate axes) and would therefore have to be called a curve. This stretches our imagination rather far, especially if we look at the points where some of these "curves" exhibit a variety of pathological behavior.

Figure 6.1.1

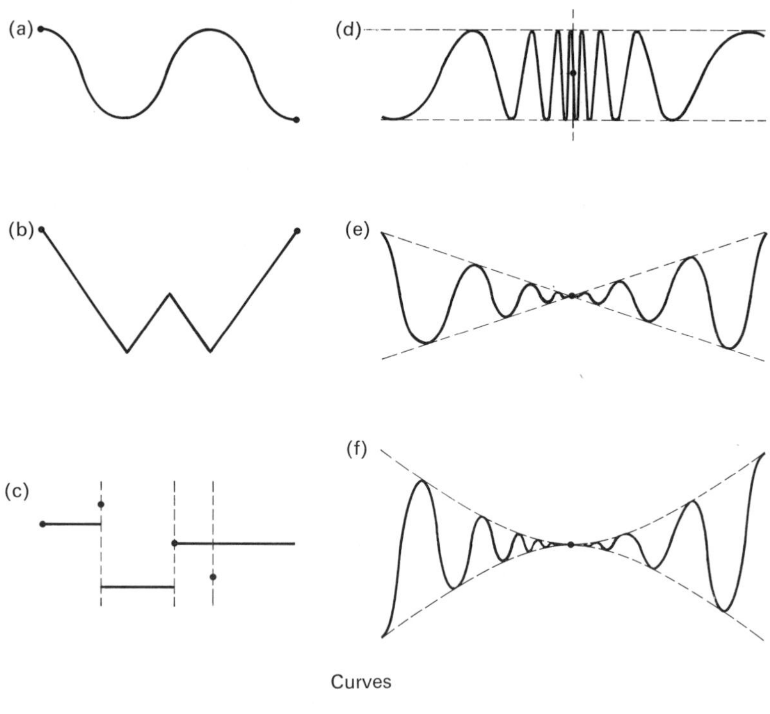

Curves

We found that the graph of a *continuous* function better conformed to our mental picture of a curve. Continuity ruled out some of the pathology evident in Figure 6.1.1. Two of the curves in this figure are ruled out by this condition. The survivors, all *continuous* curves, are shown in Figure 6.1.2.

Figure 6.1.2

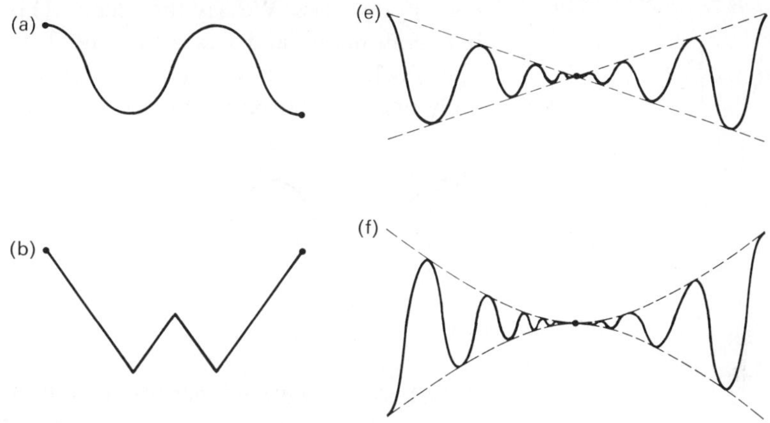

Continuous Curves

Two of the curves in Figure 6.1.2 exhibit idiosyncrasies which are still bothersome from the point of view of smoothness. Curves (b) and (e) do not have well-defined tangent lines at all points. Such an observation led us earlier to the definition of a *differentiable function*. The graph of such a function might be called a *differentiable curve*, which is characterized by the property of having tangent lines at all points. Only two of the continuous curves in Figure 6.1.2 are differentiable curves.

Figure 6.1.3

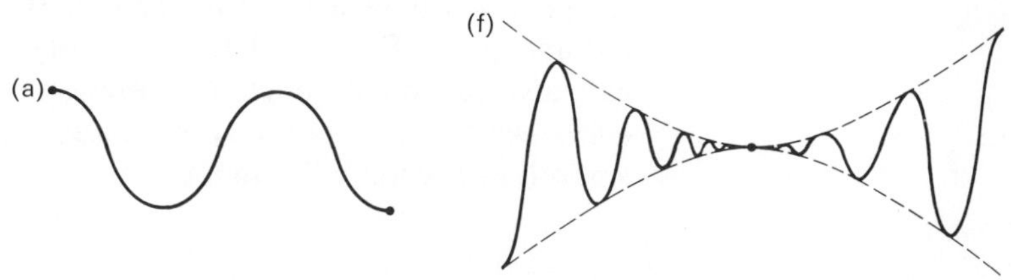

Differentiable Curves

One of the differentiable curves in Figure 6.1.3 exhibits yet another pathology which we may wish to avoid. We lack information needed to describe it precisely in terms of the geometrical figure itself, but we can suggest what the trouble is by saying that as curve (f) is traversed in a neighborhood of point P the tangent line oscillates discontinuously. This is more easily described by introducing coordinate axes and saying that the slope of the tangent to this curve is not a continuous function of x. For the function f whose graph is the curve (f),[1] this means that although $f'(x)$ exists for all x, f' is *not* continuous at the point P. This causes the wrinkle in the curve at P, which could not be erased by even the most powerful magnifying glass. We say that curve (f) is not *smooth* at P, although it *is* continuous and has a tangent. Intuitively, a smooth curve is one which possesses a *continuously varying* tangent line at all points. Only one of the original curves in Figure 6.1.1 is smooth.

[1] *Such a function might be*

$$f(x) = \begin{cases} x^2 \sin\left(\frac{1}{x}\right), & x \neq 0, \\ 0 & , x = 0. \end{cases}$$

(a) 

Smooth curve

Figure 6.1.4

Just as we have no general definition of a curve, we have no general definition of a smooth curve. But the above discussion enables us to say at least that

If f has a continuous derivative f', then the graph of $y = f(x)$ is a smooth curve.

Regarding the geometry of curves a reasonable compromise between generality and tractability is to consider only smooth curves. We shall restrict our attention to such curves in this chapter, although it should be fairly obvious how one can deal easily with curves which are *sectionally smooth*. A sectionally smooth curve is one which, although it may have discontinuities or corners, nevertheless consists of a finite number of smooth pieces (arcs). Curves (b) and (c) in Figure 6.1.1 are sectionally smooth. Moreover, one can always obtain much information about those sections of a curve which are smooth, even though the curve itself is neither smooth nor sectionally smooth.

Exercises

State whether the graph of $y = f(x)$ is a smooth curve if f is defined by

1 $f(x) = x^2 - 2x + 1.$

2 $f(x) = \sin |x|.$

3 $f(x) = \cos |x|.$

4 $f(x) = x + |x|.$

5 $f(x) = \begin{cases} x & \text{if } x \leq 0, \\ \sin x & \text{if } x > 0. \end{cases}$

6 $f(x) = \begin{cases} e^x - x^2 & \text{if } x < 0, \\ 1 + \sin x & \text{if } x \geq 0. \end{cases}$

7 $f(x) = \begin{cases} \cos x & \text{if } x \le \pi/2, \\ x - \pi/2 & \text{if } x > \pi/2. \end{cases}$ **8** $f(x) = \begin{cases} x \sin x & \text{if } x < 0, \\ x^2 + 1 & \text{if } x \ge 0. \end{cases}$

9 $f(x) = \begin{cases} \log x & \text{if } 0 < x \le 1, \\ (x-1)^2 & \text{if } x > 1. \end{cases}$ **10** $f(x) = |x - 1|^3$.

6.2

Paths; Parametric Representation of Curves

A curve is a twisted line, ray, or segment. That is all there is to curves, really, although this statement obviously raises more questions than it settles. We shall have to be a bit more precise. A twisting of a segment is a particular *mapping* of that segment into a new location. So we can say that a plane curve is determined by a line or line segment ℓ and a mapping F of ℓ (that is, of the set of points on ℓ) into the region R in which the curve lies (Figure 6.2.1). Throughout this chapter "curve" will mean "plane curve."

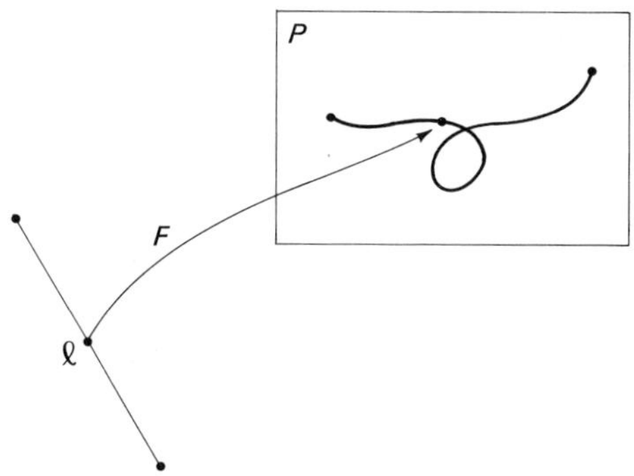

Figure 6.2.1

We may simplify a bit by considering the line or segment ℓ to be an interval I, open or closed, finite or infinite, on a coordinate line. Thus, points on ℓ may be identified with numbers t between two numbers a and b. Since we are considering only plane curves here, we may think of the region R as being a coordinate plane P. The mapping is then a function with domain I and values which are not numbers but *points* in the plane P.

Definition 6.2.1 Let $F\colon I \to P$ be a mapping of the interval I into a plane P. The mapping F is called a *path* in P. The image $F(I)$ of the interval under this mapping is called a *curve* in P.

According to Definition 6.2.1, a plane curve is a set of points in the plane, and a path is a particular way of laying off an interval I on this set of points. We may think of a path in another way. The interval I is identified with the set of all real numbers t such that

$a \le t \le b$. We may think of t as a *time variable* ranging over the time interval from $t = a$ to $t = b$. The path $F: I \to P$ is then a particular way of traversing the curve $F(I)$ during the interval of time. Thus, whereas a curve is a set of points of a special kind, a path may be considered as a *trip* along a curve. A particular curve may be associated with a large number of paths—there are many ways of traversing a given curve: forward, backward, fast, slow, etc. This fact is an obvious by-product of the fact that there are a great many different functions F, G, H, . . . on an interval I such that $F(I) = G(I) = H(I)$, ...

Since the points in a coordinate plane may be identified with ordered pairs of numbers (the coordinates of the points), we may write the mapping F in Definition 6.2.1 as follows: For each number t in I, we may write the image $F(t)$ as

$$F(t) = (g(t), f(t)),$$

where g and f are two ordinary real functions defined on I and

(6.2.1)
$$y = f(t),\quad x = g(t)$$

are the ordinate and the abscissa of the point $F(t)$ (Figure 6.2.2).

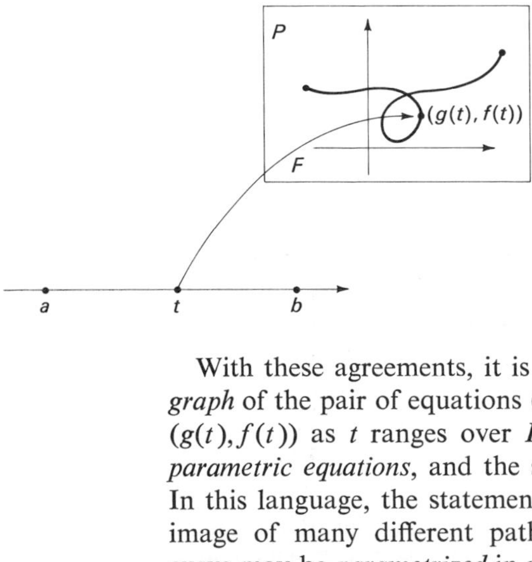

Figure 6.2.2

With these agreements, it is natural to consider the curve as the *graph* of the pair of equations (6.2.1), that is, as the set of all points $(g(t), f(t))$ as t ranges over I. These equations (6.2.1) are called *parametric equations*, and the symbol t, a *parameter* for the curve. In this language, the statement that a particular curve may be the image of many different paths may be rephrased: A particular curve may be *parametrized* in many different ways.

A curve given by a path, that is, presented as the graph of a pair of parametric equations (6.2.1), is given an orientation by that path. For the segment I is oriented (it is part of an oriented line), and we may orient the curve in the direction corresponding to increasing values of t. When we draw such a curve we usually indicate this orientation by means of an arrowhead and we also plot the images of a few values of the parameter t. This makes it easy to exhibit different paths on one given curve.

Example 1 The parametric equations

$$\begin{aligned} y &= t, \\ x &= t^2, \end{aligned} \qquad -\infty < t < \infty,$$

determine a parabola. For, eliminating t we see that

$$x = y^2.$$

Moreover, as t ranges from $-\infty$ to ∞, y assumes all real values, so all points on the indicated parabola are included in the curve. Beside two points we have indicated the value of t for which the point has coordinates $F(t) = (t^2, t)$. The arrow indicates the direction along the curve of increasing t (Figure 6.2.3).

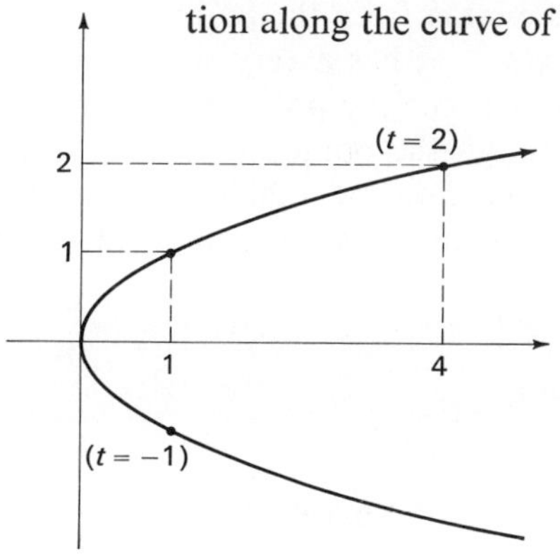

Figure 6.2.3

Example 2 The parametric equations

$$\begin{aligned} y &= \cos t, \\ x &= \sin t, \end{aligned} \qquad 0 \le t \le 2\pi,$$

determine the unit circle about the origin traversed once clockwise. For since $\sin^2 t + \cos^2 t = 1$ for all t, every point on the curve lies on the circle whose equation is

$$x^2 + y^2 = 1.$$

By inspecting the variation of $\sin t$ and $\cos t$ as t ranges from 0 to 2π we easily see that all points on the circle are represented. The point $(0, 1)$ on the circle corresponds to two values $t = 0$, $t = 2\pi$ of the parameter. See Figure 6.2.4.

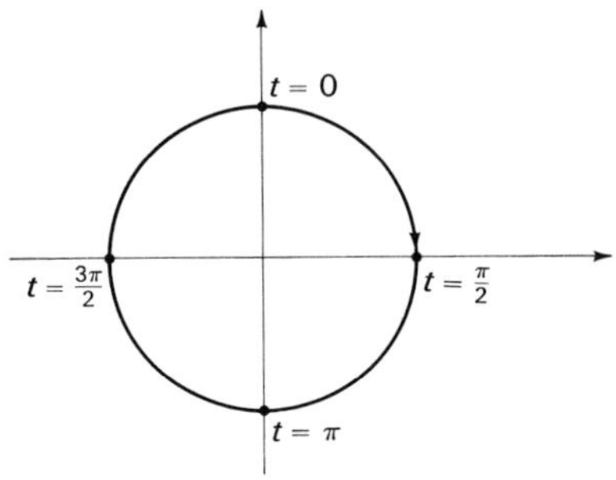

Figure 6.2.4

Example 3 The parametric equations

(6.2.2)
$$\begin{aligned} y &= \sin t, \\ x &= \cos t, \end{aligned} \qquad -\pi \leq t < \pi,$$

also describe the unit circle, but this time it is traversed once in a counterclockwise direction from the point $(-1, 0)$. In this case, the parameter t, $-\pi \leq t < \pi$, has a very simple geometric interpretation. It is the angle between the positive horizontal axis and the ray from the origin through the point $F(t)$. This is an example of the curve of Example 2 traversed by a different path, that is, of a different parametrization of the curve in Example 2 (Figure 6.2.5).

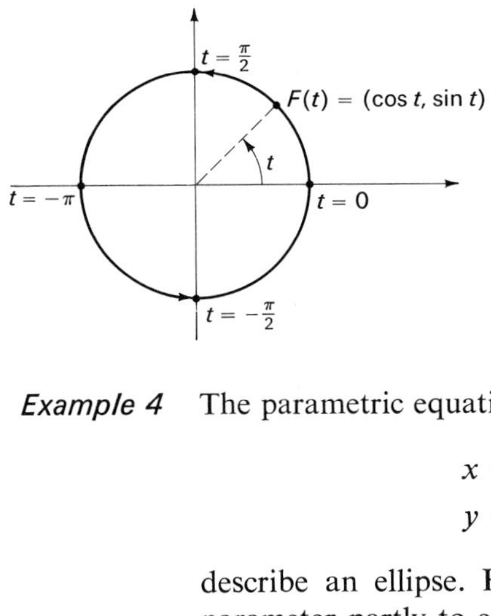

Figure 6.2.5

Example 4 The parametric equations

$$\begin{aligned} x &= a \cos \theta, \\ y &= b \sin \theta, \end{aligned} \qquad -\pi \leq \theta < \pi,$$

describe an ellipse. Here, we have used the symbol "θ" for the parameter partly to emphasize the obvious fact that symbols other than "t" may be used and partly in recognition of the geometric interpretation of this parameter, which is similar to that of the last example.

To see that the curve is an ellipse we need only observe that $x/a = \cos \theta$ and $x/b = \sin \theta$. Thus

(6.2.3) $$\frac{x^2}{a^2} + \frac{y^2}{b^2} = 1,$$

so every point on the curve lies on the ellipse with equation (6.2.3).

The geometrical significance of this parameter can be seen as follows. Draw circles about the origin of radius a and b respectively. The ray from the origin making an angle θ with the positive horizontal axis intersects these circles at points

$$P_a = (a\cos\theta,\ a\sin\theta),$$
$$P_b = (b\cos\theta,\ b\sin\theta).$$

If P is the intersection of a vertical line through P_a and a horizontal line through P_b, then P is the point on the ellipse corresponding to the value θ of the parameter. See Figure 6.2.6.

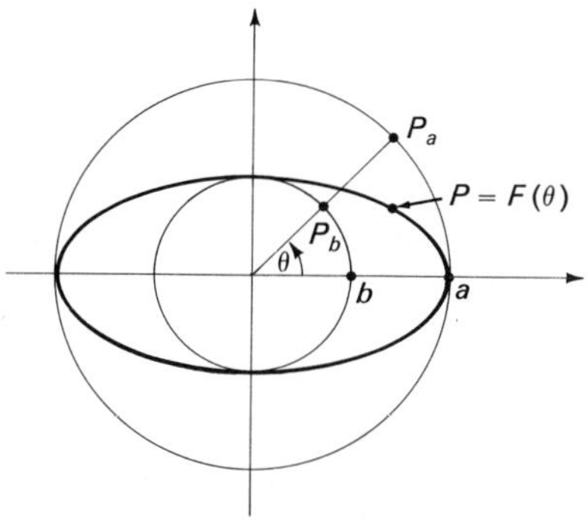

Figure 6.2.6

A curve determined by parametric equations (6.2.1) will be called a *continuous curve* if both functions f and g are continuous. Roughly speaking, such a curve has no "breaks" in it, for as $t \to t_0$ both $f(t) \to f(t_0)$ and $g(t) \to g(t_0)$. Therefore, the point $(g(t), f(t))$ approaches the point $(g(t_0), f(t_0))$, in the sense that the distance between these points approaches zero. This is all we shall say about continuous curves, because we are chiefly interested in a stronger condition, *smoothness*.

Referring to (6.2.1) again, let us suppose that both f and g are differentiable on I. Furthermore, suppose that $g'(t) \neq 0$ for all t in a closed interval $[a, b]$ contained in I. For simplicity, let us assume that $g'(t) > 0$ for $a \leq t \leq b$. Then, as we have seen, the function g has a differentiable inverse function φ defined on the interval $[g(a), g(b)]$; we write

(6.2.4) $$t = \varphi(x),\ g(a) \leq x \leq g(b).$$

Substituting (6.2.4) in the first of equations (6.2.1) we obtain

$$y = f(\varphi(x)) = [f \circ \varphi](x).$$

Thus, the curve determined by (6.2.1) is just the graph of a function $f \circ \varphi$. By the chain rule we know that $f \circ \varphi$ is differentiable, so the curve is differentiable; that is, it has a tangent at the point $(x, y) = (x, f(\varphi(x))) = (g(t), f(t))$ whose slope is

(6.2.5)
$$\frac{dy}{dx} = f'(\varphi(x)) \cdot \varphi'(x) = \frac{f'(\varphi(x))}{g'(\varphi(x))},$$

since φ is an inverse function for g. But applying (6.2.4) again, we see that (6.2.5) can be written

(6.2.6)
$$\frac{dy}{dx} = \frac{f'(t)}{g'(t)} = \frac{dy/dt}{dx/dt}.$$

Thus, we conclude that the curve has a tangent line at the point $(g(t), f(t))$ whose slope is given by (6.2.6).

The restriction that $g'(t) > 0$ for all t is too strong to be accepted. One of the primary purposes for introducing parametric equations is to handle curves which wander all about the plane with $g'(t)$ changing sign repeatedly. So let us drop the assumption that $g'(t)$ has constant sign, but add the assumption that both $f'(t)$ and $g'(t)$ are continuous functions. Then if for a certain value t_0 we have $g'(t_0) \neq 0$, we can conclude by the inertia property (Theorem 1.5.1 in *C 1*) that $g'(t) \neq 0$ for all t in some interval $[a, b]$ containing t_0. Thus, our above discussion can be applied and we conclude that the curve has a tangent at $(g(t_0), f(t_0))$ with slope given by (6.2.6).

If, on the other hand, $g'(t_0) = 0$ but $f'(t_0) \neq 0$, then we can apply the above reasoning, reversing the roles of x and y. This has the effect of reversing the coordinate axes and we conclude that *viewed from the side* the curve has a tangent line with slope

$$\frac{g'(t_0)}{f'(t_0)} = \frac{0}{f'(t_0)} = 0.$$

Having slope 0 *viewed from the side* means being parallel to the vertical axis. Thus, we conclude that in this case the curve has a vertical tangent line.

In conclusion, we have seen that if $f'(t)$ and $g'(t)$ are both continuous and the two do not vanish simultaneously at any one point, then the curve has a tangent at every point. Moreover, the slope (6.2.6) of the tangent line is also a continuous function of t whenever $g'(t) \neq 0$. The slope approaches $\pm\infty$ at a point where $g'(t) = 0$ but $f'(t) \neq 0$, which is exactly where the curve has a vertical tangent line. For this reason we say that the curve has a *continuously varying tangent line*. This is exactly what we mean by a *smooth curve*.

Theorem 6.2.1 *Let f and g be functions for which*

(i) *both $f'(t)$ and $g'(t)$ exist and are continuous for all t in an interval I and*

(ii) $[f'(t)]^2 + [g'(t)]^2 \neq 0$ *for all t in I.*[1]

Then the curve determined by the parametric equations (6.2.1) is a smooth curve. At a point where $g'(t) \neq 0$ the slope of the tangent line is $f'(t)/g'(t)$. At a point where $g'(t) = 0$ the tangent line is vertical.

[1] *This condition merely requires that $f'(t)$ and $g'(t)$ should not vanish simultaneously.*

Let us pause here to look back briefly at the Cauchy Mean Value Theorem, Theorem 4.2.1. It asserted that if f' and g' exist for all t in $[a, b]$ and $g'(t) \neq 0$ on (a, b), then there exists a $\bar{t}$ in (a, b) such that

$$\frac{f(b) - f(a)}{g(b) - g(a)} = \frac{f'(\bar{t})}{g'(\bar{t})}.$$

We now see that this condition simply asserts that for a curve defined by a pair of parametric equations (6.2.1), there is a point where the tangent line is parallel to the line joining the endpoints. Actually, it is not difficult to show that the condition $g'(t) \neq 0$ can be replaced by conditions (1) and (2) of Theorem 6.2.1 provided $g(b) \neq g(a)$. This allows the Cauchy Theorem to apply to all smooth curves, not just those which are graphs of equations of the form $y = f(x)$. Thus, although Figure 6.2.7(a) is the picture one thinks of in connection with the Mean Value Theorem, Figure 6.2.7(b) is the one more suggestive of the Cauchy Theorem.

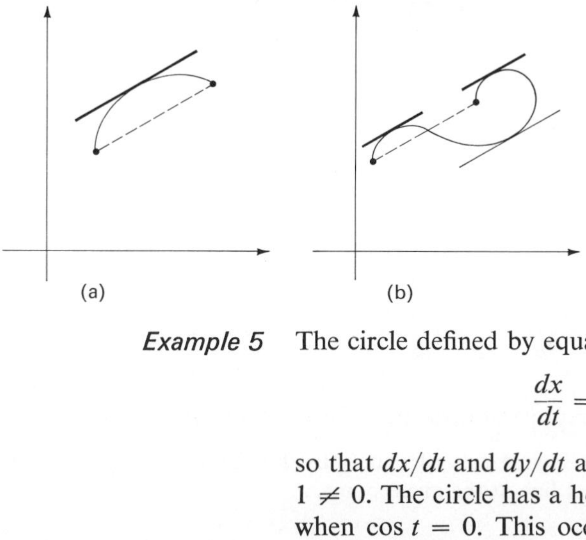

Figure 6.2.7

Example 5 The circle defined by equations (6.2.2) is a smooth curve, because

$$\frac{dx}{dt} = -\sin t, \quad \frac{dy}{dt} = \cos t,$$

so that dx/dt and dy/dt are continuous and $(dx/dt)^2 + (dy/dt)^2 = 1 \neq 0$. The circle has a horizontal tangent when $dy/dt = 0$, that is, when $\cos t = 0$. This occurs for $t = -\pi/2, \pi/2$. Thus, the circle has horizontal tangents at the points $(0, -1)$ and $(0, +1)$. Vertical tangents occur when $dx/dt = 0$, that is, when $t = -\pi, 0$. Thus, the curve has vertical tangent lines at the points $(-1, 0)$ and $(1, 0)$. The curve has a tangent of slope 1 at a point corresponding to a value of t satisfying

$$\frac{dy}{dx} = \frac{dy/dt}{dx/dt} = \frac{\cos t}{-\sin t} = -\cot t = 1.$$

For $-\pi \leq t < \pi$, this occurs for $t = -\pi/4$, $3\pi/4$. Thus, the tangents to the circle at points $(\sqrt{2}/2, -\sqrt{2}/2)$ and $(-\sqrt{2}/2, \sqrt{2}/2)$ have slope 1, and these are the only such points (Figure 6.2.8).

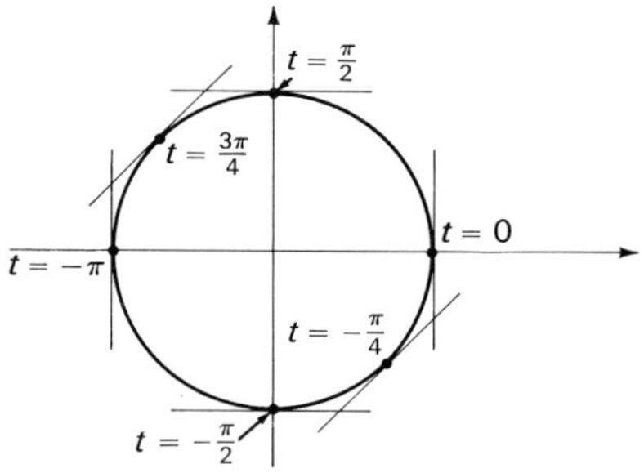

Figure 6.2.8

Exercises

1 Which of the following sets of parametric equations are parametrizations of the curve in Example 1?

(a) $x = 4t^2$, $y = 2t$, $-\infty < t < \infty$;

(b) $x = t$, $y = \sqrt{t}$, $0 \leq t < \infty$;

(c) $x = 1 - \cos^2 t$, $y = \sin t$, $-\infty < t < \infty$;

(d) $x = t^2 - 4t + 4$, $y = 2 - t$, $-\infty < t < \infty$;

(e) $x = \sec^2 t - 1$, $y = \tan t$, $-\pi/2 < t < \pi/2$;

(f) $x = t^4$, $y = t^2$, $-\infty < t < \infty$.

2 Find two different parametrizations for the ellipse of Example 4.

3 Each of the following sets of equations describes a curve. Find the slope of the tangent line at the given point.

(a) $x = t + 1$, $y = t^2 - 3$, $t = 0$;

(b) $x = 2t + 4$, $y = -3t^3 + 5$, $t = -1$;

(c) $x = 2 \sin t,\ y = 3 \cos t^2,\quad t = 0;$ (d) $x = 2 + \cos t,\ y = -3 + \sin t,\quad t = -\pi/4;$

(e) $x = (t - \sin t),\ y = (1 - \cos t),\quad t = \pi/2;$ (f) $x = \dfrac{2t}{1 + t^2},\ y = \dfrac{1 - t^2}{1 + t^2},\ t = 2.$

4 Find an equation for the tangent line to each of the following curves at the given point. Plot the graph of the curve and of its tangent line.

(a) $x = 1 + t^2,\ y = 1 - t^2,\quad t = 1;$

(b) $x = \dfrac{1}{1 + t},\ y = \dfrac{2t}{1 + t},\ t = 1;$

(c) $x = \sqrt{1 + t},\ y = \sqrt{1 - t},\quad t = 1;$

(d) $x = 25 - t^2,\ y = t^3,\quad t = 4.$

5 Find the equations of the tangent line and normal line to the curve $x = t^2$, $y = t^3$ at the point where $t = 2$.

6 Locate the points on the curve described in Exercise 5 above where the slope of the tangent line is 1.

7 Find a formula for the second derivative d^2y/dx^2 if y and x are given by parametric equations

$$y = f(t),$$
$$x = g(t).$$

$$\left(\textit{Hint: } \frac{d^2y}{dx} = \frac{d}{dx}\left(\frac{dy}{dx}\right) = \frac{d}{dt}\left(\frac{dy}{dx}\right)\frac{dt}{dx} = \frac{d}{dt}\left(\frac{dy}{dx}\right)\Big/\frac{dx}{dt}.\right)$$

8 Find the maximum and minimum points and the points of inflection for the curve

$$x = t - \sin t,$$
$$y = 1 - \cos t.$$

6.3 Curves in Polar Coordinates

To introduce coordinates in a plane we have been using the so-called Cartesian coordinate system, in which points are labeled by reference to a pair of mutually perpendicular coordinate axes. Another method of introducing coordinates makes use of the system of *polar coordinates*, in which the "reference marks" are a point, called the *pole*, and a ray from the pole, which may be considered as the positive half of a coordinate line with origin at the pole. This reference ray is called the *polar axis*.

Points are labeled in polar coordinates as follows. If P is a point, draw the line OP, where O is the pole. Establish a coordinate system (i.e. mark off units) on OP with O as the origin and with unit length the same as on the polar axis. Let θ be the angle from the polar axis to the positive ray on OP and let r be the coordinate of P in the coordinate system just established on OP (that is, $|r|$ is the distance $|OP|$, and r is positive or negative depending on whether P is in the positive or negative direction from O according to the orientation induced by the coordinates established on OP). We call r and θ polar coordinates for P and write $P \sim (r, \theta)$ (Figure 6.3.1).

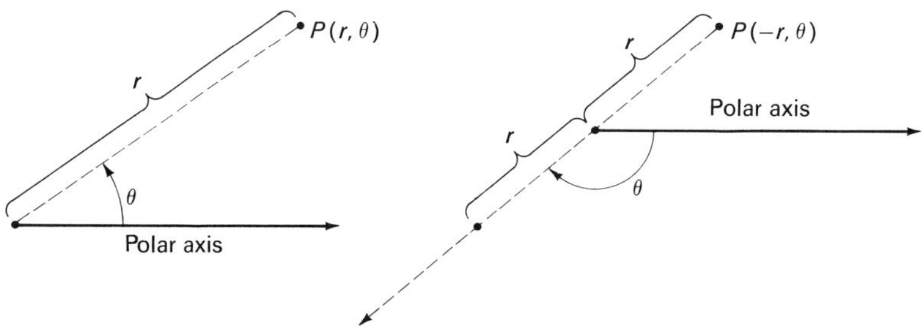

Figure 6.3.1

Clearly every point in the plane has polar coordinates. In fact, each point has infinitely many different pairs of polar coordinates, for if $P \sim (r, \theta)$, then also $P \sim (r, \theta + 2\pi)$, $P \sim (r, \theta + 4\pi)$, and $P \sim (r, \theta + 2k\pi)$, where k is any integer. Also, $P \sim (-r, \theta + \pi)$ and $P \sim (-r, \theta + 3\pi)$, etc. Thus, polar coordinates lack one desirable feature of Cartesian coordinates, namely a one-to-one correspondence between points and their Cartesian coordinates. Nevertheless, polar coordinates are frequently useful. Certain problems are more easily handled in polar coordinates than in Cartesian coordinates, and certain curves are more easily expressed by equations in polar coordinates.

Figure 6.3.2

To make efficient use of polar coordinates one must be able to transfer freely between polar and Cartesian coordinates in a given plane. This is most easily done when the axes are placed in a special position relative to each other. Let us agree to use the positive horizontal axis of a Cartesian system as the polar axis and the origin as the pole. Then the following equations which relate the Cartesian coordinates (x, y) of a point to a set of polar coordinates (r, θ) for the same point may be obtained easily from Figure 6.3.2. We leave the details to you!

(6.3.1)
$$x = r\cos\theta, \quad y = r\sin\theta, \qquad \frac{y}{x} = \tan\theta, \qquad x^2 + y^2 = r^2.$$

A curve is usually described in polar coordinates by an equation relating the two variables r and θ. Two methods are immediately available for sketching the graph of such an equation. We can use whatever information we can muster about the functions involved in the equation to find a number of pairs (r, θ) which satisfy it and to "see" how r varies with θ or vice versa. Or we can use equation (6.3.1) to find the equation in x and y of the same curve in Cartesian coordinates and then utilize our past experience to identify the graph. Both methods are illustrated in the following examples.

Example 1 Consider the equation

$$r = \theta$$

for $\theta \geq 0$. It is easy to see what the graph is. Simply imagine a ray sweeping around in a counterclockwise direction, starting at the polar axis. Follow a point on this ray for which the distance r is always equal to the angle θ. A spiral emerges (Figure 6.3.3).

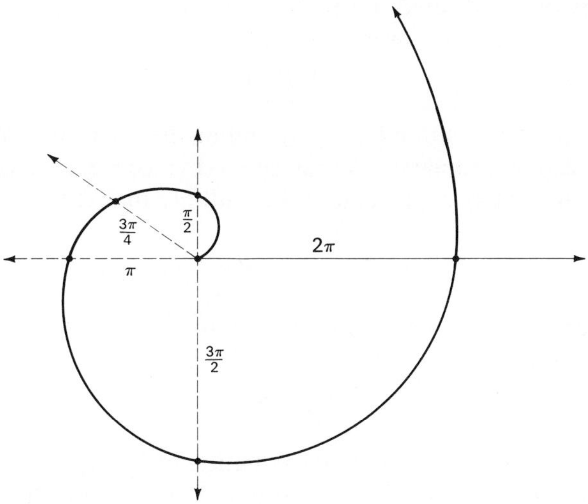

Figure 6.3.3

Example 2 Consider the equation

(6.3.2)
$$r = 3\cos\theta.$$

We know just how $\cos\theta$ varies with θ (that is, we know the graph of $r = \cos\theta$ in Cartesian r, θ coordinates). Using this knowledge and the knowledge of some specific values of $\cos\theta$ we can obtain the following information. We use the shorthand "$0 \to \theta \to \pi/2$" for the statement "θ varies monotonically from 0 to $\pi/2$."

As . . .	then correspondingly
$0 \to \theta \to \frac{\pi}{2}$	$3 \to r \to 0$
$\frac{\pi}{2} \to \theta \to \pi$	$0 \to r \to -3$
$\pi \to \theta \to \frac{3\pi}{2}$	$-3 \to r \to 0$
$\frac{3\pi}{2} \to \theta \to 2\pi$	$0 \to r \to 3$

We also have the following table of values:

θ	0	$\frac{\pi}{4}$	$\frac{\pi}{2}$	$\frac{3\pi}{4}$	π	$\frac{5\pi}{4}$	$\frac{3\pi}{2}$	$\frac{7\pi}{4}$	2π
r	3	$-\frac{3}{2}\sqrt{2}$	0	$-\frac{3}{2}\sqrt{2}$	-3	$-\frac{3}{2}\sqrt{2}$	0	$\frac{3}{2}\sqrt{2}$	3

Plotting these points and taking into consideration the monotonic variations of r indicated above, we can sketch the curve. Clearly, this curve is swept out over and over as θ varies from 2π to 4π, 4π to 6π, etc.; the same curve is also swept out for negative values of θ.

Example 3 Let us graph the equation (6.3.2) by finding the equation of the curve in Cartesian coordinates. A glance at the conversion equations (6.3.1) immediately suggests that we multiply (6.3.2) by r to obtain

(6.3.3) $$r^2 = 3r\cos\theta.$$

The equation (6.3.1) enables us to eliminate r and θ, yielding the new equation

$$x^2 + y^2 = 3x.$$

This is immediately seen to be the equation of a circle of radius 3/2 tangent to the vertical axis at the origin. We recognize the curve in Figure 6.3.4.

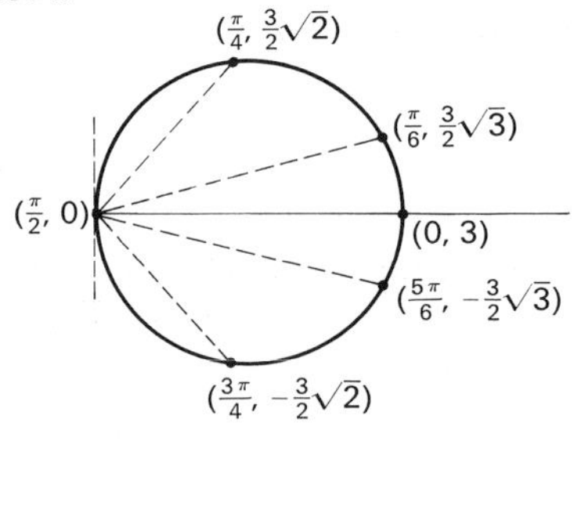

Figure 6.3.4

Example 4 Consider the equation

$$r = 2 \sin 3\theta.$$

We know the variation of sin 3θ as 3θ varies from 0 to $\pi/2$, $\pi/2$ to π, and so on, that is, as θ varies from 0 to $\pi/6$, $\pi/6$ to $\pi/3$, etc. Thus we can easily see that

As . . .	then correspondingly
$\theta \to \theta \to \frac{\pi}{6}$	$0 \to r \to 2$
$\frac{\pi}{6} \to \theta \to \frac{\pi}{3}$	$2 \to r \to 0$
$\frac{\pi}{3} \to \theta \to \frac{\pi}{2}$	$0 \to r \to -2$
$\frac{\pi}{2} \to \theta \to \frac{2\pi}{3}$	$-2 \to r \to 0$
$\frac{2\pi}{3} \to \theta \to \frac{5\pi}{6}$	$0 \to r \to 2$
$\frac{5\pi}{6} \to \theta \to \pi$	$2 \to r \to 0$

On the basis of the seven points and the variation indicated in the above table, we can construct a sketch of the curve (Figure 6.3.5). You can check that as θ continues to vary the same figure is swept out repeatedly. This curve is called the *three-leafed rose.*

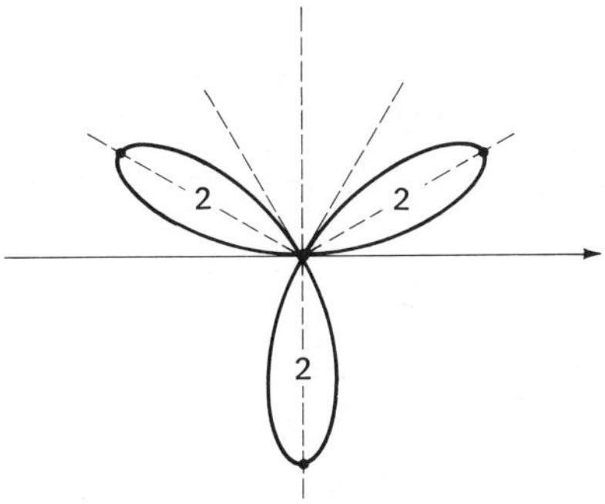

Figure 6.3.5

Each of the above equations is of the form

(6.3.4) $$r = f(\theta),$$

where f is a function with continuous derivative for all θ. It is easy to see that the graph of such a function is a smooth curve, at least if $[f(\theta)]^2 + [f'(\theta)]^2 \neq 0$ for all θ. One need only use (6.3.1) to write parametric equations for the same curve in Cartesian coordinates.

These follow readily from (6.3.4) and the relations $x = r\cos\theta$, $y = r\sin\theta$. The graph is then the graph of the parametric equations

(6.3.5)
$$x = f(\theta)\cos\theta,\qquad y = f(\theta)\sin\theta$$

with parameter θ. We then have

(6.3.6)
$$\frac{dx}{d\theta} = f'(\theta)\cos\theta - f(\theta)\sin\theta,\qquad \frac{dy}{d\theta} = f'(\theta)\sin\theta + f(\theta)\cos\theta.$$

Both the above sufficient condition for smoothness of the curve and an expression for the slope of a tangent line to the curve may be found using (6.3.6) and the information developed in the last section.

Exercises

1 Plot the following points on a polar coordinate system: $(2, \pi/4)$, $(1, 3\pi/4)$, $(-2, \pi/2)$, $(-5, \pi)$, $(3/2, -\pi/2)$, $(-\frac{1}{2}, -\pi/4)$.

2 Find another set of polar coordinates for each of the points in Exercise 1.

3 Where do all the points lie for which $r = 6$? for which $\theta = \pi/3$?

4 Find rectangular coordinates for the points with the following polar coordinates: $(3, \pi/3)$, $(10, \pi/4)$, $(5, 16\pi/3)$, $(7, -8\pi/3)$, $(-3, -\pi/6)$, $(-\frac{1}{2}, -7\pi/2)$.

5 Find a pair of polar coordinates for the points with the following rectangular coordinates: $(0, 4)$, $(3/2, 0)$, $(1, 1)$, $(-1, 1)$, $(-4\sqrt{3}, 4)$.

6 Plot the graphs of the following equations.

(a) $r = 5\cos\theta$, (b) $r = 1 - 2\sin\theta$,
(c) $r\cos\theta = 3$, (d) $r = 4/(1 + \cos\theta)$,
(e) $r = 2\sin 2\theta$, (f) $r = 3\cos 4\theta$,
(g) $r = 2(1 + \cos\theta)$, (h) $r^2 = 4\sin 2\theta$,
(i) $r = \sin^2\theta$, (j) $r = (1 - 3\cos\theta)$
(k) $r = \sin\theta + \cos\theta$, (l) $r = 2\tan\theta\sin\theta$.

7 Express the following equations in rectangular coordinates and plot their graphs.

(a) $r = 5$, (b) $r\sin\theta = 4$,
(c) $r = 2\sin\theta$, (d) $r = 3\cos\theta + 2\sin\theta$,
(e) $r(1 - \sin\theta) = 4$, (f) $r^2\cos 2\theta = 9$.

8 Use (6.3.6) to show that the slope of the tangent at a point on the curve $r = f(\theta)$ is given by

$$\frac{f'(\theta)\sin\theta + f(\theta)\cos\theta}{f'(\theta)\cos\theta - f(\theta)\sin\theta}.$$

9 Find the slope of the tangent to each of the following curves at the indicated point.

(a) $r = 3\sin\theta$, $(3/2, \pi/6)$; (b) $r = 4\cos 3\theta$, $(0, \pi/6)$;
(c) $r = 3\cos\theta + 4\sin\theta$, $(3, 0)$; (d) $r^2\sin 2\theta = 5$, at $\theta = \theta_1$;
(e) $r = 2(1 + \cos\theta)$, $(2, \pi/2)$; (f) $r = 3(1 - \sin\theta)$, $(3/2, \pi/6)$.

10 Show that the curves $r = k\sin\theta$ and $r = k\cos\theta$ intersect at right angles.

11 Draw the graphs of each of the following pairs of equations in one polar coordinate plane, and find the points where they intersect.

(a) $r = 4\cos\theta$, $r = 2$; (b) $r\cos\theta = 6$, $r = 10$;
(c) $r = \sin\theta$, $r = \cos 2\theta$; (d) $r = 1 + \cos\theta$, $r(1 + \cos\theta) = 1$.

6.4 Area and Arclength Again

In this section we consider briefly two problems: the area of a region bounded by a curve and the length of a curve, when the curve is given in either parametric or polar form. First, area:

Some Areas Let Γ be the graph of parametric equations

$$y = f(t),\quad x = g(t),\qquad t_a \le t \le t_b, \tag{6.4.1}$$

where f and g are continuous, $g'(t) > 0$, and $f(t) > 0$ for all t in the interval. Then Γ is the graph of a function $f \circ g^{-1}$ in Cartesian coordinates, where $f(g^{-1}(x))$ is defined for x in the interval $[a, b] = [g(t_a), g(t_b)]$. Since $f(t) > 0$, the graph lies above the horizontal axis, so we may consider the region R under Γ between $x = a$ and $x' = b$.

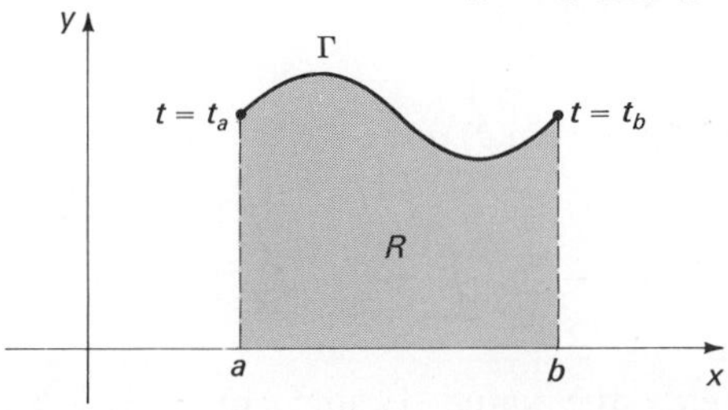

Figure 6.4.1

From earlier work we know that the area $A(R)$ of the region R is $\int_a^b y\,dx = \int_a^b f(g^{-1}(x))\,dx$. But introducing the substitution $x = g(t)$

in this integral, we find that

(6.4.2)
$$\int_a^b f(g^{-1}(x))\,dx = \int_{t_a}^{t_b} f(g^{-1}(g(t)))g'(t)\,dt.$$

Since $g^{-1}(g(t) = t$, (6.4.2) gives the simple integral expression for the area:

(6.4.3)
$$A(R) = \int_{t_a}^{t_b} f(t)g'(t)\,dt.$$

If $g'(t) < 0$, then in Figure 6.4.1, $a = g(t_b)$, $b = g(t_a)$. Following the above reasoning, then, the area $A(R)$ would be the same integral as in (6.4.3) except that $\int_{t_a}^{t_b}$ would be replaced by $\int_{t_b}^{t_a}$. Thus, the area formula (6.4.3) would have $(-)$ in front of the integral. But if $g'(t) \neq 0$ for $t_a \leq t \leq t_b$, the area of R is given correctly by the formula

(6.4.4)
$$A(R) = \left|\int_{t_a}^{t_b} f(t)g'(t)dt\right|.$$

Example 1 Consider the ellipse (Figure 6.4.2) with parametric equations

$$\begin{aligned} x &= a\cos\theta, \\ y &= b\sin\theta, \end{aligned} \quad 0 \leq \theta \leq 2\pi.$$

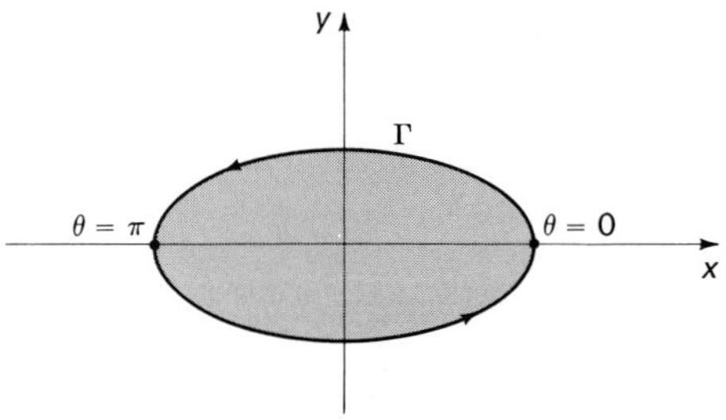

Figure 6.4.2

What is the area of the region enclosed? Clearly, the area is twice that of the region bounded by the horizontal axis and the upper half of the ellipse which is swept out as θ ranges from 0 to π. Thus, the desired area A is by (6.4.4),

$$\begin{aligned} A &= 2\left|\int_0^\pi (b\sin\theta)(-a\sin\theta)\,d\theta\right| \\ &= 2\int_0^\pi ab\sin^2\theta\,d\theta = 2ab\int_0^\pi \sin^2\theta\,d\theta. \end{aligned}$$

An integration by parts yields the value $\int_0^\pi \sin^2\theta\,d\theta = \pi/2$, so we have

(6.4.5)
$$A = \pi ab.$$

Now let γ be the graph of the equation

(6.4.6) $$r = f(\theta),\ \theta_a \le \theta \le \theta_b,$$

in polar coordinates. Suppose further that $0 \le \theta_a < \theta_b \le 2\pi$ and that $f(\theta)$ is continuous and positive in that interval. Then we may consider the region R bounded by the graph of (6.4.6) and the rays from the pole making angles θ_a and θ_b with the polar axis. Let us find the area $A(R)$ of R. See Figure 6.4.3.

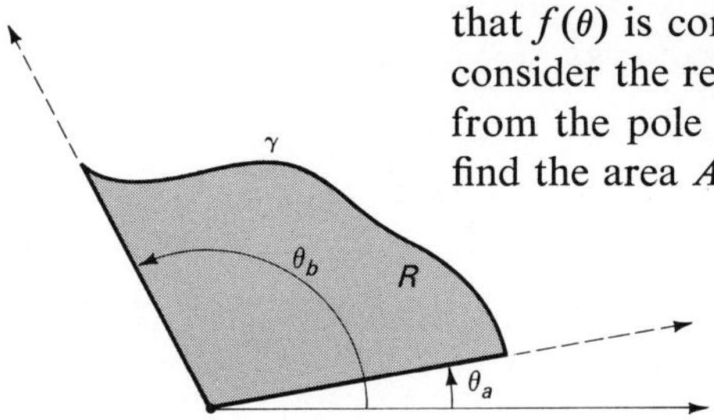

Figure 6.4.3

Form a partition $P = \{\theta_a = \theta_0, \theta_1, \theta_2, \ldots, \theta_n = \theta_b\}$ of the interval $[\theta_a, \theta_b]$. This effects a division of R into little sectors, as indicated in Figure 6.4.4.

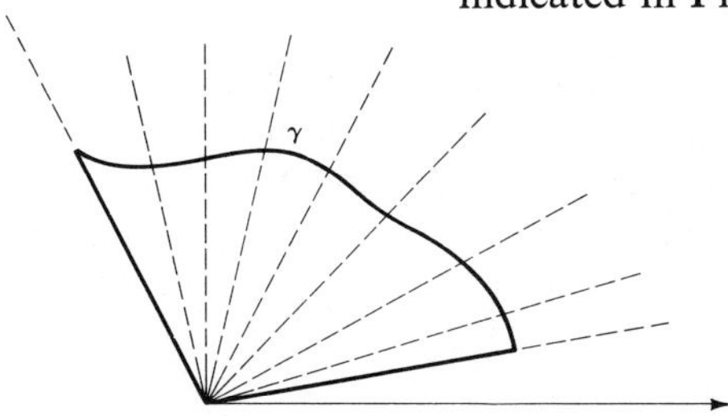

Figure 6.4.4

For each θ-subinterval $[\theta_{i-1}, \theta_i]$ let $f(\theta)$ have minimum value $f(\theta_i')$ and maximum value $f(\theta_i'')$. Then we may enclose the corresponding part of R in a circular sector of radius $f(\theta_i'')$ and central angle $\Delta\theta_i = \theta_i - \theta_{i-1}$. The circular sector with radius $f(\theta_i')$ and the same central angle is contained within this piece of R. See Figure 6.4.5.

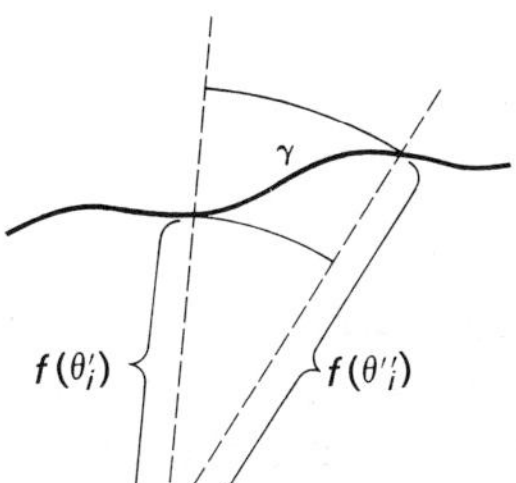

Figure 6.4.5

Since the area of a circular sector of radius r and central angle φ is $r^2\varphi/2$ (φ measured in radians), we easily construct the following estimates of $A(R)$:

(6.4.7) $$\sum_{i=1}^{n} \frac{[f(\theta_i')]^2}{2}\Delta\theta_i \le A(R) \le \sum_{i=1}^{n} \frac{[f(\theta_i'')]^2}{2}\Delta\theta_i.$$

We recognize the two sums in (6.4.7) as Riemann sums for the integral of the function $f^2/2$ over the θ-interval $[\theta_a, \theta_b]$. Thus, since $f^2/2$ is continuous and the integral is the limit of these Riemann sums (see Chapter 3), we can conclude from (6.4.7) that

$$A(R) = \int_{\theta_a}^{\theta_b} \frac{1}{2}[f(\theta)]^2\, d\theta.$$

Example 2 The area of one loop of the three-leafed rose in Example 4, Section 6.3 is

(6.4.8) $$A = \int_0^{\pi/3} \frac{1}{2}[2 \sin 3\theta]^2\, d\theta = 2\int_0^{\pi/3} [\sin 3\theta]^2\, d\theta.$$

Introducing the substitution $u = 3\theta$, we obtain

(6.4.9) $$2\int_0^{\pi/3} [\sin 3\theta]^2\, d\theta = \frac{2}{3}\int_0^{\pi} \sin^2 u\, du.$$

Once again we may use the value $\int_0^\pi \sin^2 u\, du = \pi/2$ to obtain from (6.4.8) and (6.4.9)

$$A = \frac{\pi}{3}.$$

Some Arclengths We return first to the curve Γ defined by the parametric equations (6.4.1), but now we assume that the curve is smooth. That is, we assume that both f' and g' are continuous and that $[f'(t)]^2 + [g'(t)]^2 \neq 0$.

To further simplify matters, assume that except possibly for the endpoints P_a and P_b, no point on Γ corresponds to two or more different values of t. To calculate (indeed, to *define*) the length of Γ between points $P_a = (g(t_a), f(t_a))$ and $P_b = (g(t_b), f(t_b))$ we first consider an arbitrary finite set of points $P_a = P_0, P_1, \ldots, P_n = P_b$

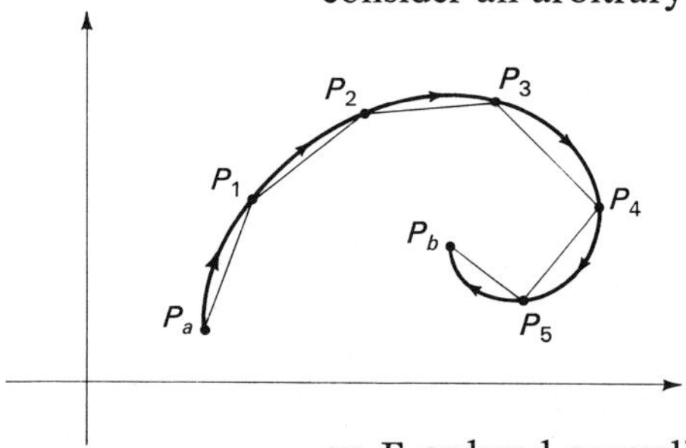

Figure 6.4.6

on Γ ordered according to the natural orientation of Γ and a corresponding partition $\{t_{a_b} = t_0, t_1, t_2, \ldots, t_n = t_b\}$ such that $P_i = (g(t_i), f(t_i))$. See Figure 6.4.6. It is natural to use the lengths $|P_{i-1}P_i|$ as approximations for the length of Γ between P_{i-1} and P_i. Using the distance formula we obtain an approximation for the length $L(\Gamma)$ of Γ:

(6.4.10) $$L(\Gamma) \sim \sum_{i=1}^{n} \sqrt{[(f(t_i) - f(t_{i-1})]^2 + [(g(t_i) - g(t_{i-1})]^2}.$$

But now we may apply the Mean Value Theorem to each function f, g on each interval $[t_{i-1}, t_i]$ to obtain

(6.4.11) $$\begin{aligned} f(t_i) - f(t_{i-1}) &= f'(t_i')\,\Delta t_i, \\ g(t_i) - g(t_{i-1}) &= g'(t_i'')\,\Delta t_i, \end{aligned}$$

where t_i' and t_i'' are numbers lying in the interval (t_{i-1}, t_i). Substituting (6.4.11) into (6.4.10) we obtain the approximation

(6.4.12)
$$L(\Gamma) \sim \sum_{i=1}^{n} \sqrt{[f'(t_i')]^2 + [g'(t_i'')]^2}\, \Delta t_i.$$

It is somewhat bothersome that the numbers t_i' and t_i'' in (6.4.12) are not the same. If they were, we would recognize the sum as a Riemann sum and we would be on our way to an integral. But the worry disappears when we note that the sum (6.4.12) is a *generalized* Riemann sum in the sense of Definition 3.6.1. Such sums were invented to handle precisely this situation, and Theorem 3.6.1 assures us that they are just as useful as ordinary Riemann sums would be, for the limit of the sum in (6.4.12) as the norm of the partition of $[t_a, t_b]$ tends to 0 is the integral

(6.4.13)
$$\int_{t_b}^{t_b} \sqrt{[f'(t)]^2 + [g'(t)]^2}\, dt.$$

Moreover, for partitions with very small norm, that is, for t_{i-1} and t_i very close for all i, continuity of f and g assure us that the points P_{i-1} and P_i on Γ will be very close. Since the approximation (6.4.10) or (6.4.12) ought to be very good when all the points P_i on Γ are close to each other in pairs and since these sums *do* approach a limit (6.4.13), it is natural to accept the inevitable and *define* the length to be

(6.4.14)
$$L(\Gamma) = \int_{t_a}^{t_b} \sqrt{[f'(t)]^2 + [g'(t)]^2}\, dt.$$

Example 3 A circle of radius r may be represented parametrically by

$$\begin{aligned} x &= r\cos t, \\ y &= r\sin t, \end{aligned} \qquad 0 \le t \le 2\pi.$$

Only the "endpoint" $(r, 0)$ is represented by two values of t, *i. e.*, 0, 2π. Clearly, the other conditions are satisfied; the circle is a smooth curve. Its length, then, is

$$L = \int_0^{2\pi} \sqrt{r^2\sin^2 t + r^2\cos^2 dt} = r\int_0^{2\pi} dt = 2\pi r,$$

a not-at-all-surprising result.

Example 4 The graph of an ordinary function f may always be interpreted as the graph of a pair of parametric equations

$$\begin{aligned} y &= f(t), \\ x &= t. \end{aligned}$$

In a sense, this amounts to making x the parameter, but it will help avoid confusion if we use the notation of (6.4.1), introducing the apparently needless new symbol "t." The length of this graph between points $(a, f(a))$ and $(b, f(b))$ can then be obtained from (6.4.14). It is

$$L = \int_a^b \sqrt{\left(\frac{dx}{dt}\right)^2 + \left(\frac{dy}{dt}\right)^2}\, dt = \int_a^b \sqrt{1 + [f'(t)]^2}\, dt.$$

This is exactly our earlier formula for the length of such a curve.

Lengths of parametric curves with a *finite number of self-intersections* may be found by splitting them into simple closed curves, that is, curves which satisfy the above conditions. But when these lengths are added together, the theorem $\int_a^b + \int_b^c = \int_a^c$ shows that formula (6.4.14) may be used as it stands for the entire curve. However, it should be recognized that strictly speaking (6.4.14) gives the length of a *path*, not a curve. When a path traces a curve, or part of the curve more than once, the length of the retraced portion emerges from (6.4.14) as many times as it is traced.

The length of a curve given in polar coordinates by an equation (6.3.4) may be obtained by means of the parametric equations (6.3.5) for the curve. If γ is the graph of (6.3.4), then its parametric equations are

$$\begin{aligned} x &= f(\theta)\cos\theta, \\ y &= f(\theta)\sin\theta, \end{aligned} \qquad \theta_a \le \theta \le \theta_b.$$

Applying (6.4.14) we obtain (you may carry out the details)

$$L(\gamma) = \int_{\theta_a}^{\theta_b} \sqrt{[f'(\theta)]^2 + [f(\theta)]^2}\, d\theta.$$

Exercises

In Exercises 1–10 find the area of the region bounded by the given curves.

1 $x = t^2$, $y = t^3$, $0 \le t \le 2$; the x-axis; and the line $x = 4$.

2 $x = \sqrt{t}$, $y = 1/t$, $1 \le t \le 4$; the x-axis; and the lines $x = 1$ and $x = 2$.

3 $x = t^2$, $y = \sin t$, $0 \le t \le \pi$; the x-axis.

4 One arch of the cycloid $x = 2(t - \sin t)$, $y = 2(1 - \cos t)$, and the x-axis.

5 $r = a\cos\theta$, $\theta = 0$, and $\theta = \pi/4$.

6 The cardioid $r = 1 + \cos\theta$.

7 The limaçon $r = 2 + \cos\theta$.

8 One leaf of the four-leafed rose $r = 3 \cos 2\theta$.

9 One loop of the lemniscate $r^2 = a^2 \cos 2\theta$.

10 Inside $r = 2 \cos \theta + 4$ and outside $r = 5$.

In Exercises 11–17 find the arclength of the arc described.

11 The total length of the cardioid $r = a(1 - \cos \theta)$.

12 $r = e^\theta, 0 \leq \theta \leq 1$.

13 The astroid $x = a \cos^3 t, y = a \sin^3 t, 0 \leq t \leq 2\pi$.

14 One arch of the cycloid $x = a(t - \sin t), y = a(1 - \cos t)$.

15 $r = 2\theta^2, 0 \leq \theta \leq 2\pi$.

16 $x = t^2, y = t^3, 0 \leq t \leq 1$.

17 The cissoid $r = 2a \tan \theta \sin \theta$ between $\theta = 0$ and $\theta = \pi/3$.

6.5

Arclength as a Parameter; Curvature

In our discussion of curves we have never ventured far from the functions which define them. We never shall, for functions are our only hold on curves. However, to introduce the next idea, let us think purely geometrically for a moment.

Figure 6.5.1

Suppose we have a smooth curve in the plane with only finitely many self-intersections (Figure 6.5.1). What should it mean to *parametrize* this curve? Parametrization of a curve should be analogous to *coordinatizing* a line; it should establish a correspondence between the numbers t in some interval $[a, b]$ and the points on the curve, a correspondence which is one-to-one except at those points which are self-intersections. This correspondence can be established in any way, the only condition being that the normal ordering of numbers ought to correspond to an *orientation* of the curve. See Figure 6.5.2.

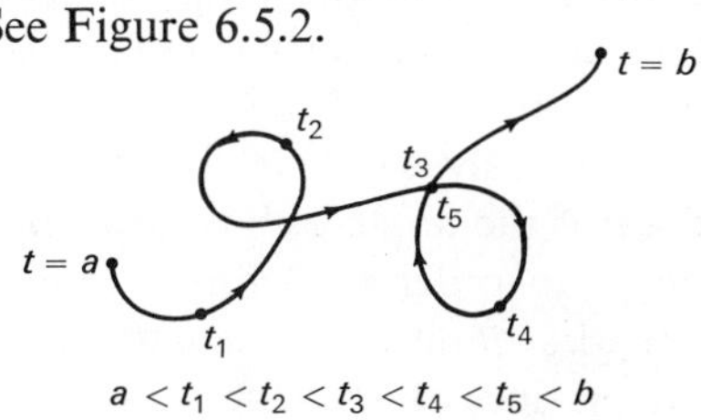

Figure 6.5.2

If we are looking for a parametrization of the curve which will have an obvious relation to the geometry at hand, it is natural to imitate that particular parametrization of a line which we have called a "coordinate system." That is, we might associate with one endpoint P_0 the number 0, and with each other point P the number s equal to the length of the curve between P_0 and P. The curve is thus parametrized by a parameter which we may call s, ranging from O to L, the total length of the curve. Moreover, this use of arclength as a parameter is a "natural" parametrization, for equal changes in s represent equal amounts of travel along the curve. See Figure 6.5.3.

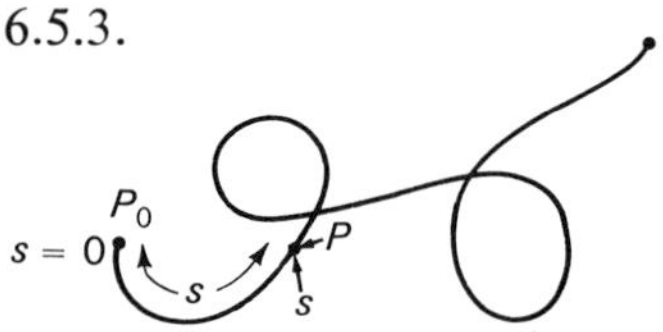

Figure 6.5.3

With the curve parametrized by arclength, it is easy to define another numerical quantity which is intimately associated with the geometry of the curve. This quantity, which we shall call *curvature*, measures how tightly curved or twisted a curve is. The quantity is closely related to the measurement of the tightness of a curve in the road which is experienced by every person who rides in an automobile. This measure involves noticing how hard one's shoulder is forced against the arm rest, or one's neighbor, as the car drives at a fixed speed around the curve. If the car travels at a predetermined speed around two different curves, the one which throws the driver hardest to one side is the sharper curve. Of course, to make such a comparison possible the car must travel at the same speed around the two curves. This dependence of the measurement on speed of travel, which in turn depends on length of curve, is the main reason for our interest in arclength as a parameter. What we are perceiving by means of our shoulder is the *rate* at which the car's direction changes *per mile of travel.*

We return now to our smooth curve parametrized by arclength, whereby we can translate this crude measurement into a calculable numerical quantity. At every point the smooth curve has a tangent line. This tangent line reflects the *direction* of the curve at the point. If we assume the existence of coordinate axes in the plane, then either the slope m of the tangent line or the angle φ between the tangent and the horizontal axis can be used as a numerical indicator of the direction of the curve.

Previously we have made more use of the slope of the tangent than of the angle φ, but now for a moment we wish to concentrate on φ. In order to define φ more precisely, let us give the tangent line at a point P on the curve an orientation in the direction of the orientation of the curve. The orientation of the curve is in the direction of increasing arclength s. Then φ will be the angle between $-\pi$ and π from the positive horizontal direction to this tangent ray (Figure 6.5.4).

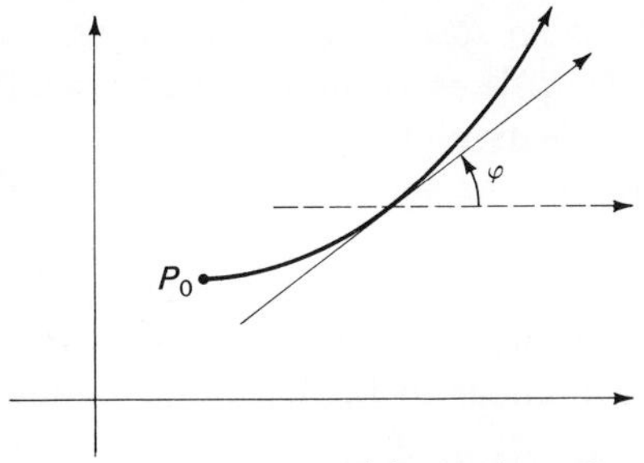

Figure 6.5.4

This angle φ fixes the direction of the curve at any point. But we are also interested in how rapidly the curve is turning at P, how tightly it is twisted. Obviously the question here is how rapidly is the tangent turning as we proceed along the curve, that is, how rapidly is φ changing with respect to s. For after all, s is precisely what measures progress along the curve.

We are talking, of course, about a derivative. For φ is a function of the arclength s, and the quantity that measures the rate of change of φ with respect to s is the derivative $d\varphi/ds$. The absolute value of this quantity is called the *curvature* of the curve at P. It measures how tightly the curve is twisted. For if $|d\varphi/ds|$ is large, the tangent is rotating rapidly per unit of arc and the curve is tightly twisted; whereas if $|d\varphi/ds|$ is small, the angle φ of inclination of the tangent changes very slowly. See Figure 6.5.5.

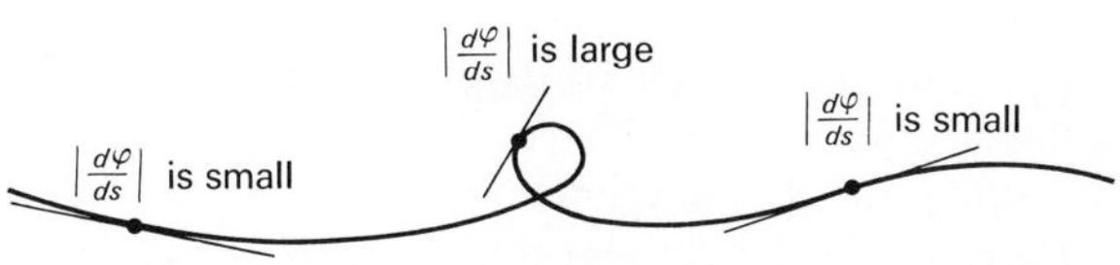

Figure 6.5.5

Definition 6.5.1 Let Γ be a smooth curve parametrized by arclength s from a fixed point P_0. Let $\varphi(s)$ be the angle between $-\pi$ and π from the positive horizontal axis to the oriented tangent at the point $P(s)$ on Γ, where Γ is oriented in the direction of increasing s. The quantity $K = |d\varphi/ds|$ is called, if it exists, the curvature of Γ at $P(s)$.

The relationship of the quantity curvature to the geometry of a curve is, we hope, obvious from the above discussion. But the definition will not be of much use unless we can actually calculate K for a given curve. In order to do this we must have a description of the curve in some manageable form, and this sends us back to the analytical methods from which we have strayed rather far.

Thus we suppose that Γ is the graph of the parametric equations

(6.5.1) $$\begin{aligned} y &= f(u), \\ x &= g(u), \end{aligned} \qquad t_a \leq u \leq t_b,$$

where for reasons that will soon be clear we now assume that $f''(u)$ and $g''(u)$ exist for all u. To insure smoothness, we continue to assume that $[f'(u)]^2 + [g'(u)]^2 \neq 0$; moreover, we shall soon see another good reason for this assumption.

We thus already have one parametrization of Γ, and our first problem is to reparametrize the curve using arclength. To do this we first select our fixed point P_0—it may as well be $(g(t_a), f(t_a))$. We have a specific formula to express arclength s from P_0 on Γ as a function of t. Let t be a number in $[t_a, t_b]$. If $s(t)$ is the arclength from P_0 to $P(t)$,

(6.5.2) $$s(t) = \int_{t_a}^{t} \sqrt{[f'(u)]^2 + [g'(u)]^2}\, du.$$

By the Fundamental Theorem of Calculus we see from (6.5.2) that

(6.5.3) $$s'(t) = \frac{ds}{dt} = \sqrt{[f'(t)]^2 + [g'(t)]^2} > 0$$

for all t. Hence the function $s(t)$ has an inverse function σ. If $s = s(t)$, we can write

(6.5.4) $$t = \sigma(s),\ 0 \leq s \leq L(\Gamma).$$

Substituting (6.5.4) into (6.5.1) we have

$$\begin{aligned} y &= f(\sigma(s)) = F(s), \\ x &= g(\sigma(s)) = G(s), \end{aligned}$$

and these are parametric equations for Γ with arclength as the parameter.

The slope of the tangent line at a point $P = (g(t), f(t)) = (G(s), F(s))$, found by differentiation with respect to either t or s, is

$$m = \frac{f'(t)}{g'(t)} = \frac{F'(s)}{G'(s)}.$$

Because φ is between $-\pi$ and π,

(6.5.5) $$\varphi = \arctan m = \arctan\left(\frac{f'(t)}{g'(t)}\right).$$

It might seem that we should use arctan $[F'(s)/G'(s)]$ for φ, since we wish to compute $d\varphi/ds$. However, using (6.5.5) the computation is simpler. Using the chain rule and the formula for the derivative of an inverse function, we have

$$K = \left|\frac{d\varphi}{ds}\right| = \left|\frac{d\varphi}{dt} \cdot \frac{dt}{ds}\right| = \left|\frac{d\varphi/dt}{ds/dt}\right|.$$

Now we may differentiate using (6.5.3) and (6.5.5) to obtain

(6.5.6)
$$K = \frac{\left|\dfrac{g'(t)f''(t) - f'(t)g''(t)}{[g'(t)]^2}\right|}{\left(1 + \dfrac{[f'(t)]^2}{[g'(t')]^2}\right)\sqrt{[f'(t)]^2 + [g'(t)]^2}},$$

or finally,

(6.5.7)
$$K = \frac{|g'(t)f''(t) - f'(t)g''(t)|}{([f'(t)]^2 + [g'(t)]^2)^{3/2}}.$$

We have thus proved:

Theorem 6.5.1 *If* $[f'(t)]^2 + [g'(t)]^2 \neq 0$ *and* $f''(t)$, $g''(t)$ *exist for all* t *in an interval, then the curvature of the graph of* (6.5.1) *at the point* $(f(t), g(t))$ is

$$K = \frac{|g'(t)f''(t) - f'(t)g''(t)|}{([f'(t)]^2 + [g'(t)]^2)^{3/2}}.$$

Theorem 6.5.1 has two useful immediate corollaries. We leave the proofs as exercises; they follow easily from the facts noted above that the graph of $y = f(x)$ in Cartesian coordinates and the graph of $r = f(\theta)$ in polar coordinates may be considered quite naturally as graphs of parametric equations.[1]

[1] *See (6.3.5) and Example 4 in Section 6.4.*

Corollary 6.5.2 *If* $f''(x)$ *exists for all* x *in* $[a, b]$, *then the curvature of the graph of* $y = f(x)$, $a \leq c \leq b$, *at the point* $(x, f(x))$ *is*

$$K = \frac{|f''(x)|}{(1 + [f'(x)]^2)^{3/2}}.$$

Corollary 6.5.3 *If* $f''(\theta)$ *exists for all* θ *in* $[\theta_a, \theta_b]$, *then the curvature of the graph of* $r = f(\theta)$ *at the point* (r, θ) *is*

$$K = \frac{|[f(\theta)]^2 - f''(\theta)f(\theta) + 2[f'(\theta)]^2|}{([f(\theta)]^2 + [f'(\theta)]^2)^{3/2}}.$$

The following examples will illustrate the calculation of the curvature and more of its properties.

Example 1 The ellipse

$$x = a\cos t, \quad y = b\sin t, \qquad 0 \leq t \leq 2\pi,$$

illustrated in Figure 6.5.6 has curvature

$$K = \frac{ab}{(a^2\sin^2 t + b^2\cos^2 t)^{3/2}},$$

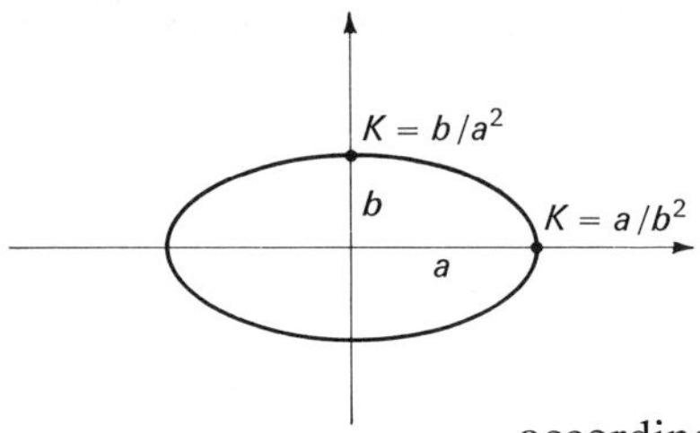

Figure 6.5.6

according to Theorem 6.5.1. At the point $(a, 0)$ (where $t = 0$) the curvature is $ab/b^3 = a/b^2$. At the point $(0, b)$ $(t = \pi/2)$, $K = ab/a^3 = b/a^2$. Note that if $a > b$, then $a/b^2 > b/a^2$, so that K is larger at $(a, 0)$ than at $(0, b)$.

Example 2 A circle of radius R can be expressed as the graph of

$$r = R > 0$$

in polar coordinates. Thus, using Corollary 6.5.3 we have (because $f(\theta)$ is constant)

$$K = \frac{|R^2|}{R^3} = \frac{1}{R}.$$

This result illustrates very clearly the role of K as an indicator of tightness of twist. For a circle of small radius R is a rapidly turning curve, whereas a circle with large radius is relatively "straight."

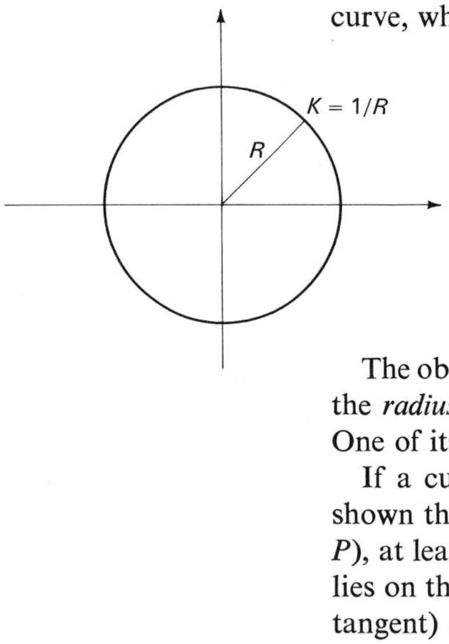

Figure 6.5.7

The observation in Example 2 leads us to call the quantity $1/K = R$ the *radius of curvature of* the curve Γ at the point P (Figure 6.5.7). One of its uses is outlined below.

If a curve Γ has curvature $K \neq 0$ at a point P, then it can be shown that the curve lies entirely on one side of its tangent line (at P), at least in the immediate vicinity of P. Let C be the point which lies on the normal line to Γ at P (the normal is perpendicular to the tangent) on the same side of the tangent line as the curve and at a distance $R = 1/K$ from P. This point C is called the *center of curvature* of Γ at P. The circle through P with center at C is called the *circle of curvature* of Γ at P (Figure 6.5.8).

Note that the circle of curvature is not only tangent to Γ at P, it also has the same curvature as Γ at P—the circle is not only going in the same direction as Γ at P; it is also turning at the same

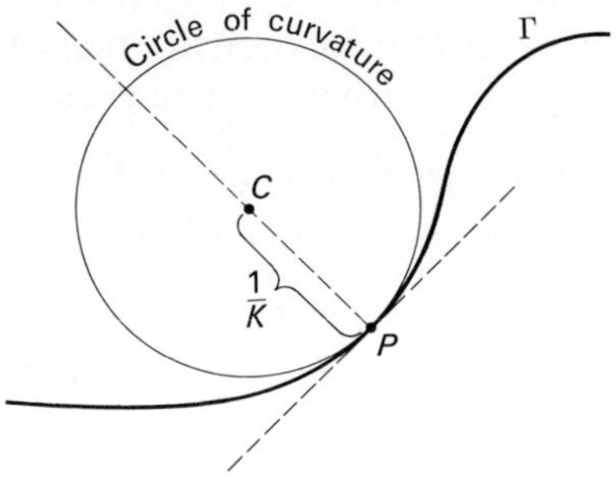

Figure 6.5.8

rate. Of all possible circles, the circle of curvature is the one which *best fits* Γ at P.

Our final example illustrates this *best-fitting* property of the circle of curvature in a much stronger way.

Example 3 Let us try to *define* a best-fitting circle for a curve Γ at a point P without using the notion of curvature.

First we must agree that to fit Γ well at P a circle must certainly have the same tangent as Γ at P. Thus, its center should be on the normal line. But this still leaves infinitely many possible circles. To determine one uniquely, we must fix a radius. We wish to do this by imposing a condition that will conform to our intuitive notion of "improving the fit." Let us consider a second point P' on Γ.

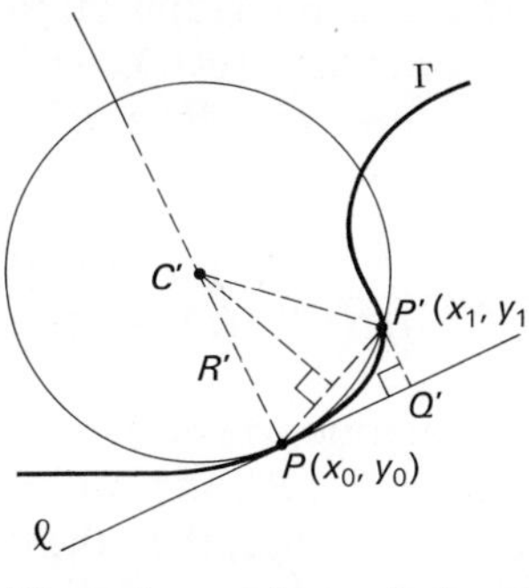

Figure 6.5.9

There is a unique circle which (i) passes through P, (ii) has its center on the normal to Γ at P, and (iii) passes through P'. Now let P' slide along Γ toward P and observe both the center C' and radius R' of this circle. The limiting position of this center C' and the limit of this radius as $P' \to P$ should be the center and radius of a circle which fits Γ best at P. From this point on, we shall leave much of the computation to you. Let ℓ be the line tangent to Γ at P and let Q' be the projection of P' on ℓ. From plane geometry using Figure 6.5.9, we can show that

(6.5.8) $$R' = \frac{1}{2}\,\frac{|PP'|^2}{|P'Q'|}\,.$$

Letting the coordinates of P and P' be (x_0, y_0) and (x_1, y_1) respectively, and denoting by m the slope of the tangent line at P, we can prove either from plane or from analytic geometry that

(6.5.9)
$$|PP'|^2 = (x_1 - x_0)^2 + (y_1 - y_0)^2,$$
$$|P'Q'| = \left|\frac{y_1 - y_0 - m(x_1 - x_0)}{\sqrt{1 + m^2}}\right|.$$

Figure 6.5.10 should be helpful in working out proofs of (6.5.9). Thus,

(6.5.10)
$$R' = \tfrac{1}{2}\sqrt{1 + m^2}\left|\frac{(x_1 - x_0)^2 + (y_1 - y_0)^2}{(y_1 - y_0) - m(x_1 - x_0)}\right|.$$

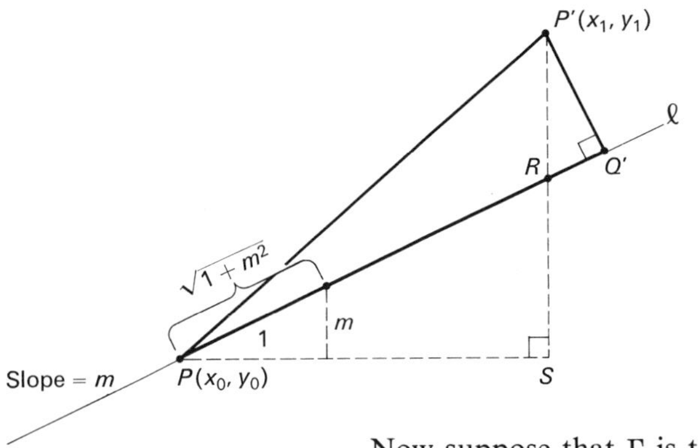

Figure 6.5.10

Now suppose that Γ is the graph of parametric equations (6.5.1) and that P has coordinates $(g(t_0), f(t_0))$ and P' has coordinates $(g(t), f(t))$. As we know, the slope m is $f'(t_0)/g'(t_0)$. Substituting into (6.5.10) we have

(6.5.11)
$$R' = \tfrac{1}{2}\sqrt{[f'(t_0)]^2 + [g'(t_0)]^2}\left|\frac{[g(t) - g(t_0)]^2 + [f(t) - f(t_0)]^2}{g'(t_0)[f(t) - f(t_0)] - f'(t_0)[g(t) - g(t_0)]}\right|.$$

(We leave the computation to you.)

We wish to find $\lim_{t\to t_0} R'$, the radius of our best-fitting circle. We use l'Hospital's Rule once, then the definitions

$$f'(t_0) = \lim_{t\to t_0}\frac{f(t) - f(t_0)}{t - t_0}, \quad f''(t_0) = \lim_{t\to t_0}\frac{f'(t) - f'(t_0)}{t - t_0},$$

and the similar ones for $g'(t_0)$, $g''(t_0)$. We obtain

(6.5.12)
$$R = \lim_{t\to t_0} R' = \frac{([f'(t_0)]^2 + [g'(t_0)]^2)^{3/2}}{|g'(t_0)f''(t_0) - f'(t_0)g''(t_0)|}.$$

But this is just the reciprocal of the curvature K of Γ at P, so we have rediscovered the center of curvature! Curvature is one of the most important quantities in the differential geometry of curves.

Exercises

1 Find the curvature of each of the following curves at the indicated points.

(a) $y = 2x^2$, $(1, 2)$;

(b) $y = x(x + 2)$, $(-2, 0)$;

(c) $y = x^2 - x^3$, $(1, 0)$;

(d) $x = t^2$, $y = 3t$, $t = 1$;

(e) $y = 1/x$, (x_1, y_1);

(f) $y = 1/x^2$, $(1, 1)$;

(g) $y = \sin x$, $(\pi/2, 1)$;

(h) $y = e^x$, $(0, 1)$;

(i) $y = \log(x + 1)$, $(0, 0)$;

(j) $y = \tan x$, $(\pi/4, 1)$;

(k) $x = t - \sin t$, $y = 1 - \cos t$, $t = \pi/2$;

(l) $r = 1 - \cos\theta$, $\theta = \pi/4$.

2 Find the radius of curvature of each of the following curves at the indicated points.

(a) $y = x^2 - x$, $(1, 0)$;

(b) $y = \dfrac{1}{\sqrt{x}}$, $(4, \frac{1}{2})$;

(c) $y = \cos x$, $(0, 1)$;

(d) $y = e^{-x}$, $(0, 1)$;

(e) $x = t^3$, $y = t^2$, $t = 2$;

(f) $y = \sec x$, $(\pi/6, 2\sqrt{3}/3)$;

(g) $r = \dfrac{2}{1 + \cos\theta}$, $\theta = 0$;

(h) $x = 3t^3$, $y = 3t - t^3$, $t = 1$.

3 (a) Let f be a twice differentiable function in $[a, b]$. Show that the equation of the normal line to the curve $y = f(x)$ at the point (x_0, y_0) is

$$(x - x_0) + f'(x_0)(y - y_0) = 0.$$

(b) Show that the coordinates (x_1, y_1) of the center of curvature of $y = f(x)$ at (x_0, y_0) are

$$x_1 = x_0 - f'(x_0)\frac{(1 + [f'(x_0)]^2)}{f''(x_0)}$$

and

$$y_1 = y_0 + \frac{1 + [f'(x_0)]^2}{f''(x_0)}.$$

(*Hint:* The center of curvature must lie on both the normal line to the curve at (x_0, y_0) and the circle at (x_0, y_0) with radius R. Note that there are two points which satisfy these conditions.)

4 Use the formulas derived in Exercise 3 to find an equation for the circle of curvature for each of the following curves at the indicated point. Make a careful sketch of the graphs of the curve and the circle of curvature in the same plane.

(a) $y = x^2$, $(0, 0)$;

(b) $y = \sin x$, $(\pi/2, 1)$;

(c) $y = x^3$, $(2, 8)$;

(d) $xy = 6$, $(2, 3)$;

(e) $y = \sqrt{x}$, $(1, 1)$;

(f) $y = \log(x + 1)$, $(0, 0)$.

5 Consider the curve described by the parametric equations

$$x = t,$$
$$y = t^2.$$

For each value of t the curve has a center of curvature given by the formulas in Exercise 3. Find parametric equations for the curve which is traced out by these centers of curvature.

6 Verify formulas (6.5.6) and (6.5.7).

7 Prove Corollary 6.5.2.

8 Prove Corollary 6.5.3.

9 Verify formula (6.5.8).

10 Verify formula (6.5.9), referring to Figure 6.5.10.

11 Verify formulas (6.5.10) and (6.5.11).

12 Verify formula (6.5.12) in two steps:

(a) Use l'Hospital's Rule to evaluate

$$\lim_{t \to t_0} \frac{(t - t_0)^2}{g'(t_0)(f(t) - f(t_0)) - f'(t_0)(g(t) - g(t_0))}.$$

(b) Multiply the numerator and denominator of the fraction in (6.5.11) by $(t - t_0)^2$, and then find $\lim_{t \to t_0} R'$.

13 Suppose that f is a twice differentiable function on $[a, b]$ and that the graph of f has zero curvature at each point. Show that the graph of f must be a straight line.

14 Suppose that f is a twice differentiable function on $[a, b]$. Prove that the graph of f has zero curvature at each of its points of inflection.

15 Show that the graph of $f(x) = x^4$ has zero curvature at the point $(0, 0)$, but does not have a point of inflection there.

6.6 Vectors

We shall conclude this chapter with a brief discussion of motions along curves in a plane, for which we shall need the notion of a vector. The introduction of this notion in an elementary text presents a special problem. On a superficial and intuitive level, vectors seem the simplest and most natural of objects, nothing more than arrows. However, when one begins to make serious use of vectors and to look carefully at their properties, one uncovers in the idea of a vector unsuspected subtlety and ultimately, unsuspected power. This process of discovery could lead us into an important branch of mathematics. Fully developed, the discussion of vectors becomes what is called *linear algebra*, a branch of algebra which has important applications in geometry, the calculus, and many parts of engineering and the physical and social sciences. Linear algebra should be an early part of the education of every student interested in any of these areas. It is the subject of another volume in this series, appropriately entitled *Linear Algebra*, designed to be studied before, or concurrently with, an introduction to the calculus of functions of several variables, also the subject of a subsequent volume in this series.

Thus, in order not to get involved in a lengthy discourse we must be content to remain on a somewhat superficial and intuitive level. We hope that at this level the notion of vector will remain for you the simplest and most uncomplicated of ideas.

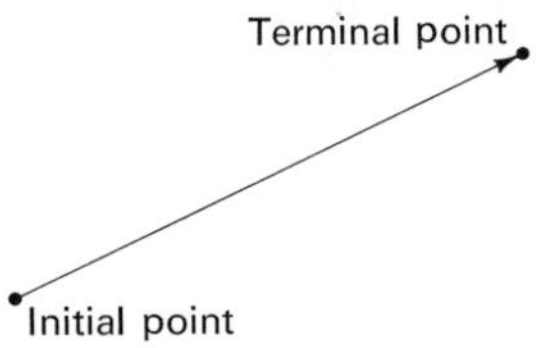

Figure 6.6.1

As suggested above, a vector is an arrow, that is, a directed line segment. It has direction, which might be specified by means of an angle or a slope, and *length*, the ordinary Euclidean length of the segment. Its orientation means that it has an *initial* point and a *terminal* point, a head and a tail.

The word *vector* encompasses more than the simple notion of arrow, however. A vector is an arrow used in a particular way: one way is to identify, or label, points in a coordinate plane. If P is a point, then there is a unique arrow with initial point at the origin and terminal point P. Conversely, each arrow with initial point at the origin uniquely determines a point in the plane. For completeness we shall say that the origin itself is "pointed out" by a special arrow with length zero. Thus, the arrows with initial points at the origin form a set of labels for the points in the plane (Figure 6.6.2).

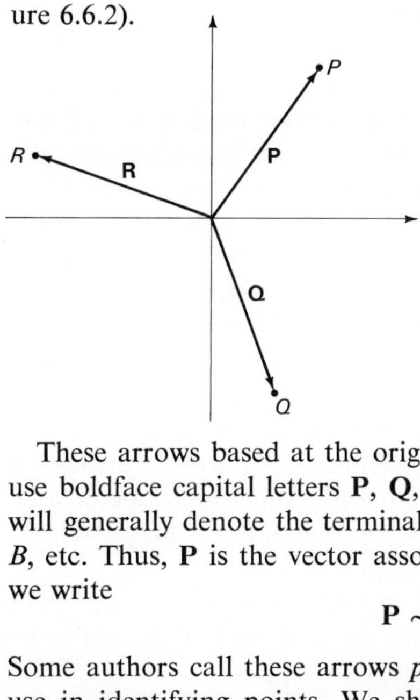

Figure 6.6.2

These arrows based at the origin will be called *vectors*. We shall use boldface capital letters **P**, **Q**, **A**, **B**, etc., to denote vectors and will generally denote the terminal points of the vectors by P, Q, A, B, etc. Thus, **P** is the vector associated with the terminal point P; we write

$$\mathbf{P} \sim P.$$

Some authors call these arrows *position vectors* to emphasize their use in identifying points. We shall use this terminology later on when we wish to emphasize this use as opposed to a number of different uses which we shall point out.

Vectors are also significant as archetypes of all arrows in the plane. Suppose Q, R are two points neither of which is the origin.

We may consider the arrow from the (initial) point Q to the (terminal) point R—we shall denote this arrow by $\overrightarrow{QR}$. Then there is a unique vector $\mathbf{P}$ which has the same length and the same direction as $\overrightarrow{QR}$. Thus this vector is closely associated with the arrow $\overrightarrow{QR}$ and with all other arrows having the same length and direction. Another way of looking at this association of a vector with $\overrightarrow{QR}$ is the following: Imagine moving the arrow $\overrightarrow{QR}$ without changing its direction or its length until its initial point lies at the origin. The arrow is now a vector $\mathbf{P}$, the vector associated with $\overrightarrow{QR}$ (Figure 6.6.3).

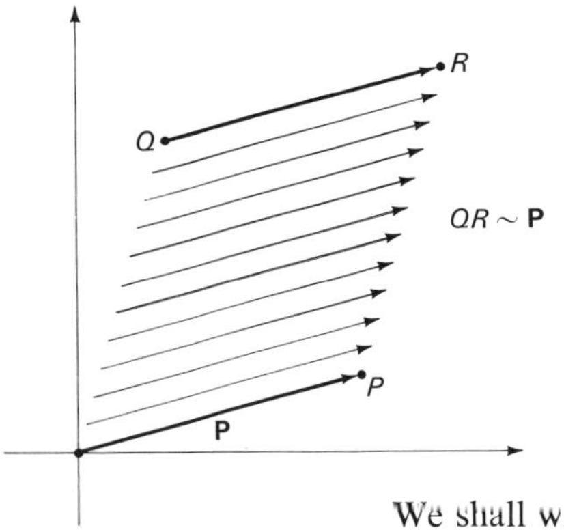

Figure 6.6.3

We shall write

$$\overrightarrow{QR} \sim \mathbf{P}$$

and say that the arrow $\overrightarrow{QR}$ is *equivalent* to $\mathbf{P}$. Any two arrows equivalent to the same vector will be called equivalent to each other.

Still another use for vectors is as indicators of *displacements*. If $\mathbf{P}$ is a vector, we may imagine a point being moved from the origin to the point P. This movement is called a displacement. But such a displacement can occur anywhere in the plane, not only at the origin. Any point Q in the plane can be given the same displacement; it can be moved as far and in the same direction as the origin was moved to reach P. If this displacement moves Q to the new position R, then the arrow $\overrightarrow{QR}$ serves as an indicator or trace of that event. Thus, a vector $\mathbf{P}$ determines a displacement which can be thought of as moving *all* points in the plane a certain distance in a certain direction. The traces of all these events are all arrows in the plane equivalent to $\mathbf{P}$.

Thus, vectors determine displacements or arrows all over the plane and *any statement about any arrow in the plane which does not depend upon its position*, as determined by its initial point, *can be interpreted as a statement about vectors*. Since a great many physical and geometrical ideas can be illustrated and interpreted in terms of arrows, you will find that vectors have an important role to play in the language of science.

So far, our discussion has been entirely geometric. But the real importance of vectors rests on some additional properties with which they are endowed. To describe some of these properties we discuss two different ways of specifying a vector numerically.

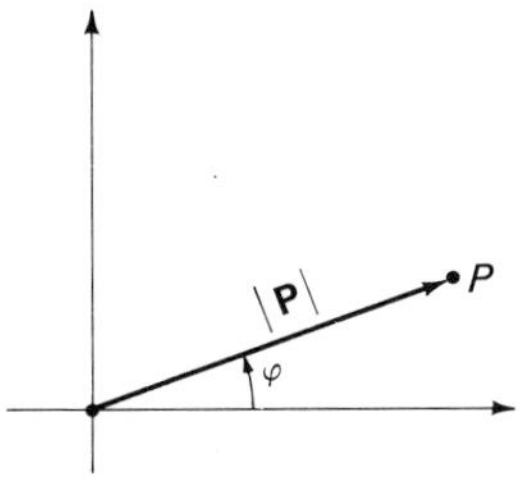

Figure 6.6.4

It should be clear enough that a vector **P**, an arrow with its initial point at the origin, is completely determined by its direction and its magnitude. Its direction can be specified by the radian measure between $-\pi$ and π of the angle from the positive x-axis to the oriented ray $\overrightarrow{OP}$. Its magnitude or length is a positive real number which we shall denote by $|\mathbf{P}|$. See Figure 6.6.4. The pair $(|\mathbf{P}|, \varphi)$ are polar coordinates of the point P.

Another method of specifying a vector **P** is to give the Cartesian coordinates (x, y) of the point P. This ordered pair of numbers uniquely determines P and hence **P** also. We shall now need to use a special terminology which has developed historically. In this terminology, real numbers are called *scalars*, and the coordinates x, y of P are called the *scalar components* of the vector **P**. In the past we have been in the habit of identifying ordered pairs (x, y) with points P in the plane. Points P are now associated with vectors **P**, though we shall want to preserve the distinction between point and vector. Number pairs are therefore associated with vectors as well as with points and we preserve the point-vector distinction by varying the terminology: *coordinates* of points, *scalar components* of vectors. We carry this distinction into our notation also. If x and y are the coordinates of P, hence the scalar components of **P**, we will write

$$(x, y) = P,$$
$$\langle x, y\rangle = \mathbf{P},$$

as illustrated in Figure 6.6.5.

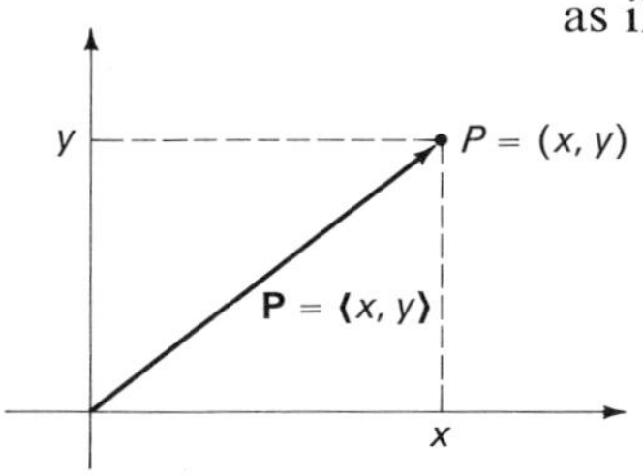

Figure 6.6.5

This connection between vectors and number pairs is of central importance in linear algebra. Let us see how we can use it to define this special kind of algebra associated with vectors. The algebra consists of two operations on vectors, *addition* and *multiplication by a scalar*. These are defined as follows: If $\mathbf{A} = \langle x, y\rangle$ and $\mathbf{B} = \langle r, s\rangle$ and if α is a scalar (real number), then

(6.6.1)
$$\mathbf{A} + \mathbf{B} = \langle x + r, y + s\rangle,$$
$$\alpha\mathbf{A} = \langle \alpha x, \alpha y\rangle.$$

These operations satisfy certain properties which are easily verified. They include the following:

If **A, B, C** are any three vectors and α, β any two scalars, then:

(6.6.2)

(i) $(\mathbf{A} + \mathbf{B}) + \mathbf{C} = \mathbf{A} + (\mathbf{B} + \mathbf{C})$,

(ii) $\mathbf{A} + \mathbf{B} = \mathbf{B} + \mathbf{A}$,

(iii) $\mathbf{A} + \mathbf{O} = \mathbf{A}$, where $\mathbf{O} = \langle 0, 0 \rangle$,

(iv) $\mathbf{A} + (-1)\mathbf{A} = \mathbf{O}$,

(v) $(\alpha + \beta)\mathbf{A} = \alpha\mathbf{A} + \beta\mathbf{A}$, $(\alpha\beta)\mathbf{A} = \alpha(\beta\mathbf{A})$,

(vi) $\alpha(\mathbf{A} + \mathbf{B}) = \alpha\mathbf{A} + \alpha\mathbf{B}$.

Any set of objects with two operations having properties (6.6.2) is called a *vector space*. Linear algebra is the study of vector spaces, of which the above is but one of many examples. Another example is the set of all real-valued continuous functions $f, g, h, \ldots$ defined on $(-\infty, \infty)$ with operations $f + g$ and αf (α constant) defined in the usual way.

We shall leave the proofs of statements (6.6.2) as an exercise. Before turning to an application of the vector concept, we pause to point out the geometrical interpretation of the operations (6.6.1) and one or two other geometric implications of the representation of a vector in terms of its scalar components. As we have said, the *direction* of **P** may be specified by the angle φ between $-\pi$ and π from the positive horizontal direction to the arrow **P**. It should be clear from Figure 6.6.6 that if $\mathbf{P} = \langle x, y \rangle$, then

$$\tan \varphi = \frac{y}{x}.$$

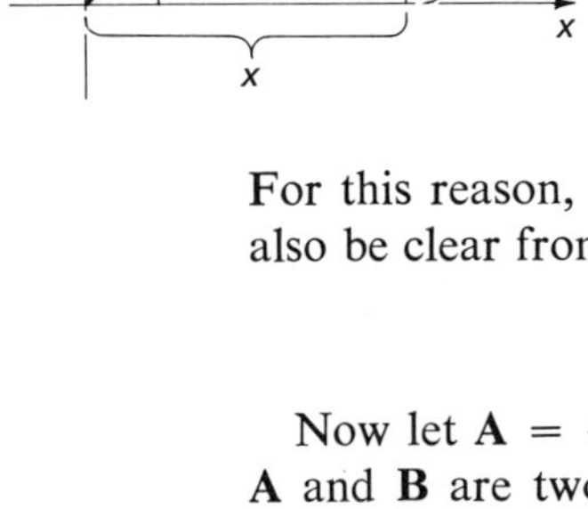

Figure 6.6.6

For this reason, we call the quantity y/x the *slope* of **P**. It should also be clear from Figure 6.6.6 that

$$|\mathbf{P}| = \sqrt{x^2 + y^2}.$$

Now let $\mathbf{A} = \langle x, y \rangle$ and $\mathbf{B} = \langle r, s \rangle$ be vectors and α be a scalar. **A** and **B** are two sides of a parallelogram with one vertex at the origin. If we complete this parallelogram, then the vector **C**, where C is the vertex of the parallelogram opposite **O**, is the vector sum $\mathbf{A} + \mathbf{B}$. This can be seen from Figure 6.6.7. A vector has many

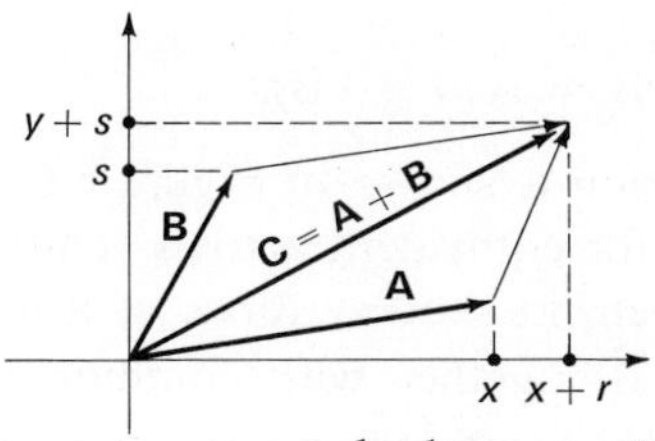

Figure 6.6.7

shadows or traces in the plane, the arrows which are equivalent to it in the sense given above. A vector addition can also be depicted in shadow form. To illustrate, it is convenient to establish one more notational convention. If **P** is a vector, then we shall use the label $\vec{\mathbf{P}}$ for *any* arrow equivalent to **P**. This notation gives us the freedom to move the vector **P** anywhere in the plane, so to speak, for any visual purpose we might have. For example, the vector addition $\mathbf{A} + \mathbf{B} = \mathbf{C}$ shown in Figure 6.6.7 might be seen in shadow form as follows. Construct arrows $\vec{\mathbf{A}}$ and $\vec{\mathbf{B}}$ in such a way that the terminal point of $\vec{\mathbf{A}}$ and the initial point of $\vec{\mathbf{B}}$ coincide. Then a glance at Figure 6.6.8 will convince you that the arrow $\vec{\mathbf{C}}$ extending from the initial point of $\vec{\mathbf{A}}$ to the terminal point of $\vec{\mathbf{B}}$ is equivalent to $\mathbf{A} + \mathbf{B}$. We could then label it $\vec{\mathbf{A}} + \vec{\mathbf{B}}$.

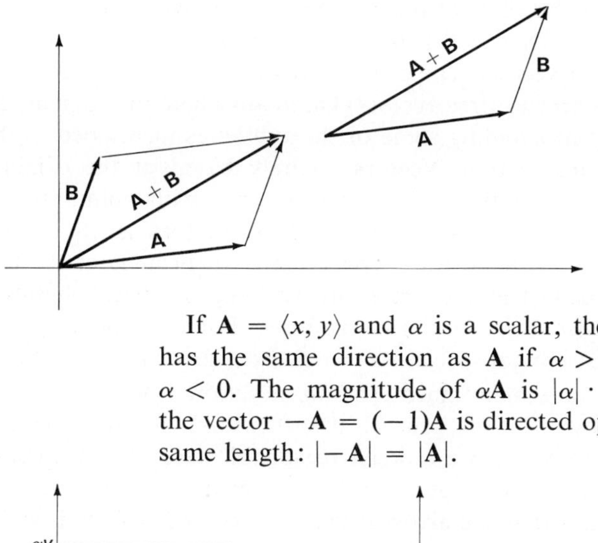

Figure 6.6.8

If $\mathbf{A} = \langle x, y\rangle$ and α is a scalar, then the vector $\alpha\mathbf{A} = \langle \alpha x, \alpha y\rangle$ has the same direction as **A** if $\alpha > 0$, the opposite direction if $\alpha < 0$. The magnitude of $\alpha\mathbf{A}$ is $|\alpha| \cdot |\mathbf{A}|$. See Figure 6.6.9. Thus, the vector $-\mathbf{A} = (-1)\mathbf{A}$ is directed oppositely from **A** and has the same length: $|-\mathbf{A}| = |\mathbf{A}|$.

Figure 6.6.9

We can define

$$\mathbf{A} - \mathbf{B} = \mathbf{A} + (-\mathbf{B}),$$

and the effect of this operation is shown in Figure 6.6.10, both for the vectors themselves and for equivalent arrows. Thus, the vector operation $\mathbf{A} - \mathbf{B}$ can be seen, in shadow form, as a manipulation which puts arrows $\vec{\mathbf{A}}$ and $\vec{\mathbf{B}}$ together with common initial point and produces $\vec{\mathbf{A}} - \vec{\mathbf{B}}$ as the arrow extending from the terminal point of $\vec{\mathbf{B}}$ to the terminal point of $\vec{\mathbf{A}}$.

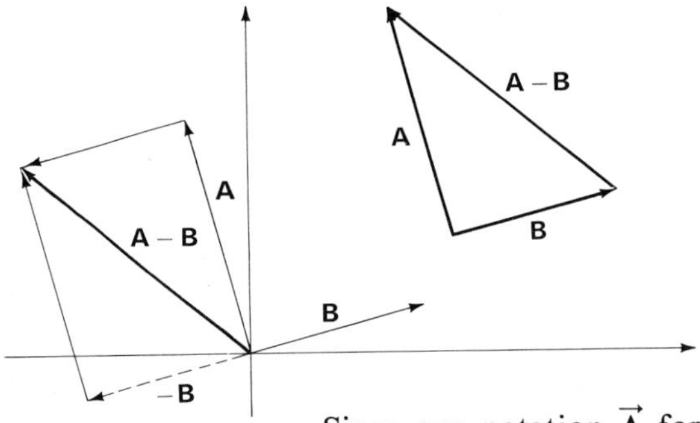

Figure 6.6.10

Since our notation $\vec{\mathbf{A}}$ for *any* arrow equivalent to the vector **A** gives the effect of our being able to move **A** all over the plane, $\vec{\mathbf{A}}$ is frequently called a *free vector* by physicists and others who make use of vectors. The distinction which we have made between vectors and arrows (free vectors) based anywhere in the plane has been our way of avoiding some of the subtleties mentioned at the beginning of this section. Vectors securely based at the origin are needed because of the clear correspondence with points in the plane and because the algebra peculiar to vectors is so easily defined in terms of them. But arrows scattered throughout the plane are also needed whenever vectors are used in geometrical physics for purposes of visualizing various quantities. For conceptual purposes vectors must remain special—as singled-out arrows—so in this introduction we have maintained separate notations: **A** for a vector, $\vec{\mathbf{A}}$ for any arrow. But in general use the dual notation becomes tedious. Moreover, if we exercise reasonable care, it is not really necessary. Referring to Figure 6.6.11, we must agree that the picture in (a) means that the arrow is *either* a vector **P** (if the initial point is the origin) *or* an arrow equivalent to the vector **P**. Then the picture (b) indicates an arrow (shadow) representation of the vector addition (c) occurring back at the origin. One is tempted to call this the "meanwhile, back at the ranch" convention.

Figure 6.6.11

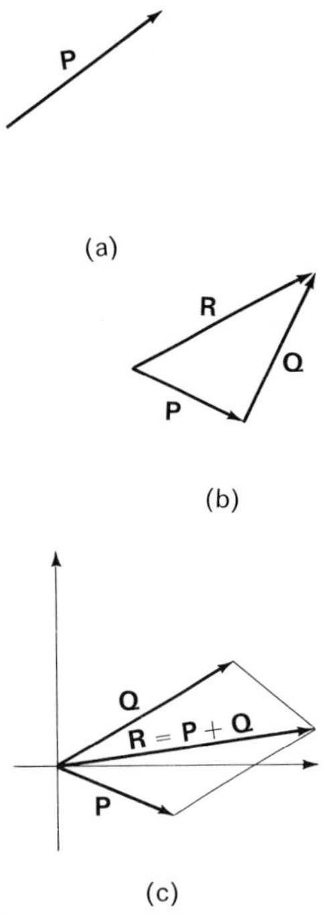

With these agreements we may dispense with the cumbersome notation $\vec{\mathbf{P}}$ in the next section as we see several applications of vectors to the motion of a particle along a curve.

Exercises

In Exercises 1–9 let $\mathbf{A} = \langle 3, 5\rangle$, $\mathbf{B} = \langle -2, 1\rangle$, $\mathbf{C} = \langle -3, -5\rangle$, $\mathbf{D} = \langle 5, -4\rangle$. In each case compute the indicated vector and then check your computation geometrically by using the parallelogram method.

1 $\mathbf{A} + \mathbf{B}$.

2 $2\mathbf{C} + \mathbf{D}$.

3 $\mathbf{A} + \mathbf{B} + \mathbf{C}$.

4 $\mathbf{C} + \mathbf{D} - \mathbf{A}$.

5 $\frac{1}{2}\mathbf{A} + \frac{1}{4}\mathbf{B}$.

6 $\mathbf{A} + 2\mathbf{D}$.

7 $2\mathbf{A} - 3\mathbf{B}$.

8 $\mathbf{A} + \mathbf{B} + \mathbf{C} + \mathbf{D}$.

9 Find $|\mathbf{A} + \mathbf{B}|$, $|\mathbf{B} - \mathbf{D}|$, $|2\mathbf{A} + 3\mathbf{C}|$.

10 Verify (6.6.2)(i)–(iii).

11 Verify (6.6.2)(iv)–(vi).

12 Prove that $\mathbf{A} + \mathbf{A} = 2\mathbf{A}$ and thus that our definitions for addition and scalar multiplication are consistent in this case.

13 Find scalars α and β so that

$$\alpha\langle 2, -1\rangle + \beta\langle 4, -3\rangle = \langle 20, -14\rangle.$$

14 Find $|\mathbf{A} - \mathbf{B}|$ if $|\mathbf{A}| = 13$, $|\mathbf{B}| = 19$, and $|\mathbf{A} + \mathbf{B}| = 24$.

15 Prove that if $\mathbf{X}$ is a vector such that $\mathbf{A} + \mathbf{X} = \mathbf{A}$ for all vectors $\mathbf{A}$, then $\mathbf{X} = \mathbf{O}$. Thus the zero vector is the only vector with property (6.6.2) (iii).

16 Find a vector $\mathbf{X}$ such that $\langle 2, 3\rangle + \mathbf{X} = \langle 3, -1\rangle$.

17 Explain the difference between a vector and an arrow.

18 In each of the following, two points are given. The first is the initial point and the second is the terminal point of an arrow. Find the vector which is equivalent to this arrow and determine its length.

(a) (12, 8), (4, 14); (b) (110, 58), (140, 98);
(c) (−5, −7), (−17, −12); (d) (−2, 1), (1, 5).

6.7 Motion along a Curve

Let Γ be a plane curve and suppose an object moves along Γ in such a way that it lies on the point $P(t)$ at an arbitrary time t. We can assume that the observed motion takes place between times $t = t_a$ and $t = t_b$. If we denote the coordinate of a point P by (x, y) and the coordinates of $P(t)$ by $(g(t), f(t))$, then this motion is governed by two functions f and g according to the equations

(6.7.1)
$$y = f(t),$$
$$x = g(t).$$

Equations (6.7.1) are, of course, just parametric equations for the curve Γ, with the parameter t interpreted as a "time variable."

Such motions along curves are, in fact, the most basic and intuitive application of parametric equations and explain our habit of using the letter t as a parameter.

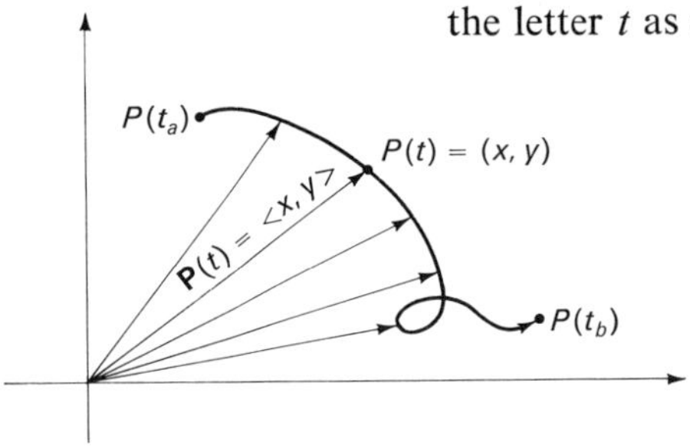

Figure 6.7.1

We can describe this motion in another somewhat more compact notation using vectors. For each time t let $\mathbf{P}(t)$ be the position vector which determines the point $P(t)$. If we denote a general position vector by $\langle x, y\rangle$, then the motion could also be described by the equation

$$\langle x, y\rangle = \mathbf{P}(t). \tag{6.7.2}$$

This means, of course, that the position vector $\langle x, y\rangle$ of the object at each time t is given by $\mathbf{P}(t)$. If we use (6.7.1) and the above notation, then we can write $\mathbf{P}(t) = \langle g(t), f(t)\rangle$. Then (6.7.2) becomes

$$\langle x, y\rangle = \langle g(t), f(t)\rangle. \tag{6.7.3}$$

See Figure 6.7.1.

Clearly, (6.7.3) is just a vector form of the pair of equations (6.7.1). Nevertheless, the recognition that motion along a curve can be interpreted as being governed by a single vector-valued function $\mathbf{P}(t)$ *exactly* as motion along a line is governed by a (real-valued) position function $s(t)$ is an important insight. It is important precisely because it enables us to continue the description of such motion in the pattern of our discussion of straight-line motion, using the newly developed algebra of vectors. In fact, we can follow the pattern of straight-line motion so closely that we can unfold the necessary definitions and conventions with little comment.

If $\mathbf{P}(t_0)$ and $\mathbf{P}(t_0 + h)$ are the position vectors of the object at times t_0 and $t_0 + h$, so that the positions are $(g(t_0), f(t_0))$ and $(g(t_0 + h), f(t_0 + h))$, then the vector

$\mathbf{P}(t_0 + h) - \mathbf{P}(t_0)$ *is called the displacement vector for the motion between times t_0 and $t_0 + h$.*

The displacement vector is, according to the algebra of vectors set up in the last section,

$$\mathbf{P}(t_0 + h) - \mathbf{P}(t_0) = \langle g(t_0 + h) - g(t_0), f(t_0 + h) - f(t_0)\rangle.$$

This vector (actually an equivalent arrow) is pictured in Figure 6.7.2.

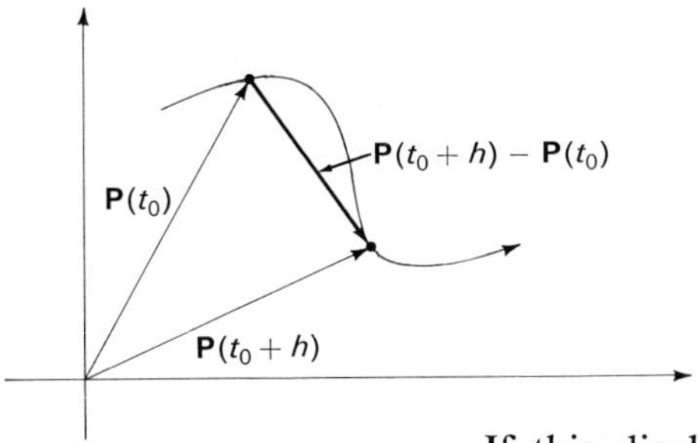

Figure 6.7.2

If this displacement vector is multiplied by the scalar $1/h$, we obtain what we call the average-velocity vector (Figure 6.7.3):

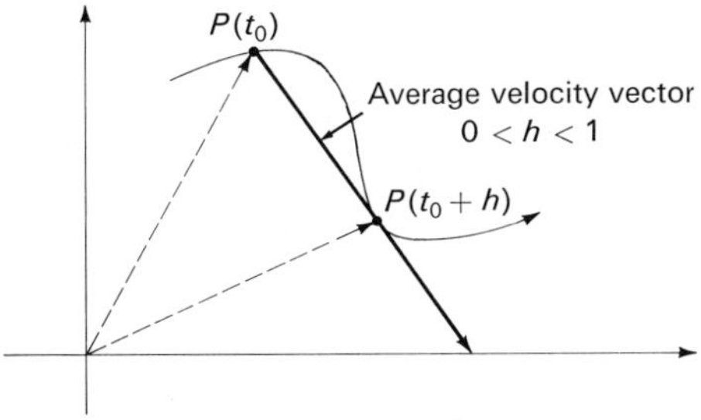

Figure 6.7.3

(6.7.4) $\frac{1}{h}[\mathbf{P}(t_0 + h) - \mathbf{P}(t_0)]$ *is the average-velocity vector for the motion between times* t_0 *and* $t_0 + h$.

The average-velocity vector is parallel to the displacement vector, but has different length if $h \neq 1$. If $0 < h < 1$, then the average-velocity vector is longer than the displacement vector.

To define what we mean by the *velocity vector* at time t_0 we must use the notion of a limit of a vector function. This presents no real problem. In summary:

When we say that a vector $\mathbf{A}$ is the limit of $\mathbf{F}(t)$ as $t \to t_0$, we mean that the length $|\mathbf{F}(t) - \mathbf{A}|$ tends to zero as $t \to t_0$. Since the length $|\mathbf{F}(t) - \mathbf{A}|$ is an ordinary real-valued function of t, our previous limit definitions apply. But if $\mathbf{F}(t) = \langle g(t), f(t)\rangle$ and $\mathbf{A} = \langle x, y\rangle$, then $\mathbf{F}(t) - \mathbf{A} = \langle g(t) - x, f(t) - y\rangle$ and $|\mathbf{F}(t) - \mathbf{A}| = [(g(t) - x)^2 + (f(t) - y)^2]^{1/2}$. It is easy to see that $|\mathbf{F}(t) - \mathbf{A}| \to 0$ if and only if $f(t) \to y$ and $g(t) \to x$, that is, if $\mathbf{A} = \langle \lim g(t), \lim f(t)\rangle$. This amounts to a definition

(6.7.5)
$$\lim_{t\to t_0} \mathbf{F}(t) = \langle \lim_{t\to t_0} g(t), \lim_{t\to t_0} f(t)\rangle.$$

If we then define the derivative of a vector function as

(6.7.6)
$$\begin{aligned}\mathbf{F}'(t_0) &= \lim_{h\to 0} \frac{1}{h}(\mathbf{F}(t_0 + h) - \mathbf{F}(t_0))\\ &= \lim_{h\to 0} \left\langle \frac{1}{h}(g(t_0 + h) - g(t_0)), \frac{1}{h}(f(t_0 + h) - f(t_0))\right\rangle,\end{aligned}$$

we see by (6.7.5) that

$$\mathbf{F}'(t_0) = \langle g'(t_0), f'(t_0)\rangle.$$

Returning now to our discussion of motion along the curve Γ and comparing (6.7.6) with (6.7.4), we see that it is natural to define velocity as follows:

The vector $\mathbf{V}(t_0) = \mathbf{P}'(t_0) = \langle g'(t_0), f'(t_0)\rangle$ *is called the velocity vector for this motion at time* t_0.

In other words, the vector $\mathbf{P}'(t_0)$ represents the rate of change of the moving vector $\mathbf{P}(t)$ at a certain instant t_0. But then $\mathbf{P}'(t)$ is also a vector which moves with time (although we do not picture this as a position vector). By analogy with the above, the vector $\mathbf{P}''(t_0) = \langle g''(t_0), f''(t_0)\rangle$, called the *acceleration vector*, represents the rate of change of the velocity vector $\mathbf{P}'(t)$:

The vector $\mathbf{A}(t_0) = \mathbf{P}''(t_0) = \langle g''(t_0), f''(t_0)\rangle$ *is the acceleration vector for the motion at time* t_0.

Before turning to specific examples we pause to point out two more geometrical properties of the velocity vector. We state these in the form of a theorem.

Theorem 6.7.1

(i) *The velocity vector* $\mathbf{V}(t_0)$ *is tangent to the curve* Γ *along which the motion takes place at the point* $P(t_0)$.

(ii) *If* $s(t)$ *represents the length of* Γ *between the initial point* $P(t_0)$ *and the point* $P(t)$, *then the magnitude* $|\mathbf{V}(t_0)|$ *of the velocity vector is* $s'(t_0)$.

This theorem states that the velocity vector indicates by its direction the *direction* of the motion of $t = t_0$ and by its magnitude the *rate of progress* along Γ. It is customary to call $|\mathbf{V}(t_0)| = s'(t)$ the *speed* of the motion at time t_0.

Proof

(i) If we think of $\mathbf{V}(t_0)$ as having its initial point at $P(t_0)$, then we need only prove that the slope of $\mathbf{V}(t_0)$ is the same as the slope of Γ at $P(t_0)$. But since $\mathbf{V}(t_0) = \mathbf{P}'(t_0) = \langle g'(t_0), f'(t_0)\rangle$, this follows immediately from Theorem 6.2.1 and the remark following Figure 6.6.6.

(ii) Since $\mathbf{V}(t_0) = \langle g'(t_0), f'(t_0)\rangle$,

$$|\mathbf{V}(t_0)| = \sqrt{[f'(t_0)]^2 + [g'(t_0)]^2}.$$

But we saw in (6.5.3) that this is just $s'(t_0)$.

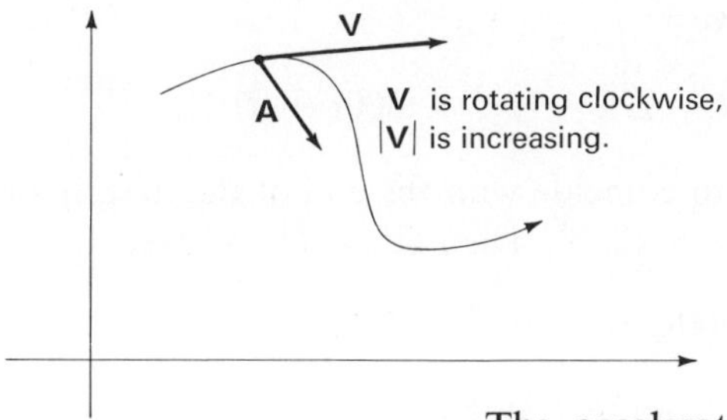

Figure 6.7.4

The acceleration vector $\mathbf{A}(t_0)$ (Figure 6.7.4) can point in any direction relative to $\mathbf{V}(t_0)$. Generally speaking, it points in the direction of rotation of $\mathbf{V}(t)$. If the angle from $\mathbf{V}(t_0)$ to $\mathbf{A}(t_0)$ is less than a right angle, then the speed $s'(t)$ is increasing at $t = t_0$. If the angle is greater than a right angle, the speed is decreasing.

Example 1 Assuming a flat earth and a uniform gravitational field above its surface (locally reasonable assumptions) and also assuming a vacuum in the space above the earth (an unreasonable but useful assumption), the motion of a projectile is governed by one condition: Its acceleration vector is vertical, pointed downward, and constant. Its magnitude is, in one system of units, 32 feet per second^2.

Suppose a shell is fired from a gun in such a way that it becomes a projectile (leaves the muzzle) traveling with a muzzle speed of 600 feet/second at an angle of 30° with the surface of the earth

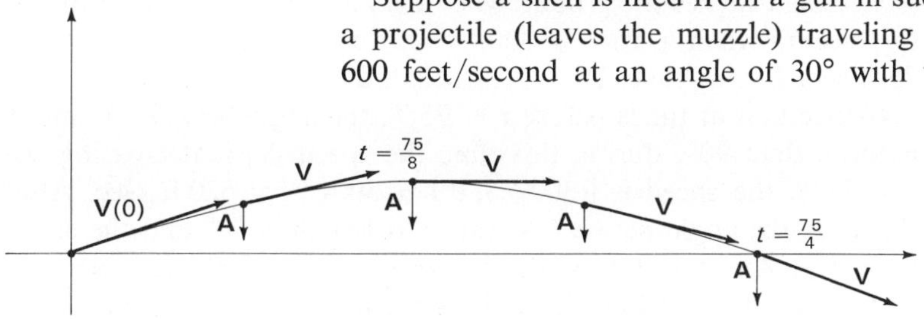

Figure 6.7.5

(Figure 6.7.5). If we use $\mathbf{P}(t)$, $\mathbf{V}(t)$, $\mathbf{A}(t)$ for the position vector, velocity, and acceleration respectively of the projectile at time t seconds after firing, then we have

$$\mathbf{A}(t) = \langle 0, -32 \rangle,$$

because an object in free fall has an acceleration directed downward with magnitude 32 feet/sec^2. Integrating once, we have

(6.7.7)
$$\mathbf{V}(t) = \langle C, -32t + D \rangle.$$

Since $\mathbf{V}(0) = \langle 300\sqrt{3}, 300 \rangle$ (why?), we see from (6.7.7) that

$$\langle 300\sqrt{3}, 300 \rangle = \langle C, D \rangle, \quad \text{or}$$
$$C = 300\sqrt{3}, \ D = 300.$$

Thus,

(6.7.8)
$$\mathbf{V}(t) = \langle 300\sqrt{3}, -32t + 300 \rangle.$$

Integrating again, we have

$$\mathbf{P}(t) = \langle 300\sqrt{3}t + C', -16t^2 + 300t + D' \rangle.$$

We may choose the origin to coincide with the end of the muzzle of the gun, in which case $\mathbf{P}(0) = \langle 0, 0 \rangle$. Then $C' = D' = 0$, so

(6.7.9) $$\mathbf{P}(t) = \langle 300\sqrt{3}t, -16t^2 + 300t \rangle.$$

Letting $\mathbf{P} = \langle x, y \rangle$, we see from (6.7.9) that

$$y = \frac{-16}{270{,}000}x^2 + \frac{\sqrt{3}}{3}x,$$

the equation of a parabola with vertical axis and passing through the origin.

The shell reaches its maximum altitude when $\mathbf{V}$ is horizontal. This occurs when its second scalar component is zero, or from (6.7.8), when $t = 300/32 = 75/8$. At this time,

$$\mathbf{P} = \left\langle \frac{\sqrt{3}}{2}(5625), \frac{1}{4}(5625) \right\rangle = 5625\left\langle \frac{\sqrt{3}}{2}, \frac{1}{4} \right\rangle.$$

The shell strikes the ground when $\mathbf{P}(t)$ is horizontal. From (6.7.9) this occurs when $-16t^2 + 300t = 0$, $t > 0$. Thus, $t = 300/16 = 75/4$ sec. At this time, $\mathbf{P} = \langle \sqrt{3}(5625), 0 \rangle = 5625\langle \sqrt{3}, 0 \rangle$.

Notice that at times before $t = 75/8$, the angle between $\mathbf{V}$ and $\mathbf{A}$ is more than 90°; during this time the speed $|\mathbf{V}|$ is decreasing. At $t = 75/8$, the speed is $|\langle 300\sqrt{3}, 0 \rangle| = 300\sqrt{3} \approx 520$ ft./sec. After this time the angle between $\mathbf{V}$ and $\mathbf{A}$ is less than 90° and the speed increases. At impact ($t = 75/4$) the speed is $|\langle 300\sqrt{3}, -300 \rangle| = 300\,\langle \sqrt{3}, -1 \rangle| = 600$ ft./sec., the same as the speed at firing.

Exercises

In Exercises 1–8, the motion of an object is described by a vector equation. Sketch the path of the object. Also sketch the velocity and acceleration vectors at the indicated time.

1 $p(t) = \langle t, 2t^2 \rangle$, $t = 1$.

2 $p(t) = \langle 2t, t^3 \rangle$, $t = 2$.

3 $p(t) = \langle 2\cos t, 2\sin t \rangle$, $t = 3\pi/4$.

4 $p(t) = \langle \sin t, 2\cos t \rangle$, $t = \pi/3$.

5 $p(t) = \langle t^2, e^t \rangle$, $t = 0$.

6 $p(t) = \langle -t, \sin 2t \rangle$, $t = 3\pi$.

7 $p(t) = \langle 2 + \cos t, 3 - \sin t \rangle$, $t = -\pi$.

8 $p(t) = \langle 2 - t, \sqrt{t} \rangle$, $t = 1$.

9 Suppose the gun in Example 1 is fired at a 45° angle of elevation. How far away does the shell hit the ground? What is the maximum altitude it reaches?

10 Prove that for a fixed initial velocity the range for a gun is obtained by firing it at a 45° angle of elevation. Obtain a formula for this range.

11 Suppose a projectile is fired with initial velocity v_0 and angle of elevation α. What is the maximum height reached by the projectile and when is it reached? What is the maximum range of the projectile and when does it strike the ground?

12 Show that doubling the initial velocity of a projectile has the effect of multiplying by 4 both the range and the maximum height of the projectile.

6.8 *Some Interesting Motions; Kepler's Laws*

Example 1 Consider the motion along a circle of radius 1 given by

$$\mathbf{P}(t) = \langle \cos t^2, \sin t^2 \rangle, \quad t > 0.$$

The velocity, speed, and acceleration are

$$\begin{aligned} \mathbf{V}(t) &= \frac{d}{dt}\mathbf{P}(t) = \langle -2t \sin t^2, 2t \cos t^2 \rangle, \\ |\mathbf{V}(t)| &= |2t| = 2t, \\ \mathbf{A}(t) &= \frac{d}{dt}\mathbf{V}(t) = \langle -2 \sin t^2 - 4t^2 \cos t^2, 2 \cos t^2 - 4t^2 \sin t^2 \rangle. \end{aligned} \tag{6.8.1}$$

Note that the speed $|\mathbf{V}|$ increases with t. The following information is easily obtained from (6.8.1):

$$\begin{aligned} \text{At } t = 1:\quad & \mathbf{P} = \langle 0.54, 0.84 \rangle, \\ & \mathbf{V} = \langle -1.68, 1.08 \rangle, \ |\mathbf{V}| \approx 2, \\ & \mathbf{A} = \langle -3.84, -2.28 \rangle. \\ \text{At } t = \sqrt{\pi/2}:\quad & \mathbf{P} = \langle 0, 1 \rangle, \\ & \mathbf{V} = \langle -\sqrt{2\pi}, 0 \rangle, \ |\mathbf{V}| = \sqrt{2\pi}, \\ & \mathbf{A} = \langle -2, -2\pi \rangle. \\ \text{At } t = \sqrt{\pi}:\quad & \mathbf{P} = \langle -1, 0 \rangle, \\ & \mathbf{V} = \langle 0, -2\sqrt{\pi} \rangle, \ |\mathbf{V}| = 2\sqrt{\pi}, \\ & \mathbf{A} = \langle 4\pi, -2 \rangle. \end{aligned}$$

This information is plotted in Figure 6.8.1 with the vectors **V** and **A** drawn to 1/4 scale.

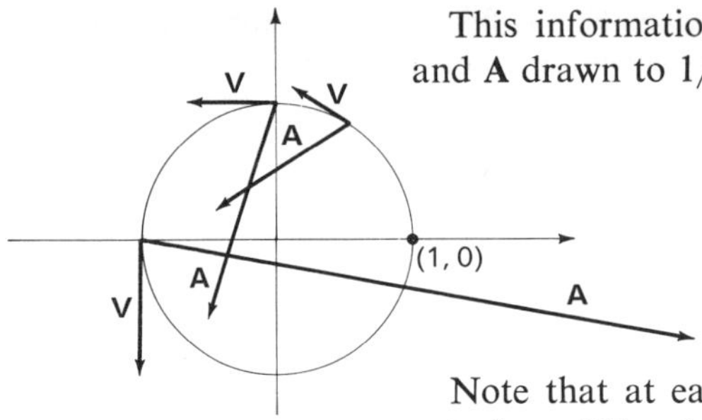

Figure 6.8.1

Note that at each point the vector **A** points in the direction of rotation of **V** and makes an angle of less than 90° with **V**.

Example 2 Any circular motion may be described by

$$\mathbf{P}(t) = \langle R\cos\theta(t), R\sin\theta(t)\rangle,$$

where R is the radius of the circle and $\theta(t)$ is the angle from the horizontal direction to P at time t. An easy exercise shows that the speed of this motion is

(6.8.2) $$|\mathbf{V}(t)| = |R\theta'(t)|.$$

We wish to discuss next a *uniform* circular motion. This is a motion in a circle at *constant speed*, and by (6.8.2) we see that $\theta'(t)$ must be constant. If we agree to let $\theta(0) = 0$, then we must have $\theta(t) = \omega t$, where ω is a constant. Let us therefore discuss the motion

(6.8.3) $$\mathbf{P}(t) = \langle R\cos(\omega t), R\sin(\omega t)\rangle = R\langle\cos(\omega t), \sin(\omega t)\rangle, \quad \omega > 0.$$

We have, immediately,

(6.8.4) $$\begin{aligned}\mathbf{V}(t) &= R\langle -\omega\sin(\omega t), \omega\cos(\omega t)\rangle,\\ |\mathbf{V}(t)| &= R\omega,\\ \mathbf{A}(t) &= R\langle -\omega^2\cos(\omega t), -\omega^2\sin(\omega t)\rangle.\end{aligned}$$

Note that ωt is the "total angle" swept out by **P** from time 0 to time t. The constant ω, the rate of change of this angle, is often called the *angular velocity* of this motion. By (6.8.4), the speed of an object moving in this way is R times the angular velocity.

Note also from (6.8.4) and (6.8.3) that

(6.8.5) $$\mathbf{A}(t) = -\omega^2\mathbf{P}(t), \ |\mathbf{A}(t)| = R\omega^2.$$

Thus, the acceleration has magnitude $R\omega^2$ and is always directed parallel but opposite to the position vector **P**. That is, **A** is directed from the point P toward the center of the circle (Figure 6.8.2).

The motion in Example 2 could be that of a ball rotating on the end of a string, an electron rotating about an atomic nucleus, a planet rotating about the sun, or the moon or a man-made satellite

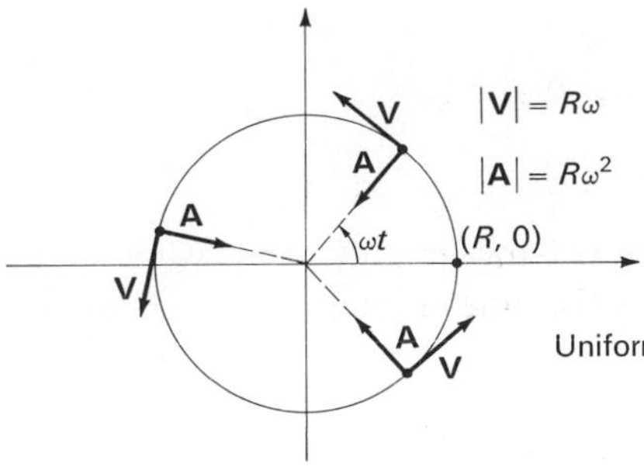

Figure 6.8.2

Uniform circular motion

rotating about the earth. Such physical motions of actual objects can be analyzed further by introducing Newton's Second Law of Motion. This law asserts that if **F** is the net force on an object, **A** is its acceleration (both vector quantities), and m is its mass (a scalar), then

(6.8.6) $$\mathbf{F} = m\mathbf{A}.$$

Thus, we conclude that if a body is rotating in a circle with constant speed (or constant angular velocity), then the net force acting on it must be directed toward the center of rotation and have constant magnitude. In fact, if the speed of the body is $v = |\mathbf{V}| = R\omega$ (see (6.8.4), where ω is the angular velocity and R is the radius of the circle, then the net force must have magnitude

(6.8.7) $$|\mathbf{F}| = mR\omega^2 = \frac{mv^2}{R}.$$

To prove (6.8.7) simply combine (6.8.5) and (6.8.6).

Example 3 Suppose a satellite of mass m rotates about a much more massive object (e.g., sun or earth) of mass M. Newton's Universal Law of Gravitation tells us that the gravitational force attracting the two objects to each other has magnitude

(6.8.8) $$G\frac{mM}{R^2},$$

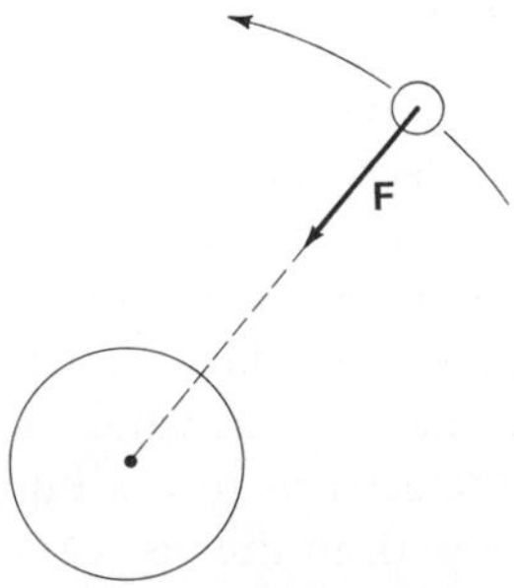

Figure 6.8.3

where R is the distance between the two centers of mass and the universal constant G can be calculated with great accuracy from known data. Let us assume that there are no other forces acting on those bodies and that the more massive body is stationary. If the smaller body is observed to be moving uniformly in a circle of radius R about the larger one, then we have exactly the situation described in Example 2 and the remarks before this example. The net force **F** acting on the smaller body is the gravitational force of attraction between the two. See Figure 6.8.3. If the angular velocity of this rotation is observed to be ω, then we have two expressions for the magnitude of **F**, given by (6.8.7) and (6.8.8). Equating them, we have

(6.8.9)
$$G\frac{mM}{R^2} = mR\omega^2, \quad \text{or}$$
$$GM = R^3\omega^2.$$

A more useful expression, involving readily observable data, results when it is noted that if T is the time required for one complete rotation about the center, then

(6.8.10)
$$T = \frac{2\pi}{\omega}, \quad \text{or} \quad \omega = \frac{2\pi}{T}.$$

From (6.8.9) and (6.8.10) we have

(6.8.11)
$$\frac{R^3}{T^2} = \frac{GM}{4\pi^2} = K.$$

K depends *only* upon the mass of the large body, and hence is constant for a large body of given mass. Equation (6.8.11) relates the period of rotation (T) to the orbital radius (R) of *any* object which rotates about this massive body in a circle, subject only to the gravitational force. This equation is the third of Johannes Kepler's famous Laws of Planetary Motion, in the special case of circular orbits.

For the case in which the central body is the earth, the constant $K = K_e$ has the approximate value 3.17×10^{10} mi.^{3}/hr.^{2}, if distance is measured in miles and time in hours. Thus, a satellite orbiting at an altitude of 100 miles above the surface of the earth (that is, at a radius $R = 100 + 3960 = 4060$ miles about the center) has a period of rotation

$$T = \left(\frac{(4060)^3}{K_e}\right)^{1/2} \approx 1.44 \text{ hours.}$$

From the observed period of rotation of the moon about the earth we may calculate its distance from the earth, since its orbit is nearly circular. This period, the lunar month, is about 656 hours, or approximately $27\frac{1}{3}$ days. Thus,

$$R = [K_e(656)^2]^{1/3} = 239{,}000 \text{ miles.}$$

Example 4 Now consider an object which travels along any curve in a plane in such a way that the net force acting on it is always directed toward a fixed point. Included of course are all the motions listed in Example 3, but also many others: noncircular planetary orbits, the travel of a ball in a flat oval path at the end of an elastic cord, the motion of an α-particle passing near an atomic nucleus which is attracted to the relatively massive nucleus by electrical forces but not trapped into a closed orbit around it. (The path in the last case is one branch of an hyperbola.)

The fixed point, the *center of attraction*, will be chosen as the origin of a coordinate system, and the position vector for the object at time t will be written, as usual,

$$\mathbf{P}(t) = \langle g(t), f(t)\rangle.$$

The condition that the net force be directed toward the origin implies by (6.8.6) that $\mathbf{A}$ and $\mathbf{P}$ are oppositely directed. Thus $\mathbf{A} = -k(t)\mathbf{P}(t)$, where $k(t)$ is a positive real number for all t. Thus,

$$\langle g''(t), f''(t)\rangle = -k(t)\langle g(t), f(t)\rangle,$$

so that

$$\frac{f''}{f} = \frac{g''}{g} = -k \quad \text{for all } t.$$

Therefore

$$f''g - fg'' = 0,$$

from which it follows by an easy exercise that

(6.8.12) $$f'g - fg' \quad \text{is constant.}$$

The discussion now takes an unexpected turn. Consider the area which is swept out by the position vector $\mathbf{P}$ between times 0 and t (Figure 6.8.4). That is the area bounded by $\mathbf{P}(0)$, $\mathbf{P}(t)$, and the path of the moving object. Call this area $\mathcal{A}(t)$—clearly it is a function of t.

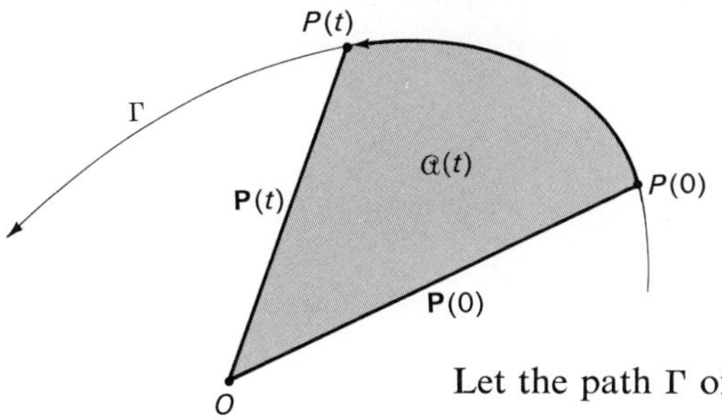

Figure 6.8.4

Let the path Γ of the object be the graph of

$$r = F(\theta)$$

in polar coordinates. If for each t the point $P(t)$ has polar coordinates $(r(t), \theta(t))$, then we can infer the following information from the above and earlier discussions:

(6.8.13) $$r(t) = F(\theta(t)) = |\mathbf{P}(t)| = \sqrt{[g(t)]^2 + [f(t)]^2},$$

(6.8.14) $$\theta(t) = \arctan\left(\frac{f(t)}{g(t)}\right),$$

(6.8.15) $$\mathcal{A}(t) = \int_{\theta(0)}^{\theta(t)} \tfrac{1}{2}[F(\theta)]^2\, d\theta.$$

Using (6.8.13) and (6.8.14), we can make a change of variable $\theta = \theta(t), d\theta = \theta'(t)\,dt$ in the integral in (6.8.15):

$$\begin{aligned}\mathcal{A}(t) &= \int_0^t \tfrac{1}{2}([f(t)]^2 + [g(t)]^2)\frac{f(t)g'(t) - g(t)f'(t)}{[f(t)]^2 + [g(t)]^2}\,dt \\ &= \tfrac{1}{2}\int_0^t [f(t)g'(t) - g(t)f'(t)]\,dt.\end{aligned}$$

Now using the Fundamental Theorem we have

$$\frac{d}{dt}\mathcal{A}(t) = \tfrac{1}{2}(f(t)g'(t) - g(t)f'(t)),$$

and by (6.8.12)

$$\frac{d}{dt}\mathcal{A}(t) = \text{constant}.$$

Thus no matter what the motion, if the net force is *centrally directed*, then the position vector **P** sweeps out area *at a constant rate*. Here is a striking degree of regularity in a situation which apparently allows great variety of motion. It was observed in the motions of the planets about the sun—literally "seen" from Tycho Brahe's great masses of astronomical data—by Johannes Kepler in the seventeenth century. It is generally referred to as Kepler's Second Law of Planetary Motion. The Law played a fundamental role in Newton's great discovery of the Law of Universal Gravitation.

simple differential equations

chapter 7

7.0

Introduction: Poohsticks and Measles

There was a game, named Poohsticks after its inventor, which Winnie-the-Pooh and his friends used to play. It involved throwing sticks into a river from the upstream side of a bridge and then speculating on which stick would emerge first from under the other side. It is quite possible that Pooh, who possessed great imagination

Figure 7.0.1

and curiosity, invented other more private games to play with the floating sticks. For example, one can imagine him wondering where the sticks go after they leave the shadow of the bridge. Where would he have to throw the stick into the water to insure that it would be carried over the little waterfall behind the island? If he threw it over here, would it negotiate a smooth path down the center channel? If he threw it there, would it get caught in the eddy behind the small point of land downstream?

In wondering about such questions and trying to answer them, this talented bear would have been trying to solve a differential equation. That is to say, he would have been trying to visualize what the path of the stick would be, depending on where it entered the water, as it was carried downstream by the current. If he thought hard enough and in a particular way about how the river singles out certain paths—these paths are often called *streamlines* by physicists—then it is quite likely that he formulated for himself the notion of *direction field.* If he had already acquired, from Owl, that repository of all knowledge in the forest, some understanding of the basic concepts of calculus, then it is almost certain that Pooh took the next logical step to the notion of *differential equation*, one of the most important ingredients in mathematics, the language of science. We shall retrace some of these thought steps in this chapter.

The scene now changes abruptly. Consider the task of a public health officer in whose community measles has just broken out. He is extremely interested in any predictions he can make concerning the growth of the infected population x as a function of time. He can construct a mathematical model which, he suspects, describes the growth of x fairly accurately.

He argues as follows: In an infinite population, the rate of increase of x should be proportional to x, since each measly person spreads the plague to others at a roughly constant rate. However, in a finite population P of people lacking measles immunity, the availability of healthy but susceptible neighbors decreases as the infected population x increases. This, of course, tends to diminish the rate of spread of the measles. It is thus reasonable to modify the rate of change of x by an additional factor $(P - x)$ which decreases toward zero as x grows toward P. He thus predicts that x will grow at a rate which is at every instant of time proportional to the product $x(P - x)$. That is,

(7.0.1) $$x'(t) = Kx(t)(P - x(t)).$$

This is a differential equation, an equation involving both a function $x(t)$ and also its derivative $x'(t)$. The health officer's problem is to solve the differential equation, to find a function which satisfies condition (7.0.1) and which therefore might be a predictor of the course of the epidemic.

The measles epidemic is closely related to Poohsticks, as we shall shortly see.

7.1 Direction Fields and First Order Differential Equations

While Pooh stood watching the river, contemplating the paths taken by twigs in the water, he might have found himself wondering how he might describe succinctly the totality of action of the current on sticks, real or imagined. It might have occurred to him that all the current does is to assign to every point in the river a direction, the direction of flow at that point. Thus, the totality of information about the river current could be tabulated by assigning to each point a little arrow marking the direction of the current at that point. It is clearly impossible to draw or otherwise record all such point-arrow pairs at one time, but the scheme can be contemplated. The resulting "map" of the current is called a *direction field.* Some of the arrows of this direction field are shown in Figure 7.1.1. Each symbol ⇢ means that at the point indicated by the dot the current is flowing in the direction of the arrow. It is understood that at each point of the unshaded region, an arrow could be drawn indicating the direction of the field (direction of the current flow) at that point. Of course, we cannot hope to draw all of these arrows, but must be content with a few representative ones as in the figure.

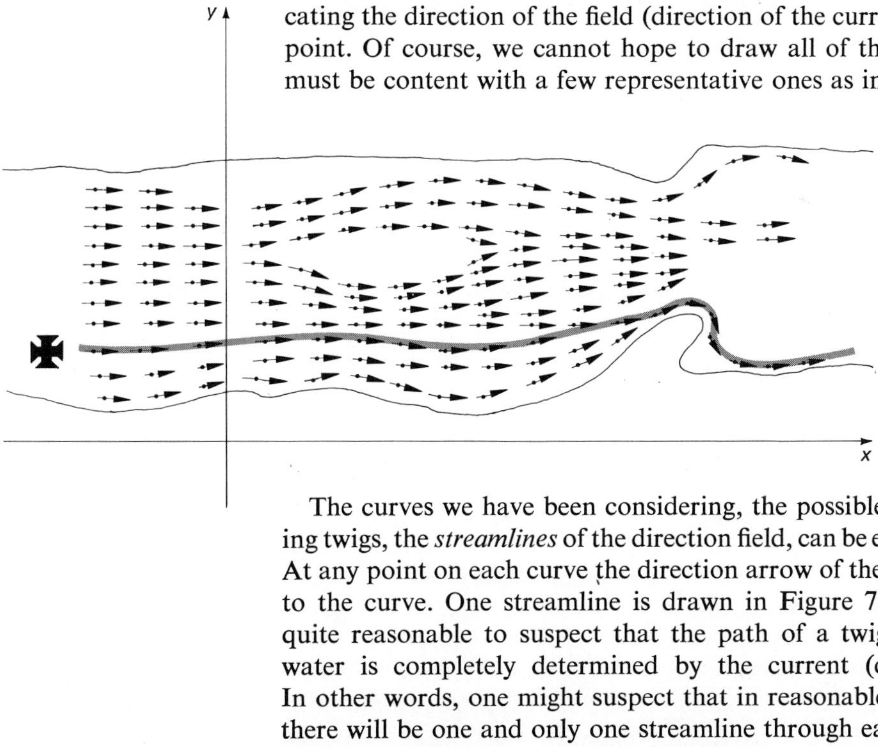

Figure 7.1.1

The curves we have been considering, the possible paths of floating twigs, the *streamlines* of the direction field, can be easily described: At any point on each curve the direction arrow of the field is tangent to the curve. One streamline is drawn in Figure 7.1.1. It appears quite reasonable to suspect that the path of a twig placed in the water is completely determined by the current (direction field). In other words, one might suspect that in reasonable circumstances there will be one and only one streamline through each point in the region of the direction field. We shall see that this will be the situation often but by no means all the time.

At this point, we hope that the idea of a direction field is quite clear to you. But you may wonder how all the information about a direction field in a region of a plane could possibly be recorded in any usable way or how any one streamline could be determined and described in a complete sense. What we have said above suggests a number of graphical techniques for constructing rough approximations to streamlines, but the precision which we have come to associate with mathematics is not yet there.

We must be both more precise and more concise: One way is to bring analysis into the picture. If we imagine coordinate axes superimposed on our map of the river—on our direction field—then we may specify each of the directions assigned by the field by a number, the *slope* of the appropriate arrow. Then, in order to describe a direction field we need only list for each point (x, y) in the desired region a number $m(x, y)$, the slope of a segment having the direction desired at that point. Thus, a direction field is determined by nothing more than a real-valued function of two variables (Figure 7.1.2).

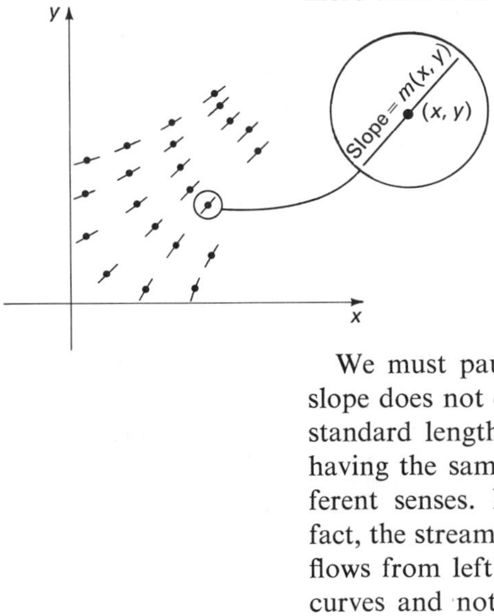

Figure 7.1.2

We must pause to admit that something has been lost here. A slope does not completely determine an arrow, even if we specify a standard length for the arrows. There are two possible directions having the same slope; a line segment can be oriented in two different senses. However, this loss does not change anything. In fact, the streamlines in our river do not depend on whether the river flows from left to right or right to left. If we care only about the curves and not about the direction in which a twig moves along one of them, then we see that only the segments are needed in Figure 7.1.1. The arrowheads can be ignored. It would be more proper to call the map in Figure 7.1.1 a *vector field.* Henceforth we shall neglect the arrowheads. When we refer to a *direction field* we are using the word "direction" as in the sentence, "The road runs in a north-south *direction.*" A vector field is an *oriented* direction field.

Now, subject to restrictions which are familiar to us by now, a curve in a region of a coordinate plane can be characterized as the graph of a function. Even when the entire curve cannot be so characterized, individual pieces often can. In the language of functions a streamline of a direction field can quite easily be described. If f is a function, then the graph of $y = f(x)$ is a streamline of the direction field $m(x, y)$ if and only if

$$f'(x) = m(x, f(x)) \tag{7.1.1}$$

for all x in the appropriate domain. This is obviously the appropriate condition, for it states that at any point $(x, f(x))$ on the curve, the slope $f'(x)$ of the tangent is just the slope $m(x, f(x))$ which the direction field assigns to that point (Figure 7.1.3).

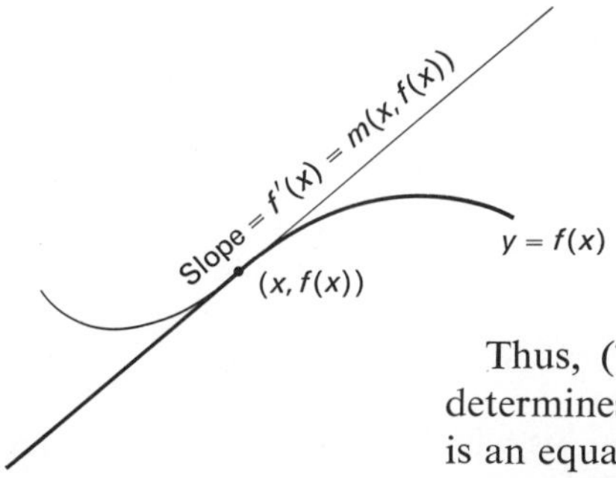

Figure 7.1.3

Thus, (7.1.1) is an analytical expression of the condition which determines streamlines of a given direction field. This expression is an equation which relates the values of a function f to the values of its derivative f'. It is called a *differential equation*. Because only the first derivative f' is involved, it is called a *first order* differential equation. Since we may also write $dy/dx = f'(x)$ if $y = f(x)$, (7.1.1) may also be written

$$\frac{dy}{dx} = m(x, y), \tag{7.1.2}$$

a more common way of writing a first order differential equation of this type. We say that a function f which satisfies (7.1.1) for all x in an interval I is a *solution* of (7.1.2) *on I*.

Notice that in the very special case where $m(x, y)$ does not vary with y, that is, when

$$m(x, y) = M(x),$$

then a solution $f(x)$ of (7.1.2) has the form

$$f(x) = \int M(x)\, dx.$$

This follows directly from the definition of an indefinite integral. For this reason, it has become customary to refer to a solution of (7.1.2) as an *integral* of the differential equation. The graph of $y = f(x)$ is called an *integral curve* of (7.1.2). For the same reason, a streamline of a direction field is also called an integral curve of the field.

Example Consider the direction field given by

$$m(x, y) = y.$$

A few of the directions given by this field are shown in Figure 7.1.4.

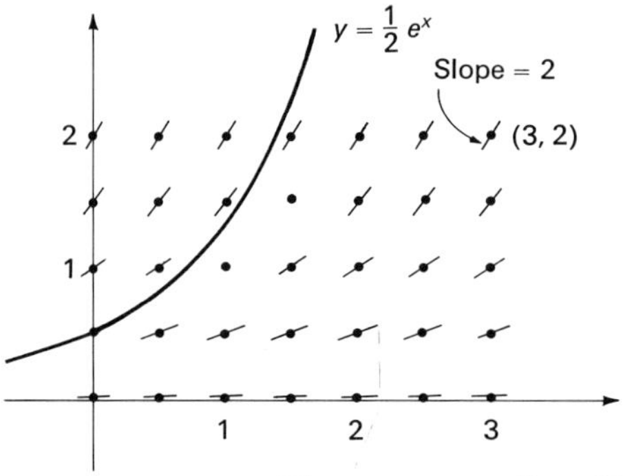

Figure 7.1.4

The corresponding differential equation is

$$\frac{dy}{dx} = y, \qquad (7.1.3)$$

the solutions of which are functions which coincide with their first derivatives. In Chapter 7 of *C 1* we saw that they are precisely the functions

$$f(x) = Ce^x$$

for arbitrary values of the constant C. (We shall see again in Section 7.4 why this is true.) Hence, the integral curves of the differential equation (7.1.3) (or the direction field pictured in Figure 7.1.4) are the graphs of

$$y = Ce^x.$$

Exercises

1 Draw a sketch of the direction field associated with the differential equation

$$\frac{dy}{dx} = x$$

by sketching a small line segment with appropriate slope through each point $(m/2, n/2)$, where m and n are integers between, say, 0 and 6. Then sketch two or three streamlines (integral curves) of this equation.

2 Repeat Exercise 1 for $dy/dx = y - 1$.

3 Repeat Exercise 1 for $dy/dx = x/y$.

4 Repeat Exercise 1 for $dy/dx = -x/y$.

5 The functions $f(x) = Ce^x + 1$, C an arbitrary constant, are all solutions of a differential equation. Find such an equation by differentiating f and eliminating C.

6 The graphs of the equations $y^2 - x^2 = k$, k an arbitrary constant, are integral curves of a differential equation $dy/dx = m(x, y)$. Find $m(x, y)$.

7 Repeat Exercise 6 for the family of graphs of the equation $x^2 + y^2 = k^2$.

7.2 Differential Equations and Their Solutions

Generally speaking, a *differential equation* is any equation of the form

(7.2.1) $$F\left(x, y, \frac{dy}{dx}, \frac{d^2y}{dx^2}, \ldots, \frac{d^ny}{dx^n}\right) = 0$$

relating an independent variable x, a function y of x, and one or more of its derivatives $dy/dx, \ldots$. Examples are

(7.2.2) $$dy/dx - 3y^{2/3} = 0,$$

(7.2.3) $$\frac{dy}{dx} - \sqrt{1 - x^2} = 0,$$

and

(7.2.4) $$\frac{d^2y}{dx^2} - \frac{dy}{dx} + 2y = x.$$

To *solve* a differential equation is to find all functions $f(x)$ which when substituted for y make the equation an identity in x. Each such function is a *solution* or *integral* of the equation. Actually we always speak of a solution of a differential equation *on a specified interval* $[a, b]$. Perhaps the equation itself makes sense only for x in some interval. Equation (7.2.3) makes sense only for x in $[-1, 1]$, for example. Or, even though the equation itself places no obvious restrictions upon x, it may happen that well-behaved solutions of the equations exist only on certain intervals.

An equation (7.2.1) is called an nth *order* differential equation, n being the order of the highest derivative occurring in the equation. Thus (7.2.2) and (7.2.3) are first order equations, and (7.2.4) is a second order differential equation. Only first order differential equations have the strong geometric interpretation given in Section 7.1 in terms of direction fields. In this very brief chapter we shall consider only first order differential equations. The general study of differential equations is a vast subject, rich and important enough to fill lengthy treatises. The study of first order differential equations is itself so large that we shall be able only to scratch its surface.

When a mathematician sees a differential equation he immediately asks himself basic questions such as the following: Does the equation have a solution? How many solutions are there? What conditions, in addition to that of satisfying the equation, are sufficient to determine a unique solution? What properties do the solutions have? How can I find all the solutions?

We observed earlier that the equation

(7.2.5) $$\frac{dy}{dx} = y$$

has solutions

(7.2.6) $$y = Ce^x.$$

All solutions of (7.2.5) are of this form. The graphs of (7.2.6) for all real values of C are the integral curves of the equation. Clearly, if we specify a point (x_0, y_0) on an integral curve, the value of the constant C can be determined. That is, only one integral curve passes through a specified point.

To say this in another way, if we specify that a solution $f(x)$ of (7.2.5) is to have a certain value y_0 at a given value x_0 of x, then the solution is uniquely determined. The condition that the solution f should satisfy,

(7.2.7) $$y_0 = f(x_0),$$

is called an *initial condition*. We have observed that solutions of (7.2.5) may be made to satisfy one initial condition of type (7.2.7) and that together with this condition the differential equation determines a function uniquely.

A contrasting situation is presented by equation (7.2.2),

$$dy/dx - 3y^{2/3} = 0.$$

It is easy to see that any function of the form

$$f(x) = (x + C)^3, \quad C \text{ a real number},$$

is a solution of (7.2.2). Such a solution may be made to satisfy any desired boundary condition of type (7.2.7). But observe the following situation.

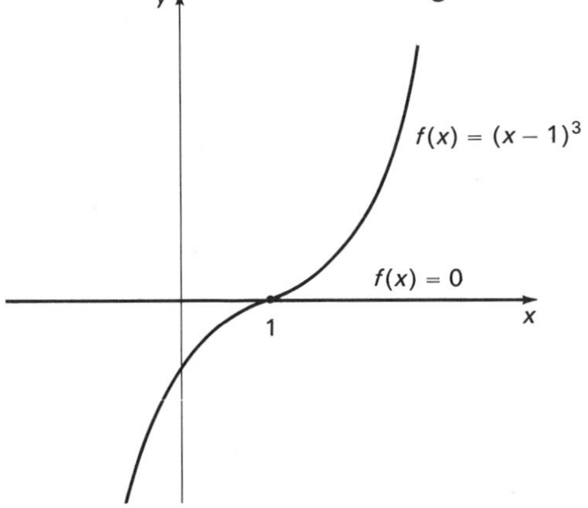

Figure 7.2.1

The function $f(x) = (x - 1)^3$ satisfies the condition $f(1) = 0$. But so does the function

$$f(x) = 0,$$

which you may quickly see is *also* a solution of (7.2.2) (Figure 7.2.1). Thus, a boundary condition (7.2.7) does *not* necessarily determine a solution of (7.2.2) uniquely.

Still more solutions f of (7.2.2) may be found such that $f(1) = 0$ by combining "pieces" of the zero function with branches of $(x + C)^3$ for other values of C. Thus

$$f(x) = \begin{cases} (x-2)^3 & \text{if } x \geq 2, \\ 0 & \text{if } 0 < x < 2, \\ x^3 & \text{if } x \leq 0 \end{cases}$$

is a third solution satisfying $f(1) = 0$ (Figure 7.2.2).

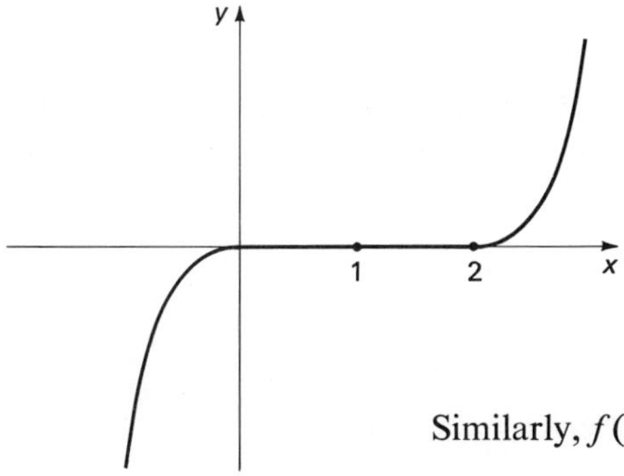

Figure 7.2.2

Similarly, $f(x) = x^3$ is a solution which satisfies the condition

$$f(1) = 1. \tag{7.2.8}$$

But the function

$$f(x) = \begin{cases} x^3 & \text{if } x \geq 0, \\ 0 & \text{if } -1 < x < 0, \\ (x+1)^3 & \text{if } x \leq -1 \end{cases}$$

is also a solution satisfying (7.2.8) (Figure 7.2.3). Infinitely many more solutions satisfying (7.2.8) may be found in the same way.

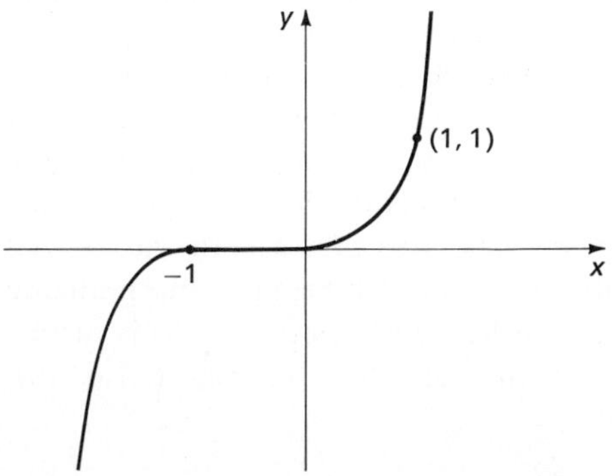

Figure 7.2.3

Notice one thing, however. All solutions which satisfy the boundary condition (7.2.8) are the same for $x \geq 0$. It appears that *on the interval* $[0, \infty)$ there is only one solution of (7.2.2) such that $f(1) = 1$. In fact, one can prove that if $y_0 \neq 0$ and x_0 is any real number, then there is some interval I of one of the forms $[a, \infty)$, $(-\infty, b]$ such that all solutions of (7.2.2) satisfying the condition (7.2.7) are the same on I.

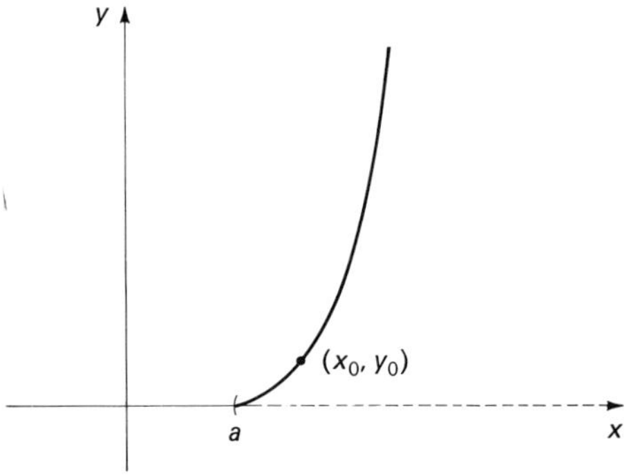

Figure 7.2.4

If, on the other hand, $y_0 = 0$, then on *any* interval I containing x_0 there are infinitely many solutions f of (7.2.2) for which $f(x_0) = y_0 = 0$.

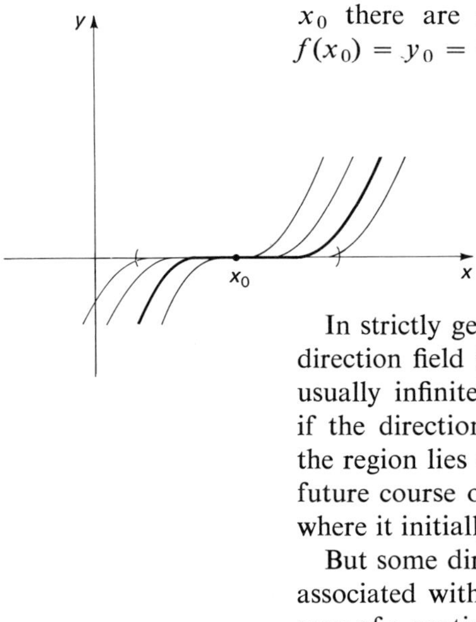

Figure 7.2.5

In strictly geometric terms, what we have said above is this. Any direction field in a region of the plane determines a great many—usually infinitely many—integral curves or streamlines. However, if the direction field is sufficiently "regular," then each point of the region lies on *one and only one* integral curve. Equivalently, the future course of the stick in the river is completely determined by where it initially lands in the water.

But some direction fields are not so regular. If the direction field associated with equation (7.2.2), shown in Figure 7.2.6(b), were a map of a portion of Pooh's river, then it would be quite possible for two sticks thrown into the water at the same spot to follow different paths. How could this be possible? Chance, perhaps? Or is it simply not possible for such a "nonregular" direction field to represent the real flow of a real river? The physicist says that it is the latter: All physical systems are *deterministic*. The mathematician, realizing that the nondeterministic direction field in Figure 7.2.6(b) is just as possible in principle as the deterministic field in Figure 7.6.2(a),

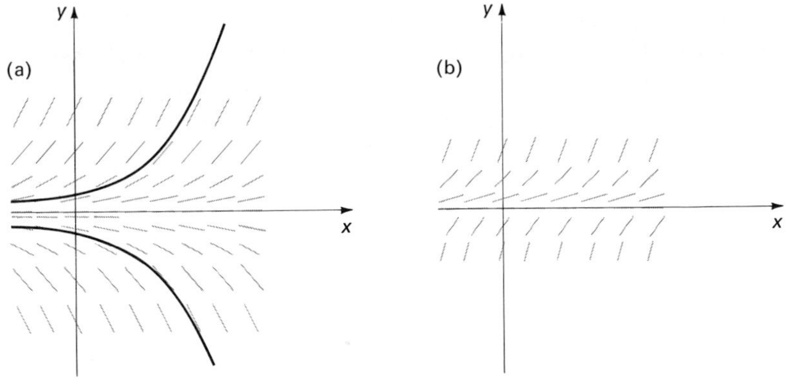

$y' = y$

$y' = 3y^{2/3}$

Figure 7.2.6

asks how he can tell one from the other. How, he asks, can one tell from the associated differential equation when a direction field will be deterministic and therefore of the type which might represent a physical situation? In other words, what conditions on a differential equation will insure that every point in the region of the direction field lies on an integral curve and that no point lies on more than one integral curve?

We shall state an important theorem which exhibits such conditions. It is usually called the Picard Existence and Uniqueness Theorem for a first order differential equation, named after the French mathematician Émile Picard (1856–1941), who did much pioneering work on differential equations. It asserts that given any admissible initial condition, a solution of (7.1.2) *exists* and is *unique*, in the sense that no other solution of (7.1.2) satisfies that initial condition.

Theorem 7.2.1 *Let the function $m(x, y)$ exist for all (x, y) in the rectangle $x_1 \le x \le x_2$, $y_1 \le y \le y_2$. If both $m(x, y)$ and the partial derivative $m_y(x, y)$ are continuous in the rectangle, then for any point (x_0, y_0) in the rectangle there exists an interval (a, b) about x_0 on the x-axis such that the following is true:*

(i) *There exists a solution f of the equation*

$$\frac{dy}{dx} = m(x, y)$$

on the interval (a, b) which satisfies the boundary condition

$$f(x_0) = y_0, \tag{7.2.9}$$

(ii) *f is the only solution of* (7.1.2) *on (a, b) which satisfies the condition* (7.2.9).

The proof of this theorem is well beyond the scope of our present discussion. However, some insight into the workings of the theorem may be obtained by looking closely at the two equations $dy/dx = y$ (7.2.5) and $dy/dx = 3y^{2/3}$ (7.2.2). For (7.2.5) $m(x, y) = y$, so $m_y(x, y) = 1$. This certainly exists and is continuous on the "infinite rectangle" consisting of the entire plane. We note also that for any (x_0, y_0) one and only one solution $f(x) = Ce^x$ exists for which $f(x_0) = y_0$. This solution corresponds to the choice $C = y_0e^{-x_0}$.

For equation (7.2.2), $m(x, y) = 3y^{2/3}$ and $m_y(x, y) = 2y^{-1/3}$. Thus, m_y exists and is continuous on any rectangle which does not intersect the x-axis. Since if $y_0 \neq 0$ then (x_0, y_0) is contained in such a rectangle, we are therefore assured that (7.2.2) has a solution $f(x)$ for which $f(x_0) = y_0$ and which is unique on some interval about x_0. On the other hand, $m_y(x, y)$ does not exist if $y = 0$; thus we expect something to go wrong in a neighborhood of $(x_0, 0)$. In this case it is the uniqueness which fails; solutions through $(x_0, 0)$ do exist, in great abundance. See Figure 7.2.7.

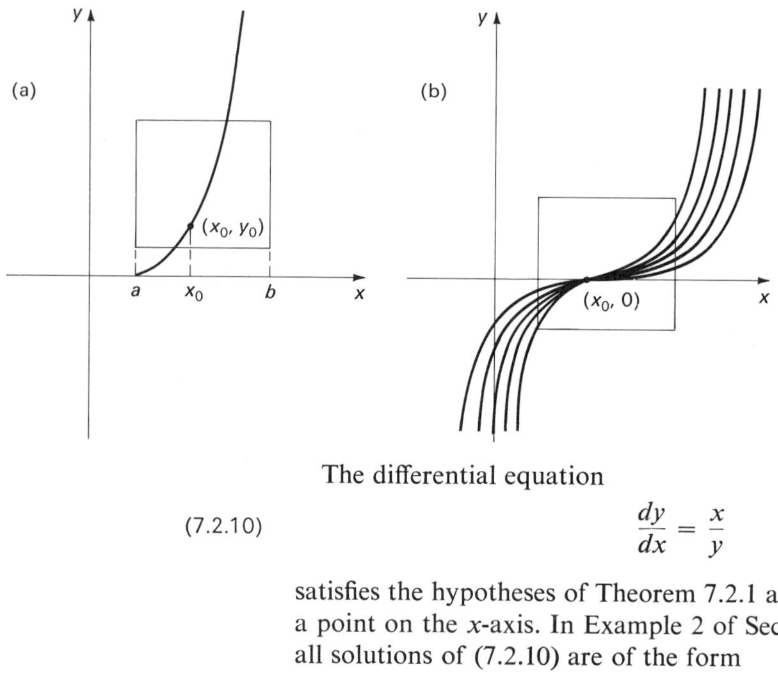

Figure 7.2.7

The differential equation

$$\frac{dy}{dx} = \frac{x}{y} \tag{7.2.10}$$

satisfies the hypotheses of Theorem 7.2.1 at any point except, again, a point on the x-axis. In Example 2 of Section 7.4 we shall see that all solutions of (7.2.10) are of the form

$$y = \pm\sqrt{x^2 + k}, \quad k \text{ constant.}$$

The only solutions which pass through $(1, 0)$ are $y = \sqrt{x^2 - 1}$, $y = -\sqrt{x^2 - 1}$. There is no uniqueness around $(1, 0)$. But here a part of the existence conclusion also fails, for no solution through $(1, 0)$ is defined for any x such that $0 < x < 1$. Thus, no solution is defined on an interval (a, b) with $a < 1 < b$. See Figure 7.2.8.

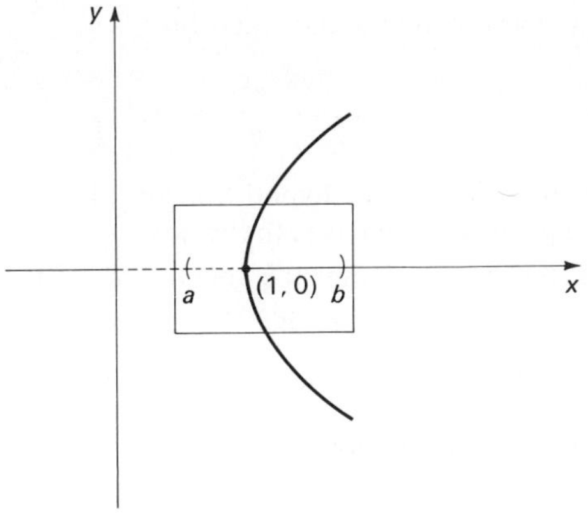

Figure 7.2.8

Exercises

1 Determine the order of the differential equations

(a) $\frac{d^2y}{dx^2} + k^2y = 0$, (b) $yy' = x$,

(c) $\frac{dy}{dx} + xy^2 = 0$, (d) $(y')^2 + y' = x^2$.

2 Verify that each of the given functions is a solution of the corresponding differential equation.

(a) $y = e^x$, $y' - y = 0$;

(b) $y = \sin 2x$, $y'' + 4y = 0$;

(c) $y = \sin x + \cos 2x$, $y^{(4)} + 5y'' + 4y = 0$;

(d) $y = e^{-3x}$, $y'' + 2y' - 3y = 0$;

(e) $y = e^x \sin 2x$, $y'' - 2y' - 5y = 0$.

3 For each of the following differential equations find the region in the xy plane where the existence and uniqueness of a solution through any specified point is guaranteed by Picard's Theorem.

(a) $\frac{dy}{dx} = \frac{x}{y}$, (b) $y' = 1 + x + y^2 + xy$,

(c) $\frac{dy}{dx} + y^2 \cos x = 0$, (d) $y' = \frac{3x - y}{4x + 5y}$,

(e) $\frac{dy}{dx} = (4 - x^2 - y^2)^{1/2}$, (f) $y' = \frac{3x^2y}{(1 + y^3)}$.

4 (a) Verify that the hypotheses of Picard's Theorem are satisfied by equation (7.2.2) on any rectangle which does not intersect the x-axis.

(b) For the point $(x_0, y_0) = (3, 8)$ find an interval (a, b) which satisfies the conclusions of that theorem.

(c) Show that the hypotheses of Picard's Theorem are not satisfied throughout any rectangle which contains any part of the x-axis.

5 (a) Show that the differential equation

$$\frac{dy}{dx} = x|y|$$

does not satisfy the hypotheses of Picard's Theorem on any rectangle which intersects the x-axis.

(b) Nevertheless, show that the conclusions of the theorem are valid throughout the entire xy plane.

6 Differential equations are frequently used to describe families of curves. For example,

(a) Show that the equation

$$x^2 - 2cx + y^2 = 1 - c^2$$

(where c is a parameter) describes a family of circles. Where are their centers and what are their radii?

(b) By differentiating, show that for each c the curve defined by the equation in (a) satisfies

$$y\frac{dy}{dx} + x = c.$$

(c) By eliminating the parameter c between the two equations in (a) and (b), show that every member of the family satisfies the single differential equation

$$y^2\left(\left(\frac{dy}{dx}\right)^2 + 1\right) = 1.$$

7 Use the method of Exercise 6 to find a differential equation which describes the following families of curves.

(a) $x^3 - 3x^2y = c$, (b) $y \cos x + xy^2 = c$,

(c) $cy^2 = x^2 + y$, (d) $y = cx + c^3 - 1$.

8 (a) Find an equation depending upon a parameter which describes the family of straight lines through the origin.

(b) Find a differential equation which also describes the family in (a).

9 The method of Exercise 6 can be extended to families of curves which are described by n parameters $c_1, c_2, \ldots, c_n$. One simply differentiates the given equation n times to produce a total of $n + 1$ equations, and then eliminates the parameters from these equations to produce a single nth order differential equation which describes the family. Thus find differential equations to describe the families

(a) $y = c_1x + c_2x^2$, (b) $y = c_1e^x + c_2e^{2x}$,

(c) $y = c_1e^x + c_2e^{-x}$, (d) $y = c_1x + c_2e^x$,

(e) $y = c_1 \sin x + c_2 \cos x$, (f) $y = c_1e^x + c_2e^{2x} + c_3e^{3x}$.

10 Find a differential equation which describes all circles with centers on the x-axis.

11 Find a differential equation which describes all circles with center at the origin.

12 Show that $y = 1/(1 - 2x)$ is a solution of the boundary value problem

$$y' = 2y^2, \; y(1) = -1.$$

In what interval is this solution valid?

13 (a) Show that $y_1(x) = 1 - x$ and $y_2(x) = -x^2/4$ are solutions of the differential equation

$$y' = \frac{-x + (x^2 + 4y)^{1/2}}{2}$$

which satisfy $y(2) = -1$. Where are these solutions valid?

(b) Why does not part (a) contradict part (ii) of the conclusion to Picard's Theorem?

7.3

Numerical Approximation of Solutions

Differential equations are among the most frequently applied mathematical objects. That is, they very often occur as a means to an end: the solutions themselves. Here Theorem 7.2.1, in spite of its great theoretical importance, is of little comfort. It is good to know that a solution satisfying a given boundary condition exists, but this knowledge is of little use if one cannot find that solution. Thus, a substantial part of the theory of differential equations is devoted to methods of actually finding solutions and studying their properties. In this section and the next we shall consider this problem for some simple first order equations.

Only in special cases is it possible to find solutions of differential equations, even first order equations. We shall investigate two of these special cases in the next section. No general method exists for finding a solution $y(x)$ of the equation

$$\frac{dy}{dx} = m(x, y) \tag{7.3.1}$$

satisfying a boundary condition

$$y(x_0) = y_0. \tag{7.3.2}$$

However, direction fields suggest a way of constructing approximations for such a solution. Suppose you were standing at a point P_0 in the plane of the direction field corresponding to (7.3.1) and wanted to walk along an integral curve from P_0. How would you proceed? You might begin by observing the indicated direction at P_0, walk a short distance in that direction to a point P_1, observe the new direction given at P_1, alter your course, walk a short distance to P_2, observe the new direction, alter course, walk a short distance to P_3, etc. That is, you might walk along a path consisting of a series of short line segments, each having slope equal to the value of $m(x, y)$ at the initial point.

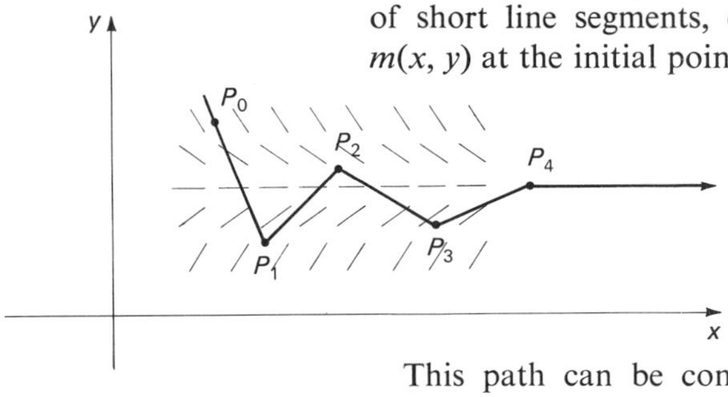

Figure 7.3.1

This path can be considered a polygonal approximation to a solution. If we assume that $m(x, y)$ is continuous so that its values change little within a small neighborhood of a point and if each segment in the polygonal path is short, then the slope at a point on that path never differs from $m(x, y)$ very much.

In order to be somewhat more systematic, let us suppose that we desire a polygonal approximation of a solution of (7.3.1) on an interval $[a, b]$ satisfying a boundary condition (7.3.2) with $x_0 = a$. We might form a partition $P = \{a = x_0, x_1, \ldots, x_n = b\}$ and construct a polygonal path as above with each segment extending from a point (x_i, y_i) to a point (x_{i+1}, y_{i+1}). The abscissas $x_0, x_1, \ldots, x_n$ are given; we need only find the ordinates $y_1, y_2, \ldots, y_n$. These can be found in order; y_0 is given by the boundary condition (7.3.2), and once y_i is known, the succeeding y_{i+1} can be found by a simple calculation. Since the segment joining (x_i, y_i) and (x_{i+1}, y_{i+1}) has slope $m(x_i, y_i)$,

$$y_{i+1} = y_i + m(x_i, y_i)\,\Delta x_i, \tag{7.3.3}$$

where $\Delta x_i = x_{i+1} - x_i$.

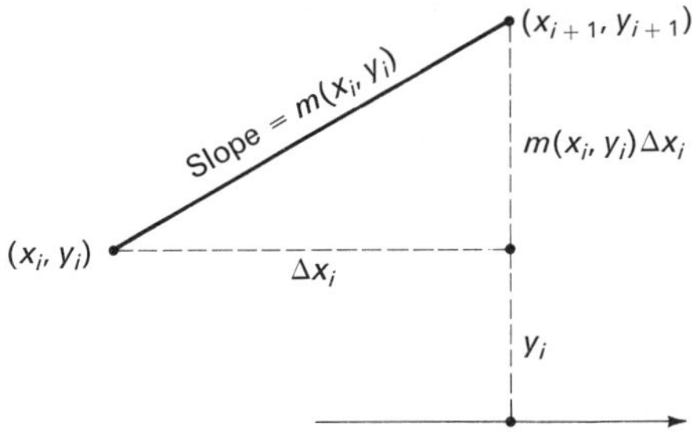

Figure 7.3.2

Thus, the calculation of the "vertices" (x_i, y_i) can be reduced to a computational routine. In fact, this routine can be turned over to a machine, and the development of the high-speed computer has made the above method and its refinements a highly practical means of obtaining precise information about solutions.

If the norm of the above partition is small, so that all of the differences Δx_i are small, then we would expect the polygonal path to be a good approximation of a solution of (7.3.1) satisfying (7.3.2). This is indeed the case, and it is possible to construct estimates of the degree of closeness of this polygon to an actual integral curve. Moreover, as you might expect, a sequence of these polygonal approximations formed from a sequence of partitions of $[a, b]$ with norms decreasing to zero can be used to construct an exact solution by a limiting process similar to that used in Chapter 3 to define a definite integral as a limit of sums. We shall not develop this process further here, but we shall illustrate with two examples.

You may recall that in earlier discussions of approximations of functions by linear functions (or approximations of curves by tangent lines) and by higher order polynomials (Taylor polynomials) we have noted that the approximations can be expected to be accurate only in the vicinity of a point where the exact value is known. This principle holds here also. The polygonal approximations mentioned above and illustrated below often diverge rapidly from the actual solution as one moves an appreciable distance from the initial point. Thus, approximation of solutions of a differential equation is a constant tug-of-war between the desire to move away from the point (x_0, y_0) and the need to use extremely fine partitions in order to keep the error within reasonable bounds. If $y(a) = y_0$ and $b - a$ is relatively large, then an enormous number of calculations may be required for a reasonably close approximation of $y(b)$, for the double reasons that $[a, b]$ is long and that Δx_i must be very small to insure accuracy. Often only a computer makes such calculations feasible.

Example 1 Consider the equation

(7.3.4) $$\frac{dy}{dx} = y$$

with the boundary condition $y(0) = 1$. We already know that the solution to this equation is $y = e^x$. Let us form a polygonal approximation on the interval $[0, 1]$ using a partition with each $\Delta x_i = 1/10$. Then, from (7.3.3) we have, for $i = 1, 2, \ldots, n$,

$$y_{i+1} = y_i + \frac{1}{10} y_i = \left(1 + \frac{1}{10}\right) y_i.$$

Thus we have

(7.3.5) $$y_0 = 1,\ y_1 = 1.1,\ y_2 = (1.1)^2,\ \ldots,\ y_n = (1.1)^n.$$

If we denote our approximation function by $\bar{y}(x)$, then we have by (7.3.5)

$$\bar{y}(0) = 1,\ \bar{y}(.1) = 1.1,\ \bar{y}(.2) = (1.1)^2,\ \ldots,\ \bar{y}(1) = (1.1)^{10},$$

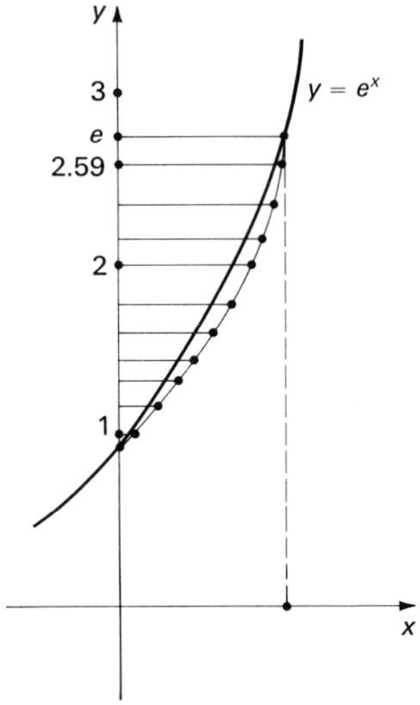

which we can compare with the values $y(x) = e^x$ of the exact solution. For example, at the last step $\bar{y}(1) = (1.1)^{10} \approx 2.593$ compared with $y(1) = e = 2.718$, a relative error of a little over 5%. See Figure 7.3.3.

Figure 7.3.3

If we were to use a finer partition with $\Delta x = 1/20$, we would then have

$$y_{i+1} = y_i + \frac{1}{20} y_i = (1.05) y_i,$$

yielding

$$y_0 = 1,\ y_1 = 1.05,\ y_2 = (1.05)^2,\ \ldots,\ y_{20} = (1.05)^{20}.$$

In this case we obtain a better approximation for $y(1) = e$, namely $(1.05)^{20} \approx 2.655$. The relative error in this approximation is about half that obtained using $\Delta x_i = 1/10$. One can show (see Exercise 7.3.5) that in general dividing Δx_i by 2 will produce about one-fourth the error at each step. However, to get to a particular value of y will now require twice as many steps, so the accumulated error in y_i will in general be about half as large. Thus the present example is more or less typical.

Example 2 Consider the differential equation

$$\frac{dy}{dx} = \frac{x}{y} \tag{7.3.6}$$

and the boundary condition

$$y(2) = 1.$$

Let us follow the above procedure to construct the first several segments of a polygonal approximation to the corresponding integral curve on an interval $[2, b]$. For each i,

$$y_{i+1} = y_i + \frac{x_i}{y_i} \Delta x_i, \quad \text{or}$$

$$y_i = y_{i-1} + \frac{x_{i-1}}{y_{i-1}} \Delta x_{i-1},\ i = 1, 2, 3, \ldots .$$

We shall again use partitions with equally spaced points. Let us consider three partitions with norms 1, $\frac{1}{2}$, $\frac{1}{4}$ respectively. The calculations for these three approximations may be done by machine. The results are summarized in a table.

i	x_i			$y_i = y_{i-1} + \frac{x_{i-1}}{y_{i-1}} \Delta_x$		
	$\Delta x = 1$	$\Delta x = \frac{1}{2}$	$\Delta x = \frac{1}{4}$	$\Delta x = 1$	$\Delta x = \frac{1}{2}$	$\Delta x = \frac{1}{4}$
0	2	2	2	1	1	1
1	3	2.5	2.25	3	2	1.5
2	4	3	2.50	4	2.63	1.88
3	5	3.5	2.75	5	3.20	2.21
4	6	4	3	6	3.74	2.52
5	7	4.5	3.25	7	4.28	2.82
			3.50			3.10

The three polygonal approximations are plotted in Figure 7.3.4. The exact solution will be found in the next section by methods we shall develop for the type of equations of which (7.3.4) and (7.3.6) are examples. In the exercises you will have a chance to construct polygonal approximations in cases where the exact solution cannot so easily be found.

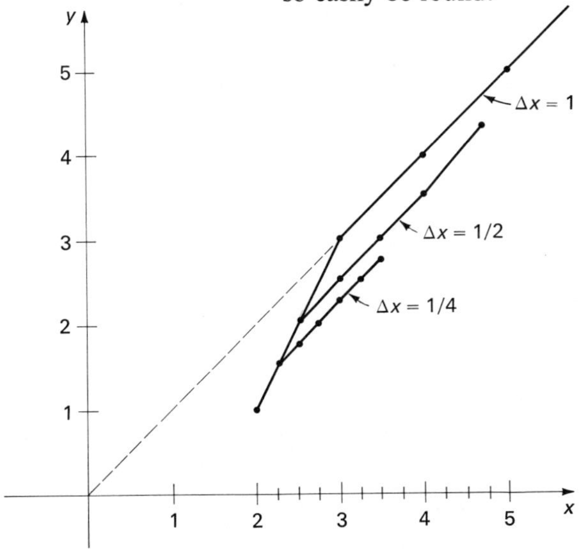

Figure 7.3.4

Exercises

1 Construct polygonal approximations to the solutions of the following differential equations using the indicated value of Δx_i.

(a) $dy/dx = 1/x$, $y(1) = 0$ over $[1, 2]$ with $\Delta x_i = 1/5$.
(b) $dy/dx = xy$, $y(0) = 1$ over $[0, 1]$ with $\Delta x_i = 1/5$.
(c) $dy/dx = 2x$, $y(0) = -1$ over $[-1, 0]$ with $\Delta x_i = -1/4$.
(d) $dy/dx = y^2$, $y(2) = 1$ over $[2, 5/2]$ with $\Delta x_i = 1/10$.
(e) $dy/dx = xy^2$, $y(-2) = 2$ over $[-2, -3/2]$ with $\Delta x_i = 1/10$.

2 Construct three polygonal approximations to the solution of

$$\frac{dy}{dx} = \tfrac{1}{2} - x + 2y, \; y(0) = 1$$

over $[0, 1]$ using $\Delta x_i = \frac{1}{2}, \frac{1}{4}$, and $\frac{1}{8}$, and compare the accuracy.

3 (a) In Example 1 show that if $\Delta x_i = 1/n$, then $\bar{y}(1) = (1 + 1/n)^n$.
(b) Hence prove that as n becomes large (and Δx_i becomes small) the approximations to $y(1) = e$ approach the actual value as a limit.

4 Suppose the solution to the differential equation

$$y' = m(x, y)$$

can be expanded in a Taylor series in an interval about $x = x_i$ which includes $x_i + \Delta x_i$.

(a) Show that

$$y(x_{i+1}) = y(x_i) + y'(x_i)\,\Delta x_i + \cdots + \frac{y^{(n)}(x_i)}{n!}(\Delta x_i)^n + \cdots.$$

(b) Which terms in the series in (a) have been neglected in our approximation of $y(x_{i+1})$?
(c) Can you suggest a procedure for approximating the solution to $y' = m(x, y)$, $y(x_0) = y_0$ which would involve, at each step, a second derivative?
(d) Use the method suggested in (c) with $\Delta x_i = 1/10$ to approximate the solution to the equation of Example 1, and compare the results.

5 (a) In view of Exercise 4(b) above, why is it reasonable to expect that replacing Δx_i by $\Delta x_i/2$ will give an estimate for $y(x_{i+1})$ with one-fourth the error?
(b) What would you expect to be the effect of replacing Δx_i by $\Delta x_i/2$ in the formula you suggested in Exercise 4(c)?

6 The methods of this section are closely related to those of Section 4.1. Since

$$\frac{dy}{dx} = m(x, y).$$

the Fundamental Theorem of Calculus says

$$y(x_{i+1}) - y(x_i) = \int_{x_i}^{x_{i+1}} m(x, y(x))\,dx.$$

Consequently, any method of approximating the integral on the right can be used to approximate $y(x_{i+1})$.

(a) Show that formula (7.3.3) corresponds to replacing $m(x, y)$ throughout the interval $[x_i, x_{i+1}]$ by its approximate value $m(x_i, y_i)$.

(b) A substantially more accurate (but more complicated) formula results from replacing $m(x, y)$ by the average of $m(x_i, y_i)$ and $m(x_{i+1}, y_i + m(x_i, y_i)\,\Delta x_i)$ throughout $[x_i, x_{i+1}]$. Derive a formula analogous to formula (7.3.3) for computing y_{i+1}.

7.4

Solutions of Some Simple First Order Differential Equations

In the first three sections we have discussed differential equations, particularly first order equations; some conditions under which solutions satisfying a boundary condition are known to exist; and a method of obtaining computable approximations of solutions. Now we shall examine two types of first order differential equations for which the solutions themselves can easily be found.

(a) Equations with Separable Variables Consider a differential equation which can be put into the following form:

(7.4.1)
$$\frac{dy}{dx} = p(x)q(y).$$

Equation (7.4.1) is said to have separable variables, because it can be written

(7.4.2)
$$\frac{1}{q(y)}\frac{dy}{dx} = p(x),$$

in which all of the y's and (y')'s appear on one side of the equation and all of the x's on the other.

If we now think of y as a function, a solution of (7.4.1), then (7.4.2) states that the two functions

$$\frac{1}{q(y(x))}y'(x) \quad \text{and} \quad p(x)$$

are identical. From what we have learned about indefinite integrals we may conclude that

(7.4.3)
$$\int \frac{1}{q(y(x))}y'(x)\,dx = \int p(x)\,dx.$$

But the first integral in (7.4.3) is precisely in a form suitable for a substitution. We may substitute simply by replacing $y(x)$ by y and $y'(x)\,dx$ by dy, treating y at that point as an independent variable.

Thus (7.4.3) becomes

(7.4.4)
$$\int \frac{1}{q(y)}\,dy = \int p(x)\,dx.$$

We may interpret the above as saying that if the variables in (7.4.1) are separated by (formal) multiplication by dx and division by $q(y)$ to yield

(7.4.5)
$$\frac{1}{q(y)}\,dy = p(x)\,dx,$$

then the equation may be "solved" by formal integration of each side of (7.4.5). The resulting equation in x and y may be solved algebraically for y, if the solution (a function, remember) $y(x)$ is desired. Since an expression of the form (7.4.4) involves a "constant of integration," the expression for $y(x)$ found by this method will contain that arbitrary constant. By carefully choosing a value for this constant, the particular solution $y(x)$ which satisfies a given boundary condition can usually be found.

Example 1 Consider the equation

(7.4.6)
$$\frac{dy}{dx} = ky.$$

We may write this as

$$\frac{1}{y}\frac{dy}{dx} = k,$$

and integrating we have

$$\int \frac{1}{y}\,dy = \int k\,dx,$$

or

$$\log y = kx + C.$$

This yields

$$y = e^{kx+C} = Ae^{kx},$$

where $A = e^C$.

We first encountered this differential equation in Chapter 7 of *C 1*. The method we used there to solve it strongly anticipated the technique we are discussing here.

Example 2 Consider the equation in Example 2 of Section 7.3,

(7.4.7)
$$\frac{dy}{dx} = \frac{x}{y}\cdot$$

We may rewrite this as

$$y\,dy = x\,dx,$$

and integrating we have

(7.4.8) $$\frac{y^2}{2} = \frac{x^2}{2} + C.$$

If we let $k = 2C$, then (7.4.8) may be written

(7.4.9) $$y^2 - x^2 = k.$$

In (7.4.9) we have the equation of a whole family of curves, one for each value of k. These are the integral curves of the differential equation (7.4.7), or of the corresponding direction field. We can find solutions (functions) of (7.4.7) by solving (7.4.9) for y in terms of x. Thus,

(7.4.10) $$y = \pm\sqrt{x^2 + k}.$$

The graphs of (7.4.9) or (7.4.10) are a family of pieces of hyperbolas, each having foci on one of the axes, and asymptotes the

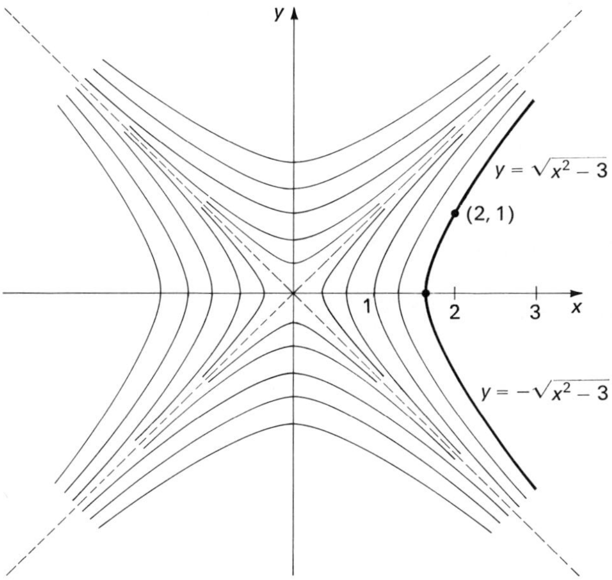

Figure 7.4.1

lines $y = \pm x$. (The last property, in case you have forgotten, means that the integral curves all approach the line $y = x$ or the line $y = -x$ as x tends either to $+\infty$ or $-\infty$.) A few of them are drawn in Figure 7.4.1. The particular solution which satisfies the boundary condition

$$y(2) = 1$$

corresponds to the value $k = -3$. For, if we set $x = 2$, $y = 1$ in (7.4.10), we obtain $1 = \sqrt{4 + k}$ or $1 = 4 + k$. Thus, $k = -3$. This solution is

(7.4.11) $$y = \sqrt{x^2 - 3}.$$

You should do two things at this point. (1) You should check by differentiation and algebraic manipulation that the function $\sqrt{x^2 - 1}$ *is* a solution of (7.4.7), and (2) you should compare the graph of (7.4.11) (see Figure 7.4.1) with the polygonal approximations found in Example 2 of Section 7.3 and pictured in Figure 7.3.4.

Example 3 Consider the differential equation

$$\frac{dy}{dx} = \frac{y}{\sqrt{1 - x^2}}.$$

We may write this as

$$\frac{dy}{y} = \frac{dx}{\sqrt{1 - x^2}}.$$

Thus, integrating, we have

$$\log y = \arcsin x + C,$$

or

$$y = Ae^{\arcsin x}, \ (A = e^C).$$

(*b*) ***First Order Linear Equations*** A differential equation which can be expressed entirely in terms of a sum of the expressions $1, y, y', y'', \ldots,$ each multiplied by a function of x, is called a *linear differential equation.* Hence a first order linear differential equation is one of the form

(7.4.12) $$y' + a(x)y = f(x),$$

where we shall assume that the functions $a(x), f(x)$ are continuous on an interval I. Such an equation is easily solved, because of the fortunate accident that the left side is easily seen to be "almost" the derivative of a recognizable function. To see this we select an episode from our recent experience which fits the pattern suggested by $y' + a(x)y$. Suppose that $A(x)$ is any antiderivative of $a(x)$, that is, $A(x) = \int a(x)\,dx$. Then we know that

$$\frac{d}{dx}e^{A(x)} = \frac{dA(x)}{dx}e^{A(x)} = a(x)e^{A(x)}.$$

No pattern resembling the differential equation is apparent yet, but if we consider the product $ye^{A(x)}$, where y is a function, then we have

(7.4.13) $$\begin{aligned}\frac{d}{dx}ye^{A(x)} &= y'e^{A(x)} + ya(x)e^{A(x)} \\ &= e^{A(x)}(y' + a(x)y).\end{aligned}$$

Here, part of (7.4.12) appears. By magic? No, for (7.4.13) is a very common derivative which a person with any appreciable experience with the exponential function could hardly have failed to encounter. Therefore, it is not at all farfetched to expect that a person interested in the differential equation (7.4.12) and willing to experiment a bit would rather quickly pull (7.4.13) out of his past experience.

In (7.4.13) we have the discovery that if the equation (7.4.12) is multiplied by the factor $e^{A(x)}$, then the left side becomes the derivative of $ye^{A(x)}$. This will enable us to find a solution easily, and for this reason $e^{A(x)}$ is often called an *integrating factor* for the differential equation. Multiplying by this factor, we change (7.4.12) to $y'e^{A(x)} + ya(x)e^{A(x)} = f(x)e^{A(x)}$, or by (7.4.13),

$$\frac{d}{dx}[ye^{A(x)}] = f(x)e^{A(x)}.$$

Integrating, we have

$$ye^{A(x)} = \int f(x)e^{A(x)}\,dx,$$

which can easily be solved for y, yielding

(7.4.14) $$y = e^{-A(x)}\int f(x)e^{A(x)}\,dx,$$

where, as we recall, $A(x) = \int a(x)\,dx$.

Thus, (7.4.14) gives explicitly a solution of (7.4.12). Since the indefinite integral $\int f(x)e^{A(x)}\,dx$ involves an arbitrary constant, a solution of this form may be found which satisfies any given initial condition of the form $y(x_0) = y_0$.

Example 4 Equation (7.4.6) is linear, for it may be written

$$y' - ky = 0.$$

Thus, using (7.4.14) with $f(x) = 0$, $a(x) = -k$, $A(x) = -kx$, we have a solution

(7.4.15) $$y = e^{kx}\int 0\,dx = Ce^{kx},$$

for, as we recall, $\int 0\,dx$ can be any constant function C. We have here the same family of solutions we found before for this differential equation.

In (7.4.14) we make use of one antiderivative of $a(x)$. What happens if two people using this formula happen to select different antiderivatives? One might use $A(x)$, the other $A(x) + K$, K a constant. In the latter case, (7.4.14) becomes

$$y = e^{-(A(x)+K)}\int f(x)e^{A(x)+K}\,dx$$

$$= e^{-(A(x)+K)}e^{K}\int f(x)e^{A(x)}\,dx$$

$$= e^{-A(x)}\int f(x)e^{A(x)}\,dx.$$

So both of these antiderivatives of $a(x)$ will produce the same function y. Thus, the function y in (7.4.14) is not affected by the choice of a representative for $\int a(x)\,dx$ as long as the same one is used each time that $e^{A(x)}$ occurs in the formula. The formula (7.4.14) can, of course, also be written

$$y = e^{-\int a(x)dx}\int f(x)e^{\int a(x)dx}\,dx,$$

which looks quite complicated but has the advantage of showing explicitly what operations are to be performed to find a solution of (7.4.12).

Example 5 Consider the equation

(7.4.16) $$y' + \frac{1}{x}y = \sin x.$$

Then $a(x) = 1/x$, so $\int a(x)\,dx = \log x$. Thus the solutions of (7.4.16) are

$$y = e^{-\log x}\int e^{\log x}\sin x\,dx.$$

Since $e^{\log x} = x$, we have

$$y = \frac{1}{x}\int x\sin x\,dx.$$

An integration by parts yields

$$y = \frac{1}{x}[-x\cos x + \sin x + C],$$

$$y = -\cos x + \frac{1}{x}\sin x + \frac{C}{x}\cdot$$

Thus, the solution of (7.4.16) which satisfies the boundary condition $y(\pi) = 0$ is

$$y = -\cos x + \frac{1}{x}\sin x - \frac{\pi}{x}\cdot$$

Exercises

Solve the following differential equations.

1 $x(3y - 2) + (x^2 + 1)\dfrac{dy}{dx} = 0.$

2 $(2x + xy)\dfrac{dy}{dx} + y = 0.$

3 $\dfrac{dy}{dx} + y = e^{2x}$.

4 $\dfrac{dy}{dx} = e^{2x-y}$.

5 $3y - xy + (x^2 + 4x)y' = 0$.

6 $e^{x/2} + y - 2y' = 0$.

7 $y(3x^2 + 7)^{1/2}\, dx + (4 - 5y^2)^{1/2}\, dy = 0$.

8 $(y \log x)\, dx + \dfrac{y-2}{x}\, dy = 0$.

9 $x\, dy + (3y - 6x^3)\, dx = 0$.

10 $x\dfrac{dy}{dx} + y = \sin x$.

11 $xe^y\, dy + \dfrac{1-x^2}{y}\, dx = 0$.

12 $(3xy - 4y - 3x)\, dx + (x^2 - 3x + 2)\, dy = 0$.

13 $2 \cos x\, (y - 3 \sin x)\, dx + \sin x\, dy = 0$.

14 $(x - 2y)\, dy + y\, dx = 0$. (*Hint:* Consider y as the independent variable.)

15 $(y^5 + 3x)\, dy - y\, dx = 0$.

16 Show that the substitution $y = vx$ reduces the equation

$$(x - 2y)\, dx + (2x + y)\, dy = 0$$

to an equation in which the variables x and v are separable. Solve the equation.

17 Apply the method suggested in Exercise 16 to the equation

$$2(2x^2 + y^2)\, dx - xy\, dy = 0.$$

7.5 Some Applications

We live in a universe crowded with moving objects. Every moment of our lives we are affected by changing quantities. Since the derivative is the basic vocabulary of the language of change, it is not surprising that differential equations should figure prominently in man's efforts to describe his physical and social surroundings. The most significant applications tend to involve differential equations of order two or higher. This can be predicted, for example, from the observation that the most frequently available information about a moving object involves its acceleration, a second derivative. Thus, in being restricted to equations of order one, we are somewhat limited in our choice of significant applications. Nevertheless, there are a few interesting applications of first order differential equations; in this section we shall examine several.

Example 1 A wide range of problems involving dilution of liquid solutions and mixtures and rates of chemical reactions lead to first order differential equations. Here is a typical one.

Suppose a 200,000-liter reservoir is initially one-quarter full of water. At a certain time, water contaminated with 2 milligrams of DDT per liter begins to flow into the reservoir from neighboring fields at a rate of 5,000 liters per hour. The liquid in the reservoir is kept perfectly mixed and the increasingly polluted mixture is pumped out at a rate of 3,000 liters per minute. What will be the concentration of DDT in the reservoir when it becomes full?

If we denote by $x(t)$ the amount of DDT in the reservoir at time t and suppose that the contaminated flow begins at time $t = 0$, then we have

$$x(0) = 0 \quad \text{and} \quad x'(t) = 10{,}000 - 3{,}000\,\frac{x(t)}{V(t)},$$

where $V(t)$ is the volume of water in the tank at time t. Clearly, $V(t) = 50{,}000 + 2{,}000t$, since the net flow of the liquid is $+2{,}000$ gallons per minute. Thus, $x(t)$ is a solution of the differential equation

(7.5.1)
$$x' + \frac{3}{2}\,\frac{x}{25 + t} = 10{,}000,$$

a linear first order equation.

We know what the solutions of (7.5.1) are like. Using (7.4.14) and noting that

$$\int \frac{3}{2(25 + t)}\,dt = \frac{3}{2}\log(25 + t),$$

we have

$$\begin{aligned} x(t) &= (25 + t)^{-3/2}\int 10^4(25 + t)^{3/2}\,dt \\ &= (25 + t)^{-3/2}[4 \cdot 10^3(25 + t)^{5/2} + C]. \end{aligned}$$

But $x(0) = 0$, so $C = -4 \cdot 5^5 \cdot 10^3$.

That is,

$$x(t) = 4 \cdot 10^3(25 + t)^{-3/2}[(25 + t)^{5/2} - 5^5].$$

The reservoir will be filled when $t = 75$, and at this time the amount of DDT contained in the reservoir will be

$$\begin{aligned} x(75) &= 4 \cdot 10^3(100)^{-3/2}(100^{5/2} - 5^5) = 4(10^5 - 5^5), \\ x(75) &= 387{,}500 \text{ mg.} \end{aligned}$$

The concentration is then 1.9375 mg. of DDT per liter of water.

Example 2

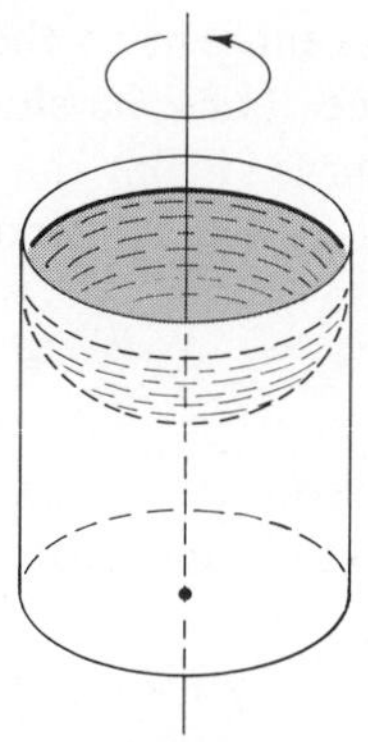

Figure 7.5.1

A cylindrical can with vertical axis containing some water is rotated at a constant angular velocity about its axis. After some time the water in the can becomes stable relative to the can, with the surface of the water taking the shape of a curved surface of revolution about the axis. We can see that this surface is a paraboloid, the surface generated by revolving a parabola about this axis, reasoning as follows: Consider a vertical cross section through the axis. Each molecule of water at the surface is acted upon by two forces, its weight and the net push of the rest of the water on it. The physicist knows that the water will be in equilibrium relative to the revolving can when and only when the second of these forces is normal to the surface at the point, i.e., when the force vector $\mathbf{F}_w$ is perpendicular to the tangent to the curve in Figure 7.5.2. The other force vector, $\mathbf{F}_{grav}$, has magnitude mg, where m is the mass of the water molecule and g is the acceleration due to gravity.

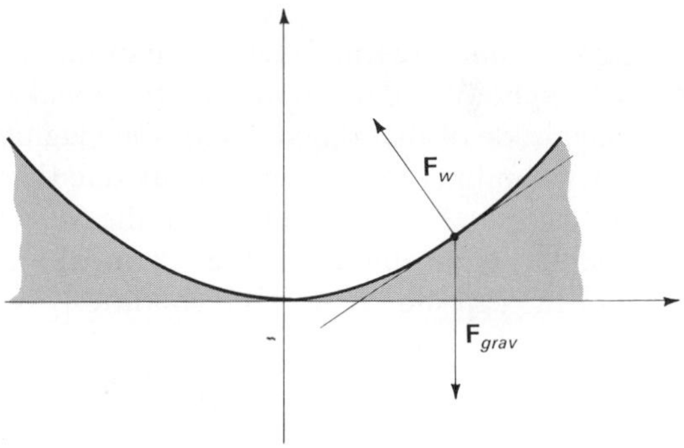

Figure 7.5.2

Since this particle is moving in a horizontal circle of radius x with constant angular speed ω, we know from an earlier discussion (see Section 6.8) that its acceleration is horizontal, directed toward the axis, and has magnitude $\omega^2 x$. Thus, the net force $\mathbf{F}_{net}$, also horizontal and directed toward the axis, has magnitude

$$m\omega^2 x.^1$$

The force vectors are as illustrated in Figure 7.5.3.

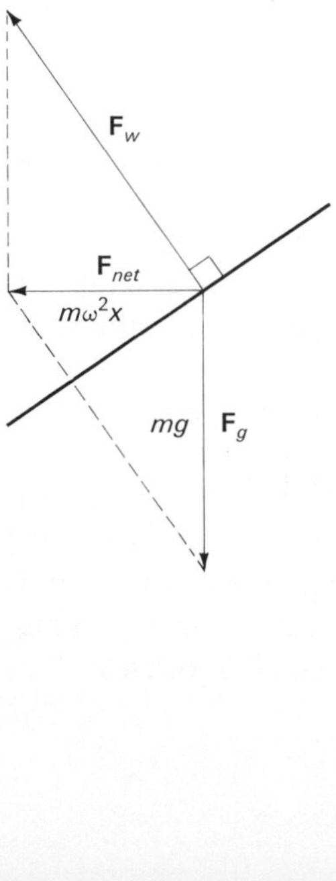

Figure 7.5.3

[1]Recall that by Newton's Second Law of Motion the net force equals the mass times the acceleration.

The heavy line is tangent to the curve which is the cross section of the water surface. Thus, its slope is $f'(x)$, where $y = f(x)$ is the equation of this curve. From the drawing we see that this slope is $m\omega^2 x/mg$. Thus, $f(x)$ is a solution of the differential equation

$$\frac{dy}{dx} = \frac{\omega^2}{g} x.$$

This is quite easily solved. If we assume that the origin is at the low point of the surface, so that $f(0) = 0$, then

$$y = f(x) = \frac{\omega^2}{2g} x^2,$$

the equation of a parabola with vertex at the origin and vertical axis. Since this curve sweeps out the surface of the water as it is rotated about the axis of the can, we see that the surface is indeed a paraboloid.

Example 3 Consider a space capsule falling through the earth's atmosphere after reentry. We shall assume that it is falling vertically, having entered the earth's atmosphere approximately in a radial direction. This is perhaps more realistic for a space vehicle returning from the moon than for one dropping out of an earth orbit. However, a nonvertical reentry represents a succeeding stage of sophistication beyond our present level. Its motion is governed by two forces, its weight **W** and the retarding frictional force **F** due to the air. The magnitude of the weight is mg; the magnitude of **F** is a very complicated combination of factors related to the density of the atmosphere, the shape of the capsule, and the velocity of the capsule. If we denote the magnitude of the frictional force by F, then the velocity v of the capsule satisfies the condition

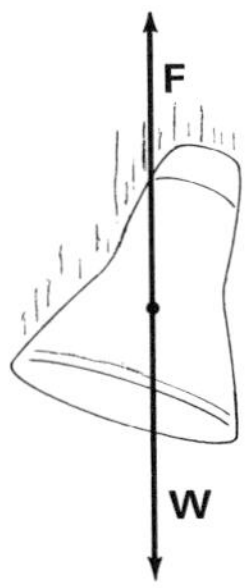

Figure 7.5.4

(7.5.2) $$m \frac{dv}{dt} = mg - F,$$

for $mg - F$ is the net force on the capsule and dv/dt is the acceleration.

The condition (7.5.2) can lead to any number of differential equations depending upon the nature of the simplifying assumption which we make about F and g. Most of them are far too complicated for us to handle here.

One set of assumptions yields an equation which is well within our experience. Moreover, the assumptions are reasonably close to experimental fact. Let us suppose that the motion of the capsule is confined to a region small enough so that the gravitational force mg is constant. Moreover, let us assume that the frictional force is proportional to the speed v of the capsule. This involves, among other things, the assumption that the region is small enough so that the atmospheric pressure does not vary appreciably within it. Then $F = Kv$, where K is a constant, and we have

$$\frac{dv}{dt} + \frac{K}{m}v = g.$$

This is a linear equation whose solution is

$$v = e^{-(K/m)t}\left[\int e^{(K/m)t}g\,dt + C\right],$$

or

$$v = Ce^{-(K/m)t} + \frac{mg}{K}.$$

Note that as t becomes large, v approaches mg/K, the so-called *terminal velocity* of the capsule. This is approximately how fast the capsule will be traveling when it strikes the earth.

Example 4 We conclude by returning to the problem of the public health officer mentioned in the introduction to this chapter. If his epidemic is proceeding according to schedule, then the epidemic function $x(t)$ = the number of infected people at time t obeys the law

(7.5.3)
$$\frac{dx}{dt} = Kx(P - x).$$

This is a differential equation whose variables can be separated. It can be written

(7.5.4)
$$\frac{dx}{x(P - x)} = K\,dt.$$

Integrating (7.5.4), after noting that

$$\frac{1}{x(P - x)} = \frac{1}{P}\left(\frac{1}{x} + \frac{1}{P - x}\right),$$

we have

$$\frac{1}{P}\log\frac{x}{P - x} = Kt + C,$$

and by an application of the exponential function,

(7.5.5)
$$\left(\frac{x}{P - x}\right)^{1/P} = e^{Kt+C} = e^{C}e^{Kt}.$$

Raising both sides of (7.5.5) to the Pth power we obtain

$$\frac{x}{P - x} = e^{PC}e^{PKt},$$

and letting $e^{PC} = C_1$ we have

(7.5.6)
$$\frac{x}{P - x} = C_1e^{PKt}.$$

Equation (7.5.6) can be solved for x:

(7.5.7) $$x = P\frac{C_1e^{PKt}}{1 + C_1e^{PKt}}.$$

Equation (7.5.7) is the health officer's prediction of the course of the epidemic. Note that $e^{PKt} \to +\infty$ as $t \to +\infty$ and that the factor $[C_1e^{PKt}/(1 + C_1e^{PKt})]$ in (7.5.7) accordingly approaches 1 as a limit. Thus, as one would expect, $x(t) \to P$ as $t \to \infty$.

Most viruses appear in cycles, so that few epidemics actually follow a path as described by (7.5.3). Many complicated factors cause a virus such as the measles virus to lose virulence with time, so that an epidemic gradually passes even though not all susceptible people have been infected. Thus, a more realistic model might include, in place of K in (7.5.3), a function, such as $e^{-\lambda t}$, $\lambda > 0$, which decays as time goes by. In Exercise 7.5.13 you will have an opportunity to pursue this model to see, for example, how many people will get the measles.

Exercises

1 What is the most general function which is equal to the square of its derivative?

2 What is the most general function which is equal to the cube of its derivative?

3 The function sin has the property that its square plus the square of its derivative equals 1. What is the most general function with this property? How does one obtain cos from its formula?

4 Find all curves such that the slope of the tangent at any point (x, y) is one-half the slope of the line from the origin to that point.

5 A particle is moving along a coordinate axis. At time $t = 0$ it is located at the origin and has a velocity in the direction of the point $P(6)$ of 30 units/min. Find an equation for the motion of the particle if

(a) its velocity is proportional to its distance from P,

(b) its velocity is proportional to the square of its distance from P,

(c) its velocity is proportional to its distance from $Q(10)$.

6 In which of the parts of Exercise 5 will the particle ever reach the point P?

7 Suppose that a 100 gallon tank is full of water. At a certain instant a brine solution containing 2 lbs. of salt per gallon begins to flow into the tank at the rate of 5 gals./min., and the perfectly mixed solution flows out at the same rate.

(a) Find the amount of salt in the tank at any time.

(b) By what time will the tank contain 10 lbs. of salt?

8 (a) Find the limiting value for the amount of salt in the tank of Exercise 7 after a long time. Why does this seem reasonable?

(b) How long will be required for the amount of salt in the tank to reach 80 per cent of this limiting value?

9 Tank A contains 100 gallons of water. Tank B contains 100 gallons of a brine solution containing 2 lbs. of salt per gallon. Both tanks are kept stirred so that their contents are uniform. The contents of tank A are pumped into tank B at the rate of 5 gal./min. and the contents of B are pumped into A at the same rate. Find the amount of salt in each tank at any time. What is the concentration of salt in each tank after a long time?

10 Consider the space capsule of Example 3. In certain situations it is more reasonable to assume that the magnitude F_f of the frictional force is proportional to the square of the velocity.

(a) Show that this leads to the differential equation

$$\frac{dv}{dt} + \frac{k}{m}v^2 = g.$$

(b) Solve this equation by separating the variables. (The notation is somewhat simplified by setting $a = \sqrt{w/k}$.)

11 A projectile is fired upward from the earth's surface at an initial velocity v_0. Assume that the only force acting on the projectile is gravity.

(a) Show that if $a = d^2r/dt^2$ is the (negative) acceleration due to gravity, then

$$a = -g\frac{R^2}{r^2},$$

where R is the radius of the earth, r is the distance from the projectile to the center of the earth, and g is the acceleration due to gravity on the surface of the earth.

(b) Show that

$$a = v\frac{dv}{dr},$$

so that

$$v\frac{dv}{dr} = -g\frac{R^2}{r_2},$$

and solve the equation. (The constant can be evaluated by considering v_0.)

(c) How large must v_0 be so that the projectile will continue upward without returning to earth? The smallest such velocity is called the *escape velocity* for the earth.

12 An insecticide applied to a colony of insects produces an instantaneous death rate of 200 per 1,000 per hour. What fraction of the colony remains at the end of 1 hour?

13 Suppose the measles virus causing the epidemic in Example 4 has a virulence factor $Ke^{-\lambda t}$. That is, the epidemic equation is $dx/dt = Ke^{-\lambda t}x(P - x)$. Find x as a function of t and show that as $t \to \infty$ $x \to (c/(1 + c))P$, where $c = e^{PK/\lambda}x(0)/(P - x(0))$.

7.6

Solutions by Power Series

We conclude with a brief discussion of one of the most useful and powerful methods of finding solutions, or approximations of solutions, of certain types of differential equations. The method derives from the fact that as we pointed out in Chapter 5, many functions can be represented as sums of convergent power series. We call such functions *analytic* functions.

Many differential equations are known to have solutions which are analytic. The most notable of such differential equations are the *linear* equations of all orders. A linear differential equation is one of the form

(7.6.1) $$a_n(x)y^{(n)} + a_{n-1}(x)y^{(n-1)} + \cdots + a_0(x)y = b(x),$$

where $a_n(x) \neq 0$. If the coefficient functions $a_i(x)$ and $b(x)$ are analytic, then it can be shown that all solutions of (7.6.1) are analytic. A general method of trying to find such a solution is the following: Replace each coefficient function by its series representation; assume that the desired solution has the form

(7.6.2) $$y = U_0 + U_1x + U_2x^2 + \cdots.$$

Then, as we know, the various derivatives $y^{(m)}$, $1 \leq m \leq n$, can also be written as power series:

$$y' = U_1 + 2U_2x + 3U_3x^3 + \cdots = \sum_{i=1}^{\infty} iU_ix^{i-1},$$

$$y'' = 2U_2 + 3 \cdot 2U_3x + 4 \cdot 3U_4x^2 + \cdots = \sum_{i=2}^{\infty} i(i-1)U_ix^{i-2}, \quad \ldots, \text{etc.}$$

All of these series for y and its derivatives and for the functions $a_i(x)$, $b(x)$ are then substituted into (7.6.1). In general, this will involve adding power series together and also multiplying power series. In more advanced treatments of power series it is shown that these operations are possible and can be carried out in the expected ways. If two power series are convergent, then they may be added together by simply adding the terms in like powers of x and summing: If $S = u_0 + u_1x + u_2x^2 + \cdots$ and $T = v_0 + v_1x + v_2x^2 + \cdots$, then $S + T = (u_0 + v_0) + (u_1 + v_1)x + (u_2 + v_2)x^2 + \cdots$. These series may also be multiplied in the obvious way, by forming all possible products of terms and collecting like powers of x:

$$S \cdot T = u_0v_0 + (u_0v_1 + u_1v_0)x + (u_0v_2 + u_1v_1 + u_2v_0)x^2 + \cdots.$$

When the substitution has been made and all of the indicated operations completed, then the entire expression (7.6.1) takes the form of a single series equation

(7.6.3) $$A_0 + A_1x + A_2x^2 + \cdots = 0.$$

Here, each coefficient A_i is itself an algebraic combination of some of the coefficients U_i from (7.6.2), involving also some of the (known) coefficients of the power series for the $a_i(x)$ and $b(x)$.

The power series representations which are being employed here can be shown to be valid in some interval $(-r, r)$. We shall assume that all are valid for the same interval. Then (7.6.3) holds also for all x in $(-r, r)$. Thus (7.6.3) represents a function which is identically zero on $(-r, r)$. In Exercise 7.6.2 you will have an opportunity to show that a power series whose sum is zero everywhere on an interval containing 0 must be the trivial series, the series *all* of whose coefficients are zero.

Thus, from (7.6.3) we conclude that

(7.6.4) $$A_i = 0 \quad \text{for } i = 0, 1, 2, \ldots.$$

Since each A_i is an expression involving certain of the coefficients U_i from (7.6.2), the conditions (7.6.4) become a sequence of *recurrence relations* which can usually be used to solve for the constants $U_0, U_1, U_2, \ldots$ successively. Once the constants U_i have been found, the series (7.6.2) is known, and hence a solution of (7.6.1) has been found.

Actually, we have been guilty of one misstatement of fact. The relations (7.6.4) do not, by themselves, determine all of the coefficients U_i. An analogous situation is seen in the lack of a unique solution for two linear algebraic equations in three variables. The value of one of the variables can generally be chosen at random, and the pair of equations then determine the values of the other two variables. Here, the recurrence relations (7.6.4) will allow values of n of the coefficients $U_0, U_1, U_2, \ldots$ to be chosen at random. The relations (7.6.4) will then uniquely determine the values of the remaining coefficients.

The last observation corresponds to the statement that the differential equation (7.6.1) has solutions which satisfy any desired set of initial conditions of the form

$$\begin{aligned} y(0) &= y_0, \\ y'(0) &= y_1, \\ y''(0) &= y_2, \\ &\ldots \\ y^{(n-1)}(0) &= y_{n-1}. \end{aligned}$$

These initial conditions and the differential equation uniquely determine the solution. If $y = \sum_{i=0}^{\infty} U_i x^i$, then $y(0) = U_0, y'(0) = U_1$, $y''(0) = 2U_2$, and in general $y^{(m)}(0) = m!\,U_m$. We have already observed that these U_i and hence the values of $y^{(i)}(0)$ can be arbitrarily chosen and then the remaining coefficients and hence the solution y are uniquely determined.

In the above discussion we have made a large number of claims with little or no proof. To verify all of these claims would require a deeper familiarity with the behavior of power series than we now have. Our purpose here is simply to introduce you to the possibility of finding solutions by power series methods and to let you explore the appropriate manipulations through several elementary exercises. We illustrate here with two examples in which power series can be used to find some solutions encountered before.

Example 1 The most common use of power series methods in this context is in finding solutions of linear differential equations with constant coefficients. Consider the general first order case:

(7.6.5)
$$y' + ay = b.$$

We seek a solution which can be expressed as a power series (7.6.2). Differentiating, we have

(7.6.6)
$$\begin{aligned} y &= \sum_{i=0}^{\infty} U_i x^i = U_0 + U_1 + U_2 x^2 + \cdots, \\ y' &= \sum_{i=1}^{\infty} i U_i x^i = U_1 + 2U_2 x + 3U_3 x^2 + \cdots, \end{aligned}$$

and substituting (7.6.6) in (7.6.5) we have

(7.6.7)
$$(U_1 + 2U_2 x + 3U_3 x^2 + \cdots) + (aU_0 + aU_1 x + aU_2 x^2 + \cdots) - b = 0.$$

Within their regions of convergence, power series can be added "termwise." That is, the expression (7.6.7) can be rewritten as

(7.6.8)
$$(U_1 + aU_0 - b) + (2U_2 + aU_1)x + (3U_3 + aU_2)x^2 + \cdots = 0.$$

The power series in (7.6.8) is identically zero, so each of its coefficients must be zero. That is,

(7.6.9)
$$\begin{aligned} U_1 + aU_0 - b &= 0, \\ 2U_2 + aU_1 &= 0, \\ 3U_3 + aU_2 &= 0, \\ &\cdots \\ iU_i + aU_{i-1} &= 0 \quad \text{for all } i. \end{aligned}$$

Let us here consider U_1 to be an arbitrary number and use (7.6.9) to find the remaining coefficients. It will be clear shortly why we prefer to let U_1 (rather than U_0) be arbitrary and solve for U_0. From the first of these conditions we have

(7.6.10)
$$U_0 = \frac{b}{a} - \frac{U_1}{a},$$

and from the remaining conditions we find that

(7.6.11) $$U_2 = -\frac{a}{2}U_1,\ U_3 = \frac{a^2}{3!}U_1,\ \ldots,\ U_i = (-1)^{i+1}\frac{a^{i-1}}{i!}U_1.$$

From (7.6.10) and (7.6.11) we see that a solution is

(7.6.12) $$\begin{aligned} y &= \frac{b}{a} - \frac{U_1}{a} + U_1x - U_1\frac{a}{2}x^2 + U_1\frac{a^2}{3!}x^3 - \cdots \\ &= \frac{b}{a} - \frac{U_1}{a}\left[1 - ax + \frac{a^2x^2}{2!} - \frac{a^3x^3}{3!} + \cdots\right].\end{aligned}$$

The series in brackets in (7.6.12) is a standard power series which we have encountered before. It converges to e^{-ax}. Thus, we have found that (7.6.5) has a solution (we let $C = U_1/a$)

(7.6.13) $$y = \frac{b}{a} + Ce^{-ax}.$$

You should compare (7.6.13) with the expressions (7.4.14), (7.4.15) of our earlier discussion of linear first order equations.

Example 2 In Chapter 5 of *C 1* we used the words *simple harmonic motion* to describe any straight line motion for which the acceleration is proportional to, but oppositely directed from, the displacement from an equilibrium point. If that equilibrium point is selected as the origin of a coordinate system on the line, then the definition says precisely that the position function $y(t)$ of a point moving with simple harmonic motion is a solution of the differential equation

(7.6.14) $$y'' + a^2y = 0.$$

We first encountered this situation in a discussion of some particular motions which turned out, after some investigation, to be simple harmonic. But now let us take the opposite point of view. Given the differential equation (7.6.14) which *defines* simple harmonic motion, what can we learn about the precise nature of this kind of motion?

One way to approach this problem is to look explicitly for solutions of (7.6.14)—one method is to look for a series solution. If

$$y = U_0 + U_1x + U_2x^2 + \cdots + U_nx^n + \cdots,$$

then

$$y'' = 2U_2 + 3\cdot 2U_3x + 4\cdot 3U_4x^2 + \cdots + n(n-1)U_nx^{n-2} + \cdots.$$

Thus, by substituting into (7.6.14), we have

$$\begin{aligned}(2U_2 + 3\cdot 2U_3x + \cdots + (n+2)(n+1)U_{n+2}x^n + \cdots)& \\ + (a^2U_0 + a^2U_1x + \cdots + a^2U_nx^n + \cdots) &= 0,\end{aligned}$$

which can be written

(7.6.15)
$$(2U_2 + a^2U_0) + (3 \cdot 2U_3 + a^2U_1)x + \cdots + ((n+2)(n+1)U_{n+2} + a^2U_n)x^n + \cdots = 0.$$

The series statement (7.6.15) yields, when the coefficients are all set equal to zero, the recurrence relations

(7.6.16)
$$\begin{aligned} 2U_2 + a^2U_0 &= 0, \\ 3 \cdot 2U_3 + a^2U_1 &= 0, \\ &\cdots \\ (n+2)(n+1)U_{n+2} + a^2U_n &= 0, \\ &\cdots. \end{aligned}$$

Since the conditions (7.6.16) relate each coefficient U_{n+2} to U_n, it is easy to see that U_0 and U_1 can be chosen arbitrarily and the rest will be determined. We have

$$\begin{aligned} &U_0, && U_1, \\ &U_2 = -\frac{a^2}{2!}U_0, && U_3 = -\frac{a^2}{3!}U_1, \\ &U_4 = \frac{a^4}{4!}U_0, && U_5 = \frac{a^4}{5!}U_1, \\ &U_6 = -\frac{a^6}{6!}U_0, && U_7 = -\frac{a^6}{7!}U_1, \quad \text{etc.} \end{aligned}$$

This gives the solution

$$y = U_0 + U_1t - \frac{a^2}{2!}U_0t^2 - \frac{a^2}{3!}U_1t^3 + \frac{a^4}{4!}U_0t^4 + \frac{a^4}{5!}U_1t^5 - \cdots.$$

Convergent power series can also be rearranged and summed in a different order without affecting their sums. Thus the solution can be rewritten

(7.6.17)
$$y = U_0\left[1 - \frac{a^2}{2!}t^2 + \frac{a^4}{4!}t^4 - \cdots\right] + U_1\left[t - \frac{a^2}{3!}t^3 + \frac{a^4}{5!}t^5 - \cdots\right], \quad \text{or}$$

$$y = U_0\left[1 - \frac{a^2}{2!}t^2 + \frac{a^4}{4!}t^4 - \cdots\right] + \frac{U_1}{a}\left[at - \frac{(at)^3}{3!} + \frac{(at)^5}{5!} - \cdots\right].$$

The two series in brackets in (7.6.17) are also standard series whose sums we recognize from Chapter 5. By (5.6.5) and (5.6.6), (7.6.17) can be written

$$y = U_0 \cos(at) + \frac{U_1}{a} \sin(at),$$

and since U_0 and U_1 are arbitrary, we may write

(7.6.18) $$y = C_1 \sin(at) + C_2 \cos(at),$$

where C_1 and C_2 are arbitrary constants. Observe that the two constants $C_1/(C_1^2 + C_2^2)^{1/2}$ and $C_2/(C_1^2 + C_2^2)^{1/2}$ are the cosine and sine respectively of an angle

$$\theta = \arccos \frac{C_1}{\sqrt{C_1^2 + C_2^2}} \cdot$$

If we let

$$A = (C_1^2 + C_2^2)^{1/2},$$

then (7.6.18) can be written

(7.6.19) $$\begin{aligned} y &= A(\cos\theta \sin(at) + \sin\theta \cos(at)), \\ y &= A \sin(\theta + at). \end{aligned}$$

A motion governed by an equation of type (7.6.19) is, for obvious reasons, called a *sinusoidal* motion. We have always known that a sinusoidal motion is simple harmonic. We need only differentiate twice to verify that the function in (7.6.19) is a solution of (7.6.14). What we have shown here is a much deeper result: that *any* simple harmonic motion is sinusoidal. Here we have an indication of the power of power series methods.

Example 3 The amount of writing necessary to solve a differential equation can often be reduced by careful use of the $\sum$ notation for a series. Consider

(7.6.20) $$y'' - 2y' + y = 0.$$

If

$$y = \sum_{i=0}^{\infty} U_i x^i,$$

then

$$y' = \sum_{i=1}^{\infty} i U_i x^{i-1},$$

$$y'' = \sum_{i=2}^{\infty} i(i-1) U_i x^{i-2}.$$

It will facilitate the algebraic manipulation of the series if the series for y, y', y'' are all written in the form $\sum(\ \ )x^i$. This requires some relabeling of the terms of the series for y' and y''. By writing out several terms you should check that

$$y = \sum_{i=0}^{\infty} U_i x^i,$$

$$y' = \sum_{i=0}^{\infty} (i+1)U_{i+1}x^i,$$

$$y'' = \sum_{i=0}^{\infty} (i+2)(i+1)U_{i+2}x^i.$$

Now substitute into (7.6.20) to obtain

$$\sum_{i=0}^{\infty} (i+2)(i+1)U_{i+2}x^i - 2\left(\sum_{i=0}^{\infty} (i+1)U_{i+1}x^i\right) + \sum_{i=0}^{\infty} U_i x^i = 0,$$

or

$$\sum_{i=0}^{\infty} [(i+2)(i+1)U_{i+2} - 2(i+1)U_{i+1} + U_i]x^i = 0.$$

Thus, we have the recurrence relations

(7.6.21) $$(i+2)(i+1)U_{i+2} - 2(i+1)U_{i+1} + U_i = 0 \quad \text{for } i = 0, 1, 2, \ldots .$$

The conditions (7.6.21) allow U_0 and U_1 to be arbitrarily chosen, and the remaining coefficients can be calculated by setting $i = 0, 1, 2, \ldots$ in the expression.

Since U_0 and U_1 can be given any values and since $y(0) = U_0$, $y'(0) = U_1$, we see that we may use series to look for a solution of (7.6.20) satisfying any given initial conditions $y(0) = y_0$, $y'(0) = y_1$. Let us look for a solution for which

$$y(0) = 0, \; y'(0) = 1.$$

Then $U_0 = 0$, $U_1 = 1$. From (7.6.21) we then obtain

$$i = 0: \quad 2U_2 - 2U_1 + U_0 = 0, \; U_2 = 1;$$

$$i = 1: \quad 6U_3 - 4U_2 + U_1 = 0, \; U_3 = \tfrac{1}{2};$$

$$i = 2: \quad 12U_4 - 6U_3 + U_2 = 0, \; U_4 = \frac{1}{3 \cdot 2};$$

$$i = 3: \quad 20U_5 - 8U_4 + U_3 = 0, \; U_5 = \frac{1}{4 \cdot 3 \cdot 2}.$$

$$\cdots$$

A pattern begins to emerge, and we may guess that $U_i = 1/(i-1)!$ for all $i \geq 1$. We know that this is true for $i = 1, 2, 3$. If for some k it is true for $i = k, k+1$, then by (7.6.21)

$$(k+2)(k+1)U_{k+2} - 2(k+1) \cdot \frac{1}{k!} + \frac{1}{(k-1)!} = 0,$$

$$(k+2)(k+1)U_{k+2} = \frac{2(k+1) - k}{k!} = \frac{k+2}{k!},$$

$$U_{k+2} = \frac{1}{k!(k+1)} = \frac{1}{(k+1)!}.$$

Thus, if our guess is true for $i = k, k + 1$, it is then true also for $i = k + 2$. This makes it possible to show by induction that $U_i = 1/(i - 1)!$ for all $i \geq 1$.

Thus

$$y = x + x^2 + \frac{x^3}{2} + \frac{x^4}{3!} + \cdots = x \sum_{i=0}^{\infty} \frac{x^i}{i!},$$

or

$$y = xe^x.$$

In Exercise 7.6.11 you will find by this method that the solution of (7.6.20) which satisfies $y(0) = 1, y'(0) = 1$ is

$$y = e^x.$$

By consulting (7.6.20) directly, we can see that for any two constants C_1, C_2,

(7.6.22) $$y = c_1e^x + c_2xe^x$$

is also a solution. In more extensive treatments of differential equations it is shown that all solutions of (7.6.20) are of the form (7.6.22).

This example is typical, in one sense, of a large number of problems in finding a solution as a power series. If one is reasonably adept at manipulating series, recurrence relations similar to (7.6.21) can be found without difficulty. Using them, one can find as many of the coefficients of the power series as one wishes. If, as in this example, a pattern is observable which makes it possible to guess the general form of U_i, it is usually not difficult to verify the guess by induction. Not difficult, that is, if the guess is correct! It must be admitted that this example is atypical in the ease with which we were able to guess the correct expression for U_i. Usually, the pattern is so complex that one must be content with finding a partial sum of the series. Even so, the resulting polynomials are useful approximations of a solution.

Exercises

1 A certain facility in manipulating power series is required for this method of solving differential equations, illustrated in Example 3. In each of the following examples the absence of one or more numbers or algebraic expressions is indicated by parentheses. Find these expressions.

(a) $\sum_{i=0}^{\infty} u_i = \sum_{i=(\ )}^{\infty} u_{i+2} = \sum_{i=-1}^{\infty} u_{(\ )},$

(b) $\sum_{n=1}^{\infty} x^n = \sum_{n=(\ )}^{\infty} x^{n+3} = \sum_{n=4}^{\infty} x^{(\ )},$

(c) $\sum_{i=1}^{\infty} i(i-1)x^{i-2} = \sum_{i=(\ )}^{\infty} (i+2)(\ )x^i$,

(d) $x \sum_{i=1}^{\infty} u_i x^i = \sum_{i=1}^{\infty} u_i x^{(\ )} = \sum_{i=(\ )}^{\infty} u_{(\ )} x^i$,

(e) $x^2 \sum_{i=0}^{\infty} i u_i x^{i-1} = \sum_{i=0}^{\infty} i u_i x^{(\ )} = \sum_{i=(\ )}^{\infty} (\ )x^i$,

(f) $\dfrac{d}{dx}\left(\sum_{i=0}^{\infty} u_i x^i\right) = \sum_{i=(\ )}^{\infty} i u_i x^{i-1} = \sum_{i=(\ )}^{\infty} (\ )x^{(\ )}$,

(g) $\sum_{i=1}^{\infty} u_i + \sum_{i=0}^{\infty} u_{i+2} = \sum_{i=1}^{\infty} u_i + \sum_{i=(\ )}^{\infty} u_i = (\ ) + 2\sum_{i=(\ )}^{\infty} u_i$,

(h) $\sum_{i=1}^{\infty} (x-1)^{2i} = \sum_{i=2}^{\infty} (x-1)^{(\ )}$.

2 Consider a power series whose sum is zero for all x in some neighborhood of 0:

$$f(x) = \sum_{i=0}^{\infty} a_i x^i = 0 \quad \text{for all } x \text{ in } (-c, c).$$

Then we know that f has derivatives of all orders each of which is zero on $(-c, c)$. Use this fact to show that $a_i = 0$ for $i = 0, 1, 2, \ldots$. (*Hint:* $f(0) = a_0, f'(0) = ?, \ldots$.)

3 (a) Find a solution in powers of x for the differential equation

$$\frac{dy}{dx} = xy.$$

(b) Find the particular solution which satisfies $y(0) = 1$.
(c) Can you identify the function represented by the series in (a)? If not, solve the equation by the method of separation of variables.

4 (a) Find a power series solution about $x = 0$ for the differential equation

$$\frac{dy}{dx} = x + y.$$

(b) Solve the equation as a first order linear equation and reconcile your answer with (a).

5 Find a series solution in powers of $(x-1)$ for the differential equation

$$\frac{dy}{dx} = y.$$

(*Hint:* Look for a solution of the form $y(x) = \sum_{i=0}^{\infty} u_i(x-1)^i$ and proceed as with series in powers of x.)

6 (a) Find a power series solution about $x = 1$ for the differential equation

$$\frac{dy}{dx} + xy = 0.$$

(*Hint:* $x = (x - 1) + 1$.)

(b) Find the particular solution $y_1(x)$ which passes through (1, 1).

(c) Use the first three nonzero terms of the series in (b) to approximate the value of $y_1(1.1)$.

7 (a) Find a power series solution in powers of x for *Airy's equation*

$$\frac{d^2y}{dx^2} = xy.$$

(b) It can be shown that the solutions to Airy's equation are oscillatory in nature (like $\sin x$ and $\cos x$) for $x < 0$, and are exponential in nature (like e^x) for $x > 0$. Why is it reasonable to expect this kind of behavior?

8 (a) Find a power series solution about $x = 0$ for the differential equation

$$\frac{d^2y}{dx^2} - x\frac{dy}{dx} - y = 0.$$

What is the recurrence relation for the coefficients?

(b) Find the particular solution which satisfies $y(0) = 0$, $y'(0) = 1$.

9 Find a power series solution about $x = 0$ for the differential equation

$$(1 - x)y'' + y = 0.$$

10 (a) Find a solution in powers of x for *Legendre's equation*

$$(1 - x^2)y'' - 2xy' + n(n + 1)y = 0.$$

(b) Show that if n is 0 or a positive integer, there is a polynomial solution of degree n. These polynomials, when multiplied by suitable constants, are the Legendre polynomials. (The constants are chosen so that if P_n is the Legendre polynomial of degree n, then $P_n(1) = 1$.)

11 By the method used in Example 3, find a solution of $y'' - 2y' + y = 0$ (7.6.20) for which $y(0) = y'(0) = 1$. Find the first five or six terms of the series, guess the form of the general term, prove from the recurrence relations that your guess is correct, and then identify from past experience the function which has this power series expansion.

solutions to selected exercises 2

chapter 1

page 307

1.1

1 (a) Suppose $|x| < \epsilon$. If $x \geq 0$, then $|x| = x < \epsilon$. If $x < 0$, then $|x| = -x < \epsilon$ implies $-\epsilon < x$. In either case, $-\epsilon < x < \epsilon$. Conversely, assume $-\epsilon < x < \epsilon$. Because $|x|$ equals either x or $-x$, clearly $|x| < \epsilon$.

(b) From (a), $|x - a| < \epsilon$ if and only if $-\epsilon < x - a < \epsilon$. By adding a or $-a$, we see that this is equivalent to $a - \epsilon < x < a + \epsilon$.

2 (a) Any $\delta > 0$ will do, (c) any $\delta < 1/30{,}000$.

3 (a) If $|x - 2| < 1$, then $-1 < x - 2 < 1$. Add 6 to get $5 < x + 4 < 7$. It follows, of course, that $-7 < x + 4 < 7$ and $|x + 4| < 7$.

(b) Let $\epsilon > 0$ and choose δ to be the smaller of the two numbers 1 and $\epsilon/7$. Then if $0 < |x - 2| < \delta$, we have $|x^2 + 2x - 8| = |x + 4|\,|x - 2| < 7 \cdot (\epsilon/7) = \epsilon$. (Here we use (a) to find $|x + 4| < 7$.)

5 (Here, as elsewhere, min (a, b) means the smaller of the numbers a and b.)

(a) Let $\epsilon > 0$ and $\delta = \epsilon/3$. Then if $0 < |x - 1| < \delta$, $|3x - 3| = 3|x - 1| < 3 \cdot (\epsilon/3) = \epsilon$.

(c) Let $\epsilon > 0$ and $\delta = \epsilon/2$. Then if $0 < |x| < \delta$, $|-2x + 7 - 7| = 2|x| < 2 \cdot (\epsilon/2) = \epsilon$.

(e) Let $\epsilon > 0$ and $\delta = \min(1, \epsilon/7)$. Then if $0 < |x + 3| < \delta$, $|x + 3| < 1$. Thus $-1 < x + 3 < 1$, or $-7 < x - 3 < -5$. Consequently, $|x - 3| < 7$. Also, $|x + 3| < \epsilon/7$, so $|x^2 - 6 - 3| = |x - 3|\,|x + 3| < 7 \cdot (\epsilon/7) = \epsilon$.

(g) Let $\epsilon > 0$ and $\delta = \min(1, \epsilon/7)$. Then if $0 < |x - 5| < \delta$, $|x - 5| < 1$ and $|x + 1| < 7$. Also, $|x - 5| < \epsilon/7$, so $|x^2 - 4x - 1 - 4| = |x + 1|\,|x - 5| < 7 \cdot (\epsilon/7) = \epsilon$.

(i) Let $\epsilon > 0$ and $\delta = \min(\frac{1}{4}, \epsilon/8)$. Then if $0 < |x + \frac{1}{2}| < \delta$, then $|x + \frac{1}{2}| < \frac{1}{4}$, or $-\frac{1}{4} < x + \frac{1}{2} < \frac{1}{4}$. So $-\frac{3}{4} < x < -\frac{1}{4}$ and $|x| > \frac{1}{4}$. Then $\left|\frac{1}{x} + 2\right| = \left|\frac{1+2x}{x}\right| = 2\frac{|x+\frac{1}{2}|}{|x|} < 2 \cdot 4 \cdot (\epsilon/8) = \epsilon$.

(k) Let $\epsilon > 0$ and $\delta = \min(1/10, 3\epsilon/5)$. Then if $0 < |x - 2| < \delta$, we have $-.1 < x - 2 < .1$, $1.9 < x < 2.1$, $(1.9)^2 < x^2 < (2.1)^2$, and $1.61 < x^2 - 2 < 2.41$. Thus $|x^2 - 2| > 1.61 > 3/2$. Moreover, we also have $3.9 < x + 2 < 4.1$, so that $|x + 2| < 4.1 < 5$. Then finally

$$\left|\frac{1}{x^2-2} - \frac{1}{2}\right| = \frac{|x+2|\,|x-2|}{2|x^2-2|} < \frac{5}{2}\cdot\frac{2}{3}\cdot\frac{3}{5}\epsilon = \epsilon.$$

(m) Let $\epsilon > 0$ and $\delta = \min(1, 10\epsilon)$. Then if $0 < |x - 4| < \delta$, $3 < x < 5$, $\sqrt{3} < \sqrt{x} < \sqrt{5}$, and $6 + 4\sqrt{3} < 4\sqrt{x} + 2x < 10 + 4\sqrt{3}$. Thus $|4\sqrt{x} + 2x| > 10$. Now

$$\frac{1}{\sqrt{x}} - \frac{1}{2} = \frac{2-\sqrt{x}}{2\sqrt{x}} = \frac{(2-\sqrt{x})(2+\sqrt{x})}{2\sqrt{x}(2+\sqrt{x})} = \frac{4-x}{4\sqrt{x}+2x}.$$

So

$$\left|\frac{1}{\sqrt{x}} - \frac{1}{2}\right| = \frac{|x-4|}{|4\sqrt{x}+2x|} < \frac{1}{10}\cdot 10\epsilon = \epsilon.$$

8 There exists an $\epsilon > 0$ such that for every $\delta > 0$ there is an x satisfying $0 < |x - a| < \delta$ but not satisfying $|f(x) - L| < \epsilon$.

1.2

page 312

1 (a) We say $\lim_{x\to c^+} f(x) = +\infty$ if for every $\epsilon > 0$ there is a $\delta > 0$ such that $f(x) > 1/\epsilon$ provided $0 < x - c < \delta$.

(c) We say $\lim_{x\to +\infty} f(x) = +\infty$ if for every $\epsilon > 0$ there is a $\delta > 0$ such that $f(x) > 1/\epsilon$ provided $x > 1/\delta$.

2 (a) 1, (c) 0, (e) 0, (g) 0, (i) 0.

3 (a) -1, (c) -1, (e) 0, (g) 1, (i) 1.

5 Suppose $\lim_{x\to c} f(x) = +\infty$ and $\lim_{x\to c} g(x) = L > 0$. Let $\epsilon > 0$, and let $\epsilon_1 = (L/2)\epsilon$. Then $\epsilon_1 > 0$, and there is a $\delta_1 > 0$ such that $f(x) > 1/\epsilon_1 = 2/\epsilon L$ provided $0 < |x - c| < \delta_1$. Let $\epsilon_2 = L/2$. There is a δ_2 such that $|g(x) - L| < L/2$, or $g(x) > L/2$, provided $0 < |x - c| < \delta_2$. It follows that if $\delta = \min(\delta_1, \delta_2)$, then for $0 < |x - c| < \delta$, $f(x)g(x) > \frac{2}{\epsilon L}\cdot\frac{L}{2} = \frac{1}{\epsilon}$. So $\lim_{x\to c} f(x)g(x) = +\infty$. A similar proof can be constructed for the other cases.

6 (a) $\frac{2}{3}$, (c) 0, (e) 0, (g) $+\infty$, (i) 0.

7 (a) $\frac{2}{3}$, (c) 0, (e) 0, (g) $+\infty$, (i) 0.

9 (a) $\lim_{x\to+\infty} e^{1/x} = e^{(\lim_{x\to+\infty} 1/x)} = e^0 = 1$,

(c) $\lim_{x\to+\infty} \sin\left(\dfrac{1}{x-2}\right) = \sin(0) = 0.$

11 It is immediate from the definitions that $\lim_{x\to c} f(x) = L$ implies both $\lim_{x\to c+} f(x) = L$ and $\lim_{x\to c-} f(x) = L$. On the other hand, if both $\lim_{x\to c+} f(x) = L$ and $\lim_{x\to c-} f(x) = L$, then for $\epsilon > 0$ there are $\delta_1, \delta_2 > 0$ such that $|f(x) - L| < \epsilon$ provided $0 < |x - c| < \delta_1$ and $x > c$, or provided $0 < |x - c| < \delta_2$ and $x < c$. Let $\delta = \min(\delta_1, \delta_2)$. Then $|f(x) - L| < \epsilon$ provided $0 < |x - c| < \delta$, and thus $\lim_{x\to c} f(x) = L$.

---

page 320

1.3

3 Let $\epsilon > 0$. Then there is n_ϵ such that if $n > n_\epsilon$, $|l - a_n| < \epsilon$. Similarly, there is an n'_ϵ such that if $n > n'_\epsilon$, $|l - b_n| < \epsilon$. Let $n''_\epsilon = \max(n_\epsilon, n'_\epsilon)$ and $n > n''$. Then $l - \epsilon < a_n < l + \epsilon$ and $l - \epsilon < b_n < l + \epsilon$. Consequently, $l - \epsilon < a_n \le c_n \le b_n < l + \epsilon$, or $l - \epsilon < c_n < l + \epsilon$. That is, $|l - c_n| < \epsilon$ and $\lim_{n\to\infty} c_n = l$.

5 Let $\epsilon > 0$. Then because $\lim_{x\to a} f(x) = L$, there is a $\delta > 0$ such that $|f(x) - L| < \epsilon$ provided $|x - a| < \delta$. But because $\lim_{n\to\infty} a_n = a$, there is an n_δ such that $|a_n - a| < \delta$ provided $n > n_\delta$. So, for $n > n_\delta$, $|f(a_n) - L| < \epsilon$. This means $\lim f(a_n) = L$.

6 (a) 0, (c) 1, (e) 0, (g) 0, (i) 0.

(k) Because $\dfrac{n!}{(2n)!} = \dfrac{1}{(n+1)(n+2)\cdots(2n)}$,

$$0 \le \frac{n!}{(2n)!} \le \frac{1}{n+1} \quad \text{and} \quad \lim \frac{n!}{(2n)!} = 0 \text{ by Theorem 1.3.2.}$$

8 (a) $$\lim\left(1 + \frac{1}{2n}\right)^n = \lim\left(1 + \frac{1}{m}\right)^{m/2} = \left(\lim\left(1 + \frac{1}{m}\right)^m\right)^{1/2} = e^{1/2},$$

(c) e^3.

11 $\lim n^{p/n} = \lim e^{(p \log n)/n} = e^{p(\lim(\log n/n))} = e^0 = 1.$

---

chapter 2

page 329

2.1

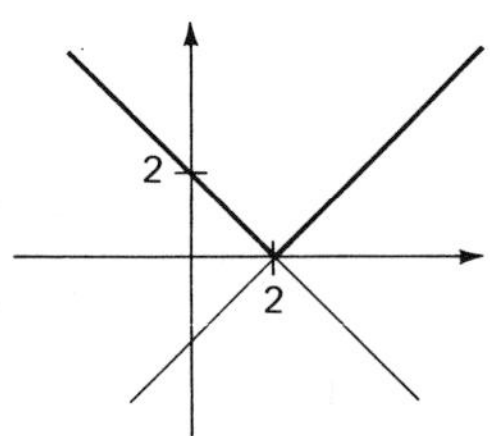

1 (a) $f'_-(2) = -1, f'_+(2) = 1$;

(c) $f'_-(-1) = -2$, $f'_-(0) = -1$, $f'_-(1) = -2$,
$f'_+(-1) = 2$, $f'_+(0) = 1$, $f'_+(1) = 2$;

(e) $g'_-(x_0) = -1$

$g'_+(x_0) = 1$

if $x_0 = k\pi$ and k is any integer;

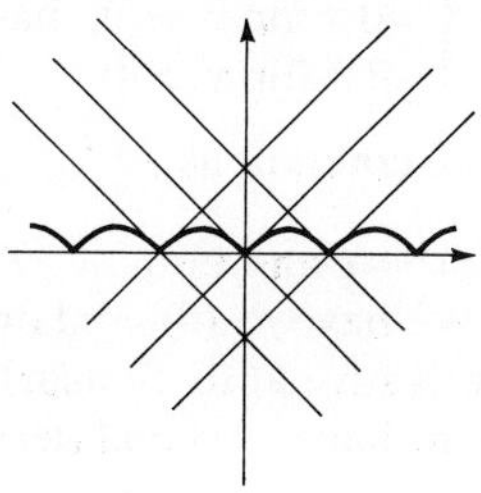

(g) $f'_-(0) = -\frac{1}{2}, f'_+(0) = \frac{1}{2}$;

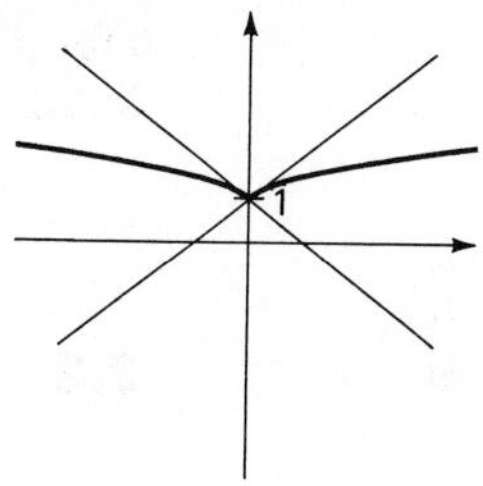

(i) $f'_-(0)$ is undefined, $f'_+(0) = 0$.

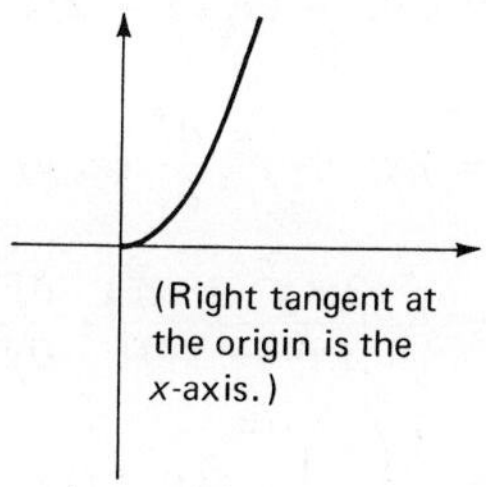

2 (a)

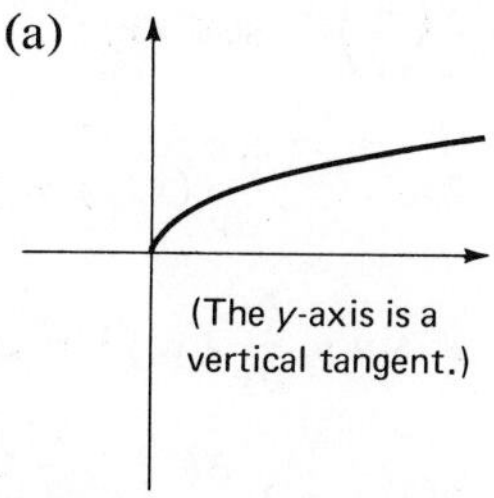

(c)

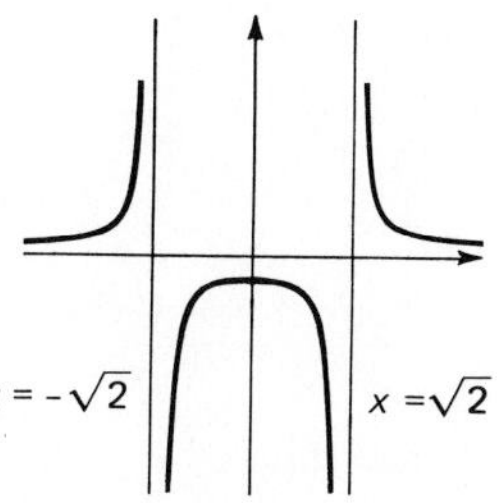

(e)

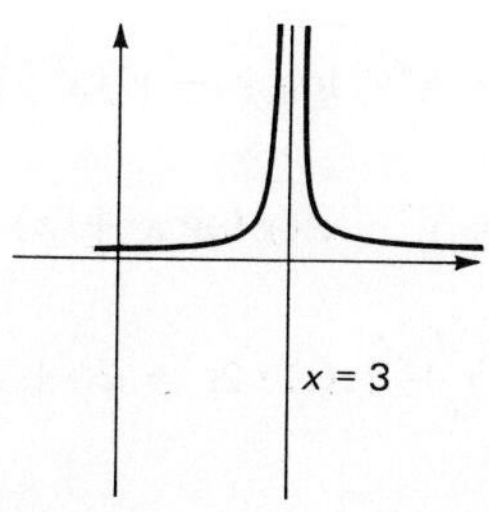

(g)

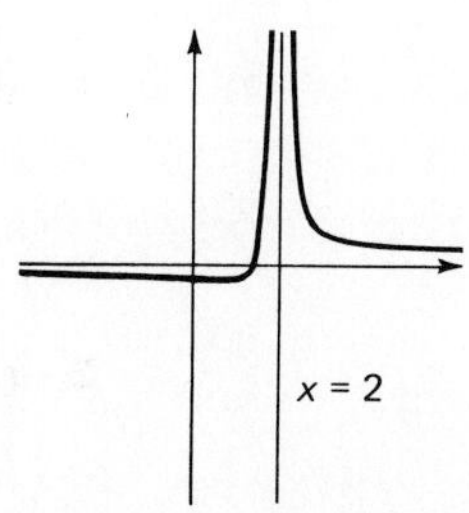

5 Without loss of generality we may assume $f'(a) < c < f'(b)$. Define $g(x) = cx - f(x)$. Then g is differentiable on $[a, b]$ and $g'(x) = c - f'(x)$. Furthermore, $g'(a) > 0$ and $g'(b) < 0$. By Theorem 4.3.2 there is a number x_0, $a < x_0 < b$, such that $g'(x_0) = c$. This means, of course, that $f'(x_0) = c$.

9 The function $f(x) = \begin{cases} -1 & \text{for } x < 0, \\ 0 & \text{for } x = 0, \\ 1 & \text{for } x > 0 \end{cases}$ has an infinite derivative at $x = 0$, where it fails to be continuous.

13 Because $\lim_{x\to 0-} f(x) = c$ and $\lim_{x\to 0+} f(x) = 0$, c must be 0 for f to be continuous at 0. We have $f'_+(0) = 0$ and $f'_-(0) = b$, so b must be 0 for f to have a first derivative at 0. Similarly, $f''_+(0) = 4$ and $f''_-(0) = 2a$, so a must be 2 for f to have a second derivative at 0.

---

page 338

2.2

1 (a) $\frac{\partial F}{\partial x}(x, y) = 2x, \frac{\partial F}{\partial y}(x, y) = 2y;$

(c) $\frac{\partial F}{\partial x}(x, y) = 2xy + y^2, \frac{\partial F}{\partial y}(x, y) = x^2 + 2xy;$

(e) $\frac{\partial F}{\partial x}(x, y) = -\frac{2y(y^2 + x^2)}{(x^2 - y^2)^2}, \frac{\partial F}{\partial y}(x, y) = \frac{2x(x^2 + y^2)}{(x^2 - y^2)^2};$

(g) $\frac{\partial F}{\partial x}(x, y) = \frac{1}{2}\left(\frac{y}{x}\right)^{1/2} \sec^2 (xy)^{1/2},$

$\frac{\partial F}{\partial y}(x, y) = \frac{1}{2}\left(\frac{x}{y}\right)^{1/2} \sec^2 (xy)^{1/2};$

(i) $\frac{\partial F}{\partial x}(x, y) = 2xe^{x^2+y^2}, \frac{\partial F}{\partial y}(x, y) = 2ye^{x^2+y^2};$

(k) $\frac{\partial F}{\partial x}(x, y) = e^x \sin y, \frac{\partial F}{\partial y}(x, y) = e^x \cos y;$

(m) $\frac{\partial F}{\partial x}(x, y) = 2^x 3^y \log 2, \frac{\partial F}{\partial y}(x, y) = 2^x 3^y \log 3;$

(o) $\frac{\partial F}{\partial x}(x, y) = x^y y^x \log y + y^x y x^{y-1} = x^{y-1} y^x (x \log y + y),$

$\frac{\partial f}{\partial y}(x, y) = y^{x-1} x^y (y \log x + x).$

2 (a) $\frac{dF}{dt} = 2x(t) + 2y(t) \cdot 2t = 2t + 4t^3,$

(c) $\dfrac{dF}{dt} = 10t^4 + 64t^3 + 132t^2 + 88t + 2.$

3 (a) $F(1.1, 2.1) = F(1, 2) + (.1)\dfrac{\partial F}{\partial x}(1, 2) + (.1)\dfrac{\partial F}{\partial y}(1, 2)$

$= 17 + .2 + 1.6 = 18.8,$

(c) $F(0.1, 92°) = 1.$

5 1/20 in.

7 $dV/dt = 19.2\pi$, i.e., the volume is increasing at the rate of 19.2π in^3/sec.

---

page 344

2.3

1 (a) Solving the equation, we find $y_1(x) = \sqrt{x+1}$ and $y_2(x) = -\sqrt{x+1}$.

(c) $y_1'(x) = 1/2\sqrt{x+1}$, $y_2'(x) = -1/2\sqrt{x+1}$.

(e) The slopes are $\frac{1}{2}$, $-\frac{1}{2}$, $\sqrt{2}/4$, $-\sqrt{2}/4$.

3 (a) $\dfrac{dy}{dx} = \dfrac{5}{2y}$, (c) $\dfrac{dy}{dx} = \dfrac{x}{2y}$,

(e) $\dfrac{dy}{dx} = -\dfrac{x}{y}$, (g) $\dfrac{dy}{dx} = \dfrac{y^2 + 4x}{3y^2 - 2xy}$,

(i) $\dfrac{dy}{dx} = \dfrac{x \cos x^2}{y \sin y^2}$, (k) $\dfrac{dy}{dx} = -\dfrac{1}{xy}$.

4 (a) $\dfrac{d^2y}{dx^2} = \dfrac{-25}{4y^3}$, (c) $\dfrac{d^2y}{dx^2} = -\dfrac{3}{4y^3}$,

(e) $\dfrac{d^2y}{dx^2} = -\dfrac{y^2 + x^2}{y^3}$, (g) $\dfrac{d^2y}{dx^2} =$

$$\frac{2y(y^2 + 4x)(3y^2 - 2xy) + 4(3y^2 - 2xy)^2 - (y^2 + 4x)^2(6y - 2x) + 2y(y^2 + 4x)}{(3y^2 - 2xy)^3},$$

(i) $\dfrac{d^2y}{dx^2} =$

$$\frac{-2x^2y^2 \sin^2 y^2 \sin x^2 + y^2 \sin^2 y^2 \cos x^2 - 2x^2y^2 \cos^2 x^2 \cos y^2 - x^2 \cos^2 x^2 \sin y^2}{(y \sin y^2)^3},$$

(k) $\dfrac{d^2y}{dx^2} = \dfrac{y^2 - 1}{x^2y^3}$.

5 (a) $\dfrac{dy}{dx} = \dfrac{\cos x}{\sin y}$, (c) $\dfrac{dy}{dx} = -\dfrac{3x^2 + 2xy - 1}{x^2 + 1}$,

(e) $\dfrac{dy}{dx} = -\sqrt{\dfrac{y}{x}}$, (g) $\dfrac{dy}{dx} = -\dfrac{y^2}{x^2}$, (i) $\dfrac{dy}{dx} = -\sqrt[3]{\dfrac{y}{x}}$.

6 (a) $x + 4y - 14 = 0$, (c) $2x - y - 1 = 0$, (e) $x + y - 2 = 0$.

7 (a) $4x - y - 5 = 0$, (c) $x + 2y + 2 = 0$, (e) $x - y = 0$.

---

page 349

2.4

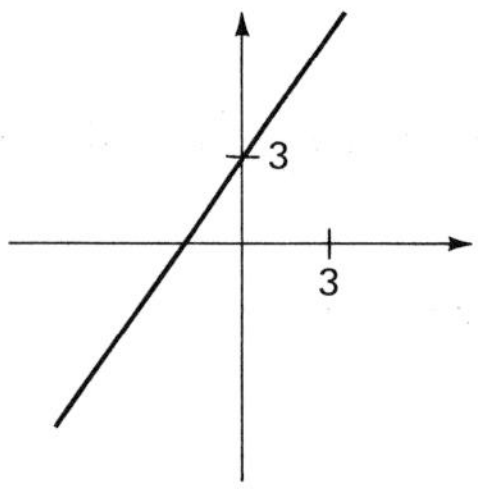

1 (a) $f'(x) = 2$, so f has an inverse on the whole set of real numbers; $f^{-1}(y) = \frac{1}{2}(y - 3)$, $df^{-1}/dy = \frac{1}{2}$.

(c) This function has no inverse on any nondegenerate interval.

(e) $f'(x) = 3x^2$, so f is increasing and has an inverse on the whole line $(-\infty, +\infty)$. The inverse of f is $f^{-1}(y) = \sqrt[3]{y - 2}$. Its derivative is $\dfrac{df^{-1}}{dy} = \dfrac{1}{3x^2} = \dfrac{1}{3(y - 2)^{2/3}}$.

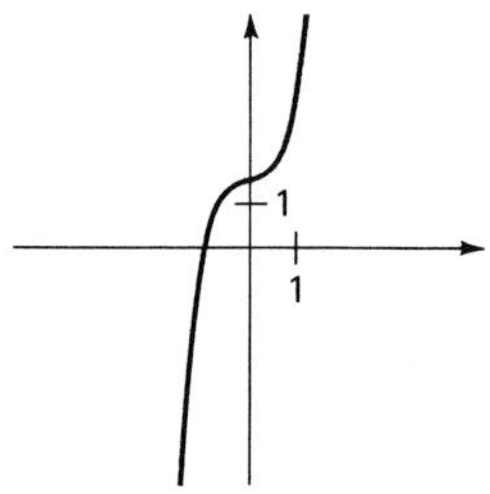

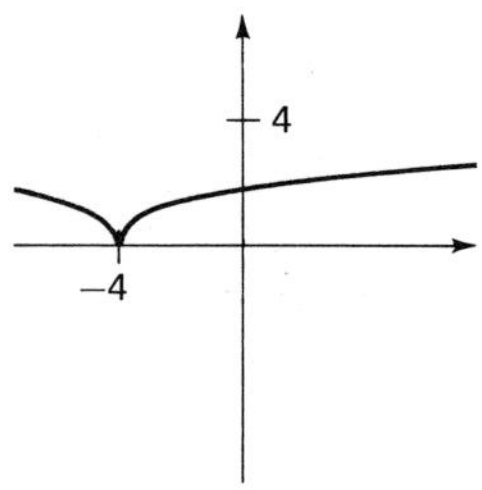

(g) $f'(x) = \frac{1}{3}(x + 4)^{-2/3}$, so f is increasing and has an inverse on $(-\infty, +\infty)$. The inverse is given by $f^{-1}(y) = y^3 - 4$, and the derivative by $df^{-1}/dy = 3(x + 4)^{2/3} = 3y^2$.

(i) $f'(x) = -1/x^2$, so f is decreasing on $(-\infty, 0)$ and $(0, +\infty)$. The inverse of f on either interval is given by $f^{-1}(y) = 1/y$. The derivative of f^{-1} is $df^{-1}/dy = -x^2 = -1/y^2$.

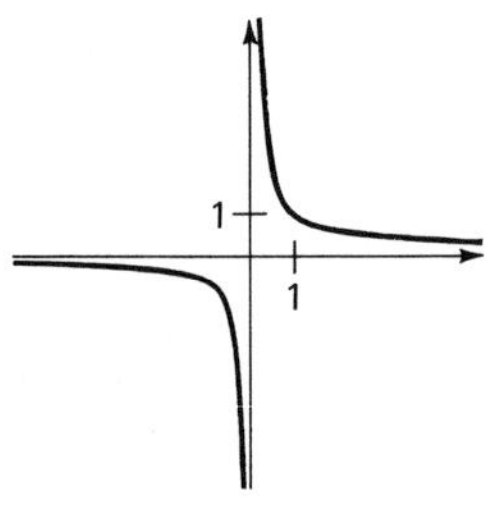

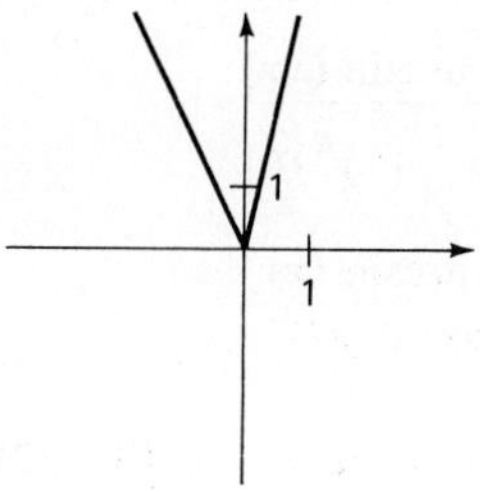

(k) f is decreasing in $(-\infty, 0]$ and is increasing in $[0, +\infty)$. For the first interval, $f^{-1}(y) = -\frac{1}{2}y$ $(y \geq 0)$ and $df^{-1}/dy = -\frac{1}{2}$. For the second interval, $f^{-1}(y) = \frac{1}{4}y$ $(y \geq 0)$ and $df^{-1}/dy = \frac{1}{4}$.

3 It is sufficient to show that one of the conditions in formula (2.4.1) is satisfied. For example, let x be in the domain of f. Then $[f^{-1} \circ g^{-1} \circ g \circ f](x) = f^{-1}(g^{-1}(g(f(x)))) = f^{-1}(f(x)) = x$. This shows that $f^{-1} \circ g^{-1}$ is the inverse of $g \circ f$.

5 Yes. A simple example is the inverse of $f(x) = \tan x$. This inverse function is defined and increasing on $(-\infty, +\infty)$ and assumes values between -1 and 1.

9 $\dfrac{df^{-1}}{dy} = e^{x^2}$.

---

page 356

2.5

1 (a) $\pi/3$, (c) $-\pi/4$, (e) $\pi/3$,
(g) $\pi/4$, (i) π, (k) $5\pi/6$.

3 (a) $-2\sqrt{5}/5$, (b) $\sqrt{5}/5$, (c) $-1/2$,
(d) $\sqrt{5}$, (e) $-\sqrt{5}/2$.

4 (a) 1, (c) $-\sqrt{2}$, (e) $|x|/\sqrt{1 + x^2}$.

5 (a) $f'(x) = \dfrac{3}{\sqrt{1 - 9x^2}}$, (c) $F'(x) = \dfrac{-1}{2\sqrt{x - x^2}}$,

(e) $f'(x) = \dfrac{1 - x^2}{x^4 - x^2 + 1}$,

(g) $F'(y) = -\dfrac{\sqrt{y} + 2\sqrt{1 - y}\arccos(\sqrt{y})}{2y^2\sqrt{1 - y}}$,

(i) $f'(x) = \dfrac{2\sqrt{4 - x^2} + \sqrt{1 - 4x^2}}{\sqrt{(1 - 4x^2)(4 - x^2)}}$.

6 (a) $\dfrac{\partial F}{\partial x}(x, y) = \arcsin(2y)$, $\dfrac{\partial F}{\partial y}(x, y) = \dfrac{2x}{\sqrt{1 - 4y^2}}$;

(c) $\dfrac{\partial F}{\partial x}(x, y) = \dfrac{-\tan y}{\sqrt{1 - x^2 \tan^2 y}}$, $\dfrac{\partial F}{\partial y}(x, y) = \dfrac{-x \sec^2 y}{\sqrt{1 - x^2 \tan^2 y}}$;

$$\text{(e)}\ \frac{\partial F}{\partial x}(x, y) = \frac{y}{(x^2 + y^2)\sqrt{1 - x^2y^2}} + \frac{2x \arcsin (xy)}{(x^2 + y^2)^2},$$

$$\frac{\partial F}{\partial y}(x, y) = \frac{x}{(x^2 + y^2)\sqrt{1 - x^2y^2}} + \frac{2y \arcsin (xy)}{(x^2 + y^2)^2}.$$

7 Define the restricted secant function as $f(x) = \sec x$, $x \in [0, \pi/2) \cup [\pi, 3\pi/2)$. Then the domain of the inverse function, $f^{-1}(y) = \operatorname{arcsec}(y)$, is $(-\infty, -1] \cup [1, +\infty)$. By Theorem 2.4.3

$$\frac{d}{dy} \operatorname{arcsec}(y) = \frac{1}{\sec x \tan x} = \frac{1}{y\sqrt{y^2 - 1}}.$$

The selection of this domain for the restricted secant function is motivated by the single simple formula which results for the derivative of the arcsecant. If one were to use instead $[0, \pi/2) \cup (\pi/2, \pi]$, the formula for the derivative would have different signs in different parts of the domain.

9 Because

$$f'(x) = \frac{1}{\sqrt{1 - x^2}} + \frac{-1}{\sqrt{1 - x^2}} = 0,$$

f must be a constant function. The constant can be identified; since $f(0) = \pi/2$, we have $f(x) = \pi/2$.

12 $x = \sqrt{b(a + b)}$ feet.

---

chapter 3

page 366

3.1

1 $\frac{1}{2}[x\sqrt{9 - x^2} + 9 \arcsin (x/3)]$.

3 $\frac{1}{2}[x\sqrt{x^2 - 16} - 16 \log (x + \sqrt{x^2 - 16})]$.

5 $\log (x + 2 + \sqrt{1 + (x + 2)^2})$. **7** $25\pi/16$.

9 $\frac{1}{3} \arccos (3/x) = \frac{1}{3} \operatorname{arcsec} (x/3)$. **11** $\dfrac{-x}{4\sqrt{x^2 - 4}}$.

13 $\sqrt{x^2 - a^2} - a \cdot \arccos \left(\dfrac{a}{x}\right)$. **15** $\dfrac{(2x - 1)}{8\sqrt{4x^2 - 4x + 5}}$.

page 372

3.2

1 $x^2 + \log(x - 2)$.

3 $\log((x - 1)(x - 2)^2)$.

5 $x^2 + 11x + \log((x - 1)^4/(x - 3)^3)$.

7 $\frac{5}{2}\log(x + 1) - \frac{8}{3}\log(x + 2) + \frac{1}{6}\log(x - 1)$.

9 $\log(x + 1) - \dfrac{2}{(x + 1)} - \dfrac{3}{2(x + 1)^2}$.

11 $\frac{1}{2}x^2 + \frac{3}{2}\log(x^2 + 2)$.

13 $\arctan(x + 1)$.

15 $\frac{1}{2}\log(x^2 + 3) + \frac{3}{2}\arctan\left(\dfrac{x}{2}\right)$.

17 $\frac{1}{25}\left(\log\left(\dfrac{(x^2 + 4)^2}{(x - 1)^4}\right) - \dfrac{10}{x - 1} + \dfrac{19}{2}\left(\arctan\left(\dfrac{x}{2}\right)\right)\right)$.

19 $\log(x - 1)^3 - \arctan\left(\dfrac{x - 1}{2}\right)$.

---

page 379

3.3

1 (a) $\displaystyle\lim_{n\to\infty} \sum_{i=1}^{n} \left(\frac{i}{n}\right)\frac{1}{n} = \int_0^1 x\,dx = \frac{1}{2}$,

(c) $\displaystyle\lim_{n\to\infty} \frac{1}{n^6} \sum_{i=0}^{n-1} i^5 = \int_0^1 x^5\,dx = \frac{1}{6}$,

(e) $\displaystyle\lim_{n\to\infty} \frac{1}{n} \sum_{i=1}^{n} \left(\frac{i + n}{n}\right)^{1/2} = \int_0^1 \sqrt{1 + x}\,dx = \frac{2}{3}(2^{3/2} - 1)$.

2 (a) $\displaystyle\int_0^1 x^3\,dx = \lim_{n\to\infty} \frac{1}{n} \sum_{i=1}^{n} \left(\frac{i}{n}\right)^3$,

(c) $\displaystyle\int_0^{\pi/2} x\cos x\,dx = \lim_{n\to\infty} \left(\frac{\pi}{2n}\right)^2 \sum_{i=1}^{n} i\cos\left(\frac{i\pi}{2n}\right)$,

(e) $\displaystyle\int_{-2}^{-1} \frac{1}{x^2}\,dx = \lim_{n\to\infty} \frac{1}{n} \sum_{i=1}^{n} \frac{1}{(-2 + i/n)^2}$.

---

page 382

3.4

1 64/27. **3** $\log(21/5) - \frac{1}{2}$. **5** $\frac{1}{2}a(e - 1/e)$.

7 $f'(x) = e^x \cdot L_0^1(f) = \int_0^1 \sqrt{1 + e^{2x}}\,dx = \int_0^1 e^x\sqrt{1 + e^{-2x}}\,dx$

$$= \int_1^e \frac{\sqrt{u^2+1}}{u}\,du = \sqrt{1+u^2} - \log\left(\frac{1+\sqrt{1+u^2}}{u}\right)\Bigg|_1^e$$

$$= \sqrt{1+e^2} + 1 - \sqrt{2} - \log(1 + \sqrt{1+e^2}) + \log(1 + \sqrt{2}).$$

9 (b) The graph of s will be concave upward at $(t_0, s(t_0))$ if

$$f'(t_0)f''(t_0) > 0.$$

(c) As an easy example, consider $f(t) = t^2$, $-1 \le t \le 1$.

---

page 388

3.5

1 (a) $\pi^2/2$, (c) $(\pi/4)(2 + \pi)$.

2 (a) $28\pi/5$, (c) $\pi(4 - 4\log 2 + 2(\log 2)^2)$.

3 (a) $2\pi^2$, (c) $\pi \log 2$.

4 (a) $64\pi/15$.

5 (a) 64π, (c) $32^2\pi/35$.

6 (a) $3\pi/10$, (c) $\pi^2/12$.

7 $4\pi a^2 b/3$.

9 The material removed can be thought of as a cylinder of radius a and length $2\sqrt{r^2 - a^2}$, plus 2 caps in the form of spherical segments. The total volume can be expressed as

$$2\pi a^2\sqrt{r^2 - a^2} + 2\pi \int_{(r^2-a^2)^{1/2}}^{r} (r^2 - x^2)\,dx = \tfrac{4}{3}\pi(r^3 - (r^2 - a^2)^{3/2}).$$

---

page 393

3.7

1 $50{,}000\pi$ ft.-lbs. **3** $100{,}000\pi$ ft.-lbs.

5 $62.5\pi \int_0^6 (x + 2)(36 - x^2)\,dx = 38{,}250\pi$ ft.-lbs.

7 The parabola is $y = \frac{1}{2}x^2$. The work required is $62.5\pi\int_0^8 x(16 - x)\,dx = 21{,}333\pi$ ft.-lbs.

---

page 397

3.8

1 $\displaystyle\int_0^1 \frac{dx}{\sqrt[3]{x}} = \lim_{u\to 0+} \tfrac{3}{2}x^{2/3}\Big|_u^1 = \tfrac{3}{2}.$ **3** $-\infty.$

5 $7\sqrt{2}.$ **7** $+\infty.$

9 $\log(2+\sqrt{3}).$ **11** $-\infty.$

13 $\pi/2.$ **15** $+\infty.$

17 $1.$ **19** $\tfrac{1}{4}.$

21 $\tfrac{1}{2}(\arctan(\tfrac{1}{2})) + \pi/4.$ **23** $\pi/3.$

25 $+\infty.$

---

chapter 4

page 401

4.1

1 (a) $\displaystyle\int_a^b f(x)\,dx = \frac{b-a}{n}(y_0 + y_1 + \cdots + y_{n-1}).$

(c) The average is given by

$$\frac{b-a}{2n}[(y_0 + y_1 + \cdots + y_{n-1}) + (y_1 + y_2 + \cdots + y_n)]$$

$$= \frac{b-a}{2n}[y_0 + 2y_1 + 2y_2 + \cdots + 2y_{n-1} + y_n],$$

which is the trapezoid method.

2 (a)

x	y
1	1.00000
2	0.50000
3	0.33333
4	0.25000
5	0.20000
6	0.16667
7	0.14286
8	0.12500
9	0.11111

Approximation by (4.1.2) is 1.82897.
Approximation by (4.1.4) is 2.27341.
Approximation by (4.1.8) is 2.21005.

(b)

x	y
1	1.00000
1.5	0.44444
2	0.25000
2.5	0.16000
3	0.11111
3.5	0.08163
4	0.0625
4.5	0.04938
5	0.04000

(c)

x	y
0.0000	0.00000
0.39270	0.61861
0.78540	0.84090
1.17810	0.96119
1.57080	1.00000
1.96350	0.96119
2.35619	0.84090
2.74889	0.61861
3.14159	0.00000

Approximation by (4.1.2) is 2.29391.
Approximation by (4.1.4) is 2.29391.
Approximation by (4.1.8) is 2.35646.

(d)

x	y
0.00000	0.00000
0.78540	0.50000
1.57080	1.00000
2.35619	0.50000
3.14159	0.00000
3.92699	0.50000
4.71239	1.00000
5.49779	0.50000
6.28319	0.00000

3 (a)

x	y
1.0	1.00000
1.1	0.90909
1.2	0.83333
1.3	0.76923
1.4	0.71428
1.5	0.66667
1.6	0.62500
1.7	0.58824
1.8	0.55556
1.9	0.52632
2.0	0.50000

The approximation by Simpson's method is 0.693150.

(b) $|E_{10}| \leq \dfrac{24(.1)^4}{3080} \approx .00000078.$

5 (a) $y = \cos x$, $y' = -\sin x$, so the arclength is $\int_0^{\pi/2} \sqrt{1 + \sin^2 x}\, dx$.

(b)

x	y
0.0000	1.00000
0.3927	1.07072
0.7854	1.22474
1.1781	1.36145
1.5708	1.41421

Approximation by trapezoid method is 1.91010.
Approximation by Simpson's rule is 1.91014.

7

x	y
0	6.00000
1	5.91608
2	5.65685
3	5.19615
4	4.47214
5	3.31662
6	0.00000

Approximation using trapezoid method is 27.55785.
Actual value by integration is $9\pi \approx$ 28.27431.

9 15.73067 square units.

10

x	πy^2
2.0	28.84262
2.4	66.76538
2.8	105.68309
3.2	136.43328
3.6	189.17901
4.0	224.84862
4.4	265.32644

Approximation of $\int_2^{4.4} \pi y^2\, dx$ by Simpson's rule is 346.14432 cubic units.

11

x	$2\pi xy$
2.0	38.07607
2.4	69.51710
2.8	102.03884
3.2	132.49970
3.6	175.52692
4.0	212.62281
4.4	254.06667

Approximation of $\int_2^{4.4} 2\pi xy\, dx$ by Simpson's rule is 334.11103 cubic units.

---

page 418

4.2

1 Let n be a positive integer. Then

$$(4.2.9)\quad \lim_{x \to c} \frac{f(x)}{(x-c)^{n-1}} = \lim_{x \to c} \frac{f'(x)}{(n-1)(x-c)^{n-2}}$$

$$= \lim_{x \to c} \frac{f''(x)}{(n-1)(n-2)(x-c)^{n-3}}$$

$$= \lim_{x \to c} \frac{f^{(n-1)}(x)}{(n-1)!} = \frac{f^{(n-1)}(c)}{(n-1)!} = 0.$$

Note that the hypotheses are used to justify each application of l'Hospital's Rule.

5 (a) $\dfrac{f(2) - f(1)}{g(2) - g(1)} = \dfrac{1}{3} \dfrac{f'(x)}{g'(x)} = \dfrac{1}{2x}$, so $\bar{t}$ can be chosen as $\bar{t} = 3/2$.

(c) $\bar{t} = \pi/2$ can be chosen.

6 In each case consider x fixed and apply Theorem 4.2.1 as indicated.
(a) $f(z) = \sin z$, $g(z) = z$ for z in $[0, x]$,
(c) $f(z) = \tan z$, $g(z) = z$ for z in $[0, x]$.

7 (*Hint:* If $x > 0$, set $f(z) = e^z$, $g(z) = z$ for z in $[0, x]$.)

9 (*Hint:* $\lim_{k\to\infty} k^{1/k} = \lim_{k\to\infty} e^{(\log k)/k} = e^{\lim ((\log k)/k)}$.)

10 $\displaystyle\lim_{x\to 3} \frac{x^2 - 9}{x - 3} = \lim_{x\to 3} \frac{2x}{1} = 6.$

12 $\displaystyle\lim_{x\to 0} \frac{\sin x - x}{x} = \lim_{x\to 0} \frac{\cos x - 1}{1} = 0.$

14 -1. **16** $\frac{1}{2}$. **18** 0. **20** 0.

22 0. **24** $+\infty$. **26** $+\infty$. **28** 1.

30 $\displaystyle\lim_{x\to+\infty} \frac{1}{x} \log(1 + x^2) = \lim_{x\to+\infty} \frac{2x}{1 + x^2} = \lim_{x\to+\infty} \frac{2}{2x} = 0.$ Therefore, $\displaystyle\lim_{x\to+\infty} (1 + x^2)^{1/x} = 1.$

32 1. **34** 1.

36 Clearly, $\displaystyle\lim_{x\to\infty} \frac{x - \sin x}{x + \sin x} = \lim_{x\to\infty} \left(1 - \frac{2 \sin x}{x + \sin x}\right) = 1.$

But $\displaystyle\lim_{x\to\infty} \frac{1 - \cos x}{1 + \cos x}$ does not exist, because $\dfrac{1 - \cos x}{1 + \cos x}$ is 0 for every even multiple of π and becomes arbitrarily large for values near every odd multiple of π.

---

page 426

4.3

1 $\cos x = \dfrac{\sqrt{2}}{2} - \dfrac{\sqrt{2}}{2}\left(x - \dfrac{\pi}{4}\right) - \dfrac{\sqrt{2}}{2}\left(x - \dfrac{\pi}{4}\right)^2 + \dfrac{\sqrt{2}}{12}\left(x - \dfrac{\pi}{4}\right)^3$

$$+\frac{\sqrt{2}}{48}\left(x-\frac{\pi}{4}\right)^4-\frac{\sin(\bar{x})}{120}\left(x-\frac{\pi}{4}\right)^5, \text{ where } \bar{x} \text{ is between } \pi/4 \text{ and } x.$$

3 $\log x = x - \frac{x^2}{2} + \frac{x^3}{3} - \frac{x^4}{4} + \frac{1}{\bar{x}^5}\frac{x^5}{5}$, where $\bar{x}$ is between 1 and x.

5 $e^{-x} = 1 - x + \frac{x^2}{2!} - \frac{x^3}{3!} + \frac{x^4}{4!} - \frac{x^5}{5!} + \frac{e^{-\bar{x}}}{6!}x^6$, where $\bar{x}$ is between 0 and x.

7 $$\tan x = 1 + 2\left(x-\frac{\pi}{4}\right) + 2\left(x-\frac{\pi}{4}\right)^2 + 4\left(x-\frac{\pi}{4}\right)^3 + \frac{\sec^2\bar{x}[5\tan^3\bar{x} + 3\tan\bar{x}]}{3}\left(x-\frac{\pi}{4}\right)^4.$$

9 $\arcsin(x) = x + \frac{x^3}{3!} + \frac{3\bar{x}(2\bar{x}^2+3)}{4!(1-\bar{x}^2)^{7/2}}x^4$, where $\bar{x}$ is between 0 and x.

11 $$\arctan(x) = \frac{\pi}{4} + \frac{1}{2}(x-1) - \tfrac{1}{4}(x-1)^2 + \frac{1}{12}(x-1)^3 + \frac{(\ \ -\bar{x}^2)}{(1+\bar{x}^2)^4}x^4.$$

13 $$\sin 61° \approx \frac{\sqrt{3}}{2} + \frac{1}{2}\left(\frac{\pi}{180}\right) - \frac{\sqrt{3}}{4}\left(\frac{\pi}{180}\right)^2 - \frac{1}{12}\left(\frac{\pi}{180}\right)^3 + \frac{\sqrt{3}}{48}\left(\frac{\pi}{180}\right)^4$$
$$= 0.874619,\ |R_4| \le \frac{1}{120}\left(\frac{\pi}{180}\right)^5.$$

16 $\sqrt[3]{28} \approx 3 + \frac{1}{27} \approx 3.037$.

---

chapter 5

5.1

page 436

1 (a) diverges, (c) converges, (e) diverges.

2 (a) 5/4, (c) 3/5, (e) $4/(1 + 3^{2/3})$.

3 14/99. **4** (a) $1/(1 - r^2)$, **5** (b) 1.

6 3/4. **12** The series diverges.

14 (a) Consider the partial sum $S_n = (a_2 - a_1) + (a_3 - a_2) + \cdots + (a_{n+1} - a_n) = a_{n+1} - a_1$. Because $\lim_{n\to\infty}(a_{n+1} - a_1) = \infty$, the series diverges.

page 442

5.2

1 converges. **3** diverges. **5** diverges.

7 diverges. **9** converges. **11** diverges.

13 converges. **15** converges. **17** diverges.

19 The nth term can be written as

$$a_n = \frac{n}{(2n-1)(2n+1)},$$

from which the series is easily seen to diverge. For example, $n/(4n^2 - 1) > 1/4n$, so the series dominates $\frac{1}{4}\sum_{n=1}^{\infty} 1/n$ and is divergent.

25 (a) Since the convergence or divergence of a series is unaffected by deleting a finite number of terms, we may assume $N = 1$. Suppose first there is an $r < 1$ with the stated property. Then for each k, $a_k \leq r^k$, and the series $\sum_{k=1}^{\infty} a_k$ can be compared with the convergent geometric series $\sum_{k=1}^{\infty} r^k$. It is convergent. If there is, on the other hand, an $s > 1$ such that $\sqrt[k]{a_k} \geq s$, then $a_k \geq s^k$ for each k and the series $\sum_{k=1}^{\infty} a_k$ dominates the divergent geometric series $\sum_{k=1}^{\infty} s^k$. It therefore diverges.

27 The error is bounded by $\displaystyle\int_{100}^{\infty} \frac{1}{x^2}\,dx = \lim_{t\to\infty}\left(\frac{-1}{t} + \frac{1}{100}\right) = \frac{1}{100}.$

Thus the sum is approximated to within 0.01.

---

page 448

5.3

1 absolutely convergent. **3** absolutely convergent.

5 divergent. **7** absolutely convergent.

9 conditionally convergent. **11** absolutely convergent.

13 absolutely convergent. **15** 999.

18 There are many such rearrangements possible; here is a means of constructing one. The series of positive terms $1 + \frac{1}{3} + \frac{1}{5} + \cdots$ is divergent. It is therefore possible by inserting parentheses to group the series into blocks of terms each of whose sum exceeds $\frac{1}{2}$:

$$(1) + (\tfrac{1}{3} + \tfrac{1}{5}) + (\tfrac{1}{7} + \tfrac{1}{9} + \tfrac{1}{11} + \cdots + \tfrac{1}{21}) + (\tfrac{1}{23} + \cdots) + \cdots.$$

(The blocks of course become longer and longer.) Now construct a rearrangement of the original series by taking alternately a *block* of positive terms and a *single* negative term (in order):

$$1 - \tfrac{1}{2} + \tfrac{1}{3} + \tfrac{1}{5} - \tfrac{1}{4} + \tfrac{1}{7} + \tfrac{1}{9} + \cdots + \tfrac{1}{21} - \tfrac{1}{6} + \tfrac{1}{23} + \cdots.$$

It is clear that this is a rearrangement, i.e., each term of the original series appears exactly once. In addition, the series must diverge, as one can see by considering those partial sums S_{n_k} whose last term is the kth negative term, i.e.,

$$S_{n_k} = 1 - \tfrac{1}{2} + \tfrac{1}{3} + \tfrac{1}{5} - \tfrac{1}{4} + \cdots - \frac{1}{2k} > \left(\tfrac{1}{2} - \tfrac{1}{2}\right) + \left(\tfrac{1}{2} - \tfrac{1}{4}\right) + \left(\tfrac{1}{2} - \tfrac{1}{6}\right) + \cdots + \left(\tfrac{1}{2} - \tfrac{1}{2^k}\right).$$

It is easy to see that the S_{n_k} form an unbounded sequence, which implies that the rearranged series diverges.

---

page 453

5.4

1 (–2, 2). **3** [1, 3]. **5** (–4, –2].

7 (–1, 1). **9** [–1, 1]. **11** [–4/3, 4/3].

13 [–1, 1].

14 We note that $\lim\limits_{k\to\infty} \dfrac{|a_k|}{|a_{k+1}|}$ does not exist because the quotient oscillates between 2 and 3. Thus Theorem 5.4.3 cannot be applied. However, observe that the series of even and odd powers

$$1 + \frac{1}{2 \cdot 3} x^2 + \frac{1}{2^2 \cdot 3^2} x^4 + \cdots = \frac{1}{1 - (x^2/6)}$$

and

$$\tfrac{1}{2}x + \frac{1}{2^2 \cdot 3} x^3 + \frac{1}{2^3 \cdot 3^2} x^5 + \cdots = \frac{\tfrac{1}{2}x}{1 - (x^2/6)}$$

are both geometric series with sums $6/(6 - x^2)$ and $3x/(6 - x^2)$ respectively. Moreover they both converge for $x^2/6 < 1$, i.e., in the interval $(-\sqrt{6}, \sqrt{6})$. Thus the sum of the original series is

$$(3x + 6)/(6 - x^2).$$

---

page 458

5.5

2 *Theorem* Let $f(x) = \sum_{k=0}^{\infty} a_k(x - c)^k$ for $|x - c| < r$. Then

$$\int_c^x f(t)\,dt = \sum_{k=0}^{\infty} \frac{a_k}{k+1} x^{k+1} \quad \text{for} \quad |x - c| < r.$$

Proof Let $g(z) = \sum_{k=0}^{\infty} a_k z^k$ for $|z| < r$. Then by substitution and Theorem 5.5.1, $\int_c^x f(t)\,dt = \int_0^{x-c} g(z)\,dz = \sum_{k=0}^{\infty} a_k(x - c)^{k+1}$.

3 (a) $\log 1/(1 - x)$,

(b) No. Part (i) fails. For example, if $f(x) = \sum_{k=1}^{\infty} \frac{x^k}{k}$, then it is easy to see that the interval of convergence is $[-1, 1)$. But the series $f'(x) = \sum_{k=1}^{\infty} x^{k-1}$ does not converge for $x = -1$. (Part (ii) can be extended.)

5 (a) $f(x) = \dfrac{2}{1+x} = 2 - 2x + 2x^2 - \cdots + (-1)^n 2x^n + \cdots;$

(c) $f(x) = \dfrac{1}{3-x} = \dfrac{1}{3}\dfrac{1}{1-(x/3)} = \dfrac{1}{3} + \dfrac{1}{9}x + \dfrac{1}{27}x^2$
$$+ \cdots + \frac{x^n}{3^{n+1}} + \cdots;$$

(e) $\dfrac{1}{1+x} + \dfrac{1}{1-x} = 2(1 + x^2 + x^4 + \cdots + x^{2n} + \cdots).$

6 (a) $(-1, 1)$, (c) $(-3, 3)$, (e) $(-1, 1)$.

8 $\arctan(x) = x - \dfrac{1}{3}x^3 + \dfrac{1}{5}x^5 - \cdots + (-1)^n \dfrac{1}{2n+1}x^{2n+1} + \cdots.$

The radius of convergence is 1.

10 (a) $\log(1-x) = -x - \dfrac{1}{2}x^2 - \dfrac{1}{3}x^3 - \cdots - \dfrac{1}{n+1}x^{n+1} - \cdots;$

(c) $\log\left(\dfrac{1}{1+x}\right) = -\log(1+x)$
$$= -x + \frac{1}{2}x^2 - \frac{1}{3}x^3 + \cdots + (-1)^{n+1}\frac{1}{n+1}x^{n+1} + \cdots.$$

11 .1192.

12 $x - \dfrac{1}{4}x^2 + \dfrac{1}{9}x^3 - \cdots + (-1)^n \dfrac{1}{(n+1)^2}x^{n+1} + \cdots.$

---

page 459

5.6

2 (a) $2x - \dfrac{8}{3!}x^3 + \dfrac{32}{5!}x^5 - \cdots + (-1)^n \dfrac{2^{2n+1}}{(2n+1)!}x^{2n+1} + \cdots;$

(c) $1 - x + \dfrac{x^2}{2!} - \cdots + \dfrac{(-1)^n x^n}{n!} + \cdots;$

(e) $1 - \dfrac{x^2}{3!} + \dfrac{x^4}{5!} - \dfrac{x^6}{7!} + \cdots + (-1)^n \dfrac{x^{2n}}{(2n+1)!} + \cdots.$

3 $1 + \dfrac{2x^2}{2!} + \dfrac{8x^4}{4!} + \cdots + (-1)^n \dfrac{2^{2n-1}x^{2n}}{(2n)!} + \cdots.$

6 (a) $1 - (x - 1) + (x - 1)^2 - (x - 1)^3 + \cdots + (-1)^n(x - 1)^n + \cdots;$

(c) $$1 - \frac{\left(x - \frac{\pi}{2}\right)^2}{2!} + \frac{\left(x - \frac{\pi}{2}\right)^4}{4!} - \cdots + (-1)^n \frac{\left(x - \frac{\pi}{2}\right)^{2n}}{(2n)!} + \cdots.$$

7 .38942.

8 (a) 0.8830, (c) 1.1052.

10 $$\frac{\sqrt{2}}{2}\left(1 - \left(x - \frac{\pi}{4}\right) - \frac{\left(x - \frac{\pi}{4}\right)^2}{2!} + \frac{\left(x - \frac{\pi}{4}\right)^3}{3!} + \frac{\left(x - \frac{\pi}{4}\right)^4}{4!} - \cdots\right).$$

12 $$-\frac{x}{2!} + \frac{x^3}{4!} - \frac{x^5}{6!} + \cdots + \frac{(-1)^n x^{2n-1}}{(2n)!} + \cdots.$$

---

page 468

5.7

1 (a) $$1 + \frac{1}{2}x - \frac{1}{8}x^2 + \frac{1}{16}x^3 - \cdots + (-1)^{k+1}\frac{3 \cdot 5 \cdot 7 \cdots (2k - 3)}{2^k k!}x^k + \cdots;$$

(c) $$2^{1/3}\left(1 + \frac{1}{6}x - \frac{1}{36}x^2 + \frac{5}{648}x^3 - \cdots + (-1)^{k+1}\frac{2 \cdot 5 \cdot 8 \cdots (3k + 2)}{3^k \cdot 2^k \cdot k!}x^k + \cdots\right);$$

(e) $1 - 6x + 27x^2 - 108x^3 + \cdots + (-1)^k(k + 1)3^k x^k + \cdots;$

(g) $$\frac{1}{8}\left(1 + \frac{1}{2}x^2 + \frac{1}{2}x^4 + \cdots + \frac{1 \cdot 4 \cdot 7 \cdots (3k - 2)}{2^k k!}x^{2k} + \cdots\right).$$

2 (a) We use the series for $\sqrt{1 + x} = 1 + \frac{1}{2}x - \frac{1}{8}x^2 + \frac{1}{16}x^3 - \cdots$ with $x = 0.02$. Because the series is an alternating series, the error in summing a finite number of terms does not exceed the first omitted term. In particular, $\frac{1}{16}(.02)^3 < 0.000005$, so

$$\sqrt{1.02} \approx 1 + \tfrac{1}{2}(0.02) - \tfrac{1}{8}(0.02)^2 = 1.00995,$$

an estimate correct to five decimal places.

(c) -0.02020.

3 0.5077. **4** 0.2000. **6** 0.0976.

9 $x + \frac{1}{2}x^2 + \frac{1}{3}x^3 + \frac{3}{40}x^5 + \cdots.$

11 Although we have shown only that the series (5.7.3) converges for $-1 < x < 1$, it can be shown that it also converges for $x = -1$, giving

$$\log 2 = \log (1 - (-1)) = 1 - \tfrac{1}{2} + \tfrac{1}{3} - \cdots,$$

which identifies the sum of the alternating harmonic series. From (5.7.4) we have, on taking $x = 1/3$,

$$\log 2 = 2(\tfrac{1}{3} + \tfrac{1}{81} + \tfrac{1}{1215} + \cdots).$$

It is clear that the second series converges much more rapidly than the first.

---

chapter 6

page 472

6.1

1 Yes. **3** Yes. **5** Yes. **7** No. **9** No.

---

page 480

6.2

1 (a), (d), and (e).

3 (a) 0, (c) 0, (e) 1.

4 (a) tangent line: $x + y - 2 = 0$,

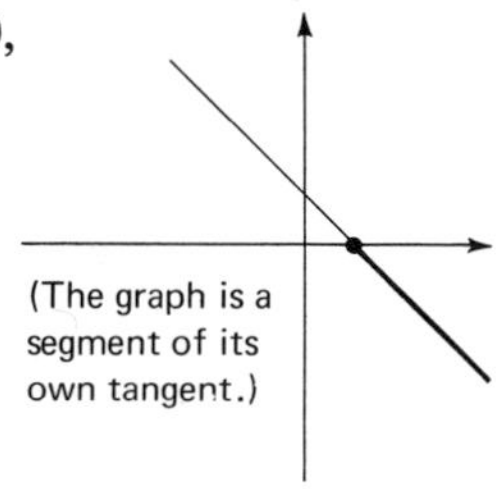

(c) $\dfrac{dy}{dx} = -\left(\dfrac{1+t}{1-t}\right)^{1/2}$ for $t \neq 1$. Moreover, $\lim\limits_{t\to 1} \dfrac{dy}{dx} = \infty$, so the curve has a vertical tangent $x = \sqrt{2}$ at the point $(\sqrt{2}, 0)$.

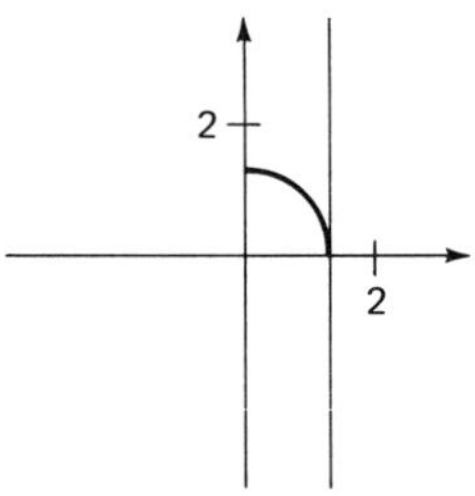

6 (4/9, 8/27).

8 Because $dy/dx = \sin t/(1 - \cos t)$, any local extreme values must occur where $t = k\pi$, k an integer. Moreover, one can easily see that $0 \leq y \leq 2$. Because $y = 0$ for $t = k\pi$, k an even integer, the curve has a local minimum at each of these points. On the other hand, $y = 2$ for $t = k\pi$, k an odd integer, so each of these points is a local maximum. Now,

$$\frac{d^2y}{dx^2} = \frac{d}{dx}\left(\frac{dy}{dx}\right) = \frac{\dfrac{d}{dt}\left(\dfrac{dy}{dx}\right)}{\dfrac{dx}{dt}} = \frac{-1}{(1 - \cos t)^2}.$$

Thus the only candidates for points of inflection are points $t = k\pi$, k an even integer, where the second derivative does not exist. But the first derivative does not exist at these points either, so there cannot be any extreme values there. The curve thus has no points of inflection.

---

page 486

6.3

3 on a circle of radius 6 centered at the pole; on a line through the pole making an angle of $\pi/3$ with the polar axis.

7 (a)

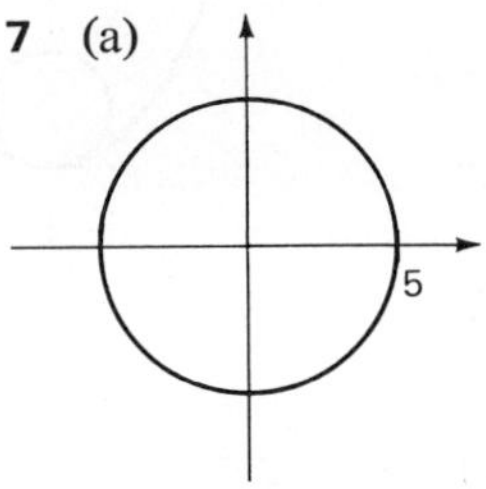

(c)

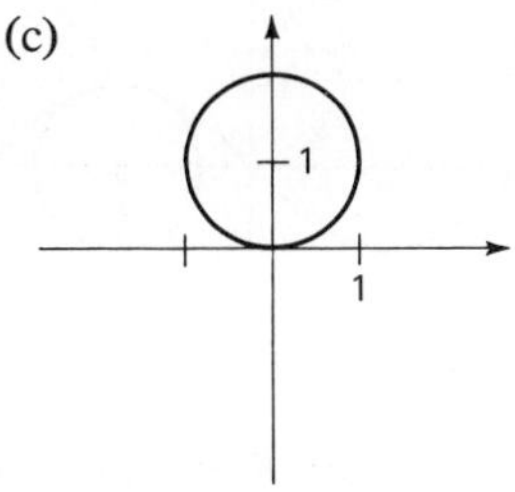

(e)

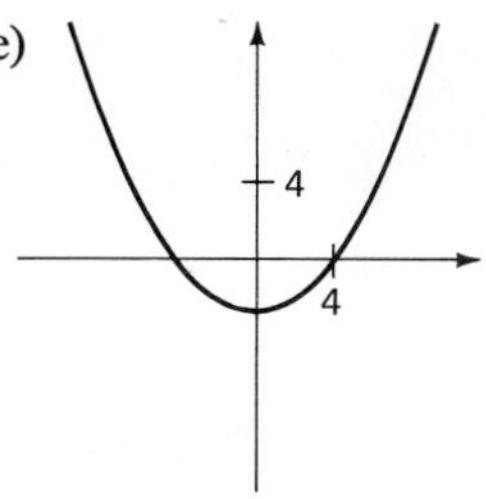

9 (a) $\sqrt{3}$, (c) $\frac{3}{4}$, (e) 0.

11 Points of intersection are
(a) $(2, \pi/3)$ and $(2, 5\pi/3)$;

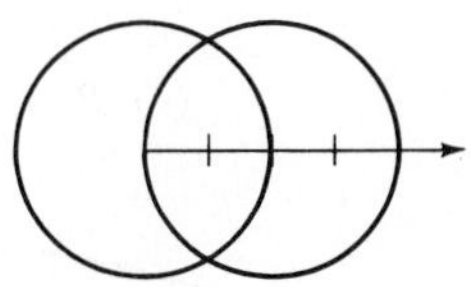

(c) $(\frac{1}{2}, \pi/6)$, $(\frac{1}{2}, 5\pi/6)$, $(-1, 3\pi/2)$, $(0, 0)$.

---

page 492

6.4

1 64/5. **3** 2π. **5** $\frac{a^2}{8}\left(\frac{\pi}{2}+1\right)$.

7 $9\pi/2$. **9** $a^2/2$. **11** $8a$.

13 $6a$. **15** $\frac{2}{3}(4+4\pi^2)^{3/2} - 16/3$.

17 $2a\sqrt{3}\left[\frac{\sqrt{7}-2}{\sqrt{3}} + \log\frac{2(2+\sqrt{3})}{\sqrt{3}+\sqrt{7}}\right]$.

---

page 501

6.5

1 (a) $4/17^{3/2}$, (c) $4/2^{3/2}$, (e) $2|x_1^3|/(1+x_1^4)^{3/2}$,
(g) 1, (i) $\sqrt{2}/4$, (k) $\sqrt{2}/4$.

2 (a) $\sqrt{2}$, (c) 1, (e) $\frac{8\cdot 10^{3/2}}{3}$.

4 (a) $R=\frac{1}{2}$, $x^2+(y-\frac{1}{2})^2=\frac{1}{4}$;

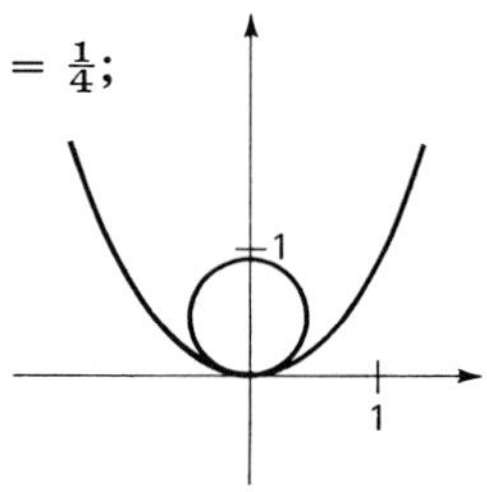

(b)

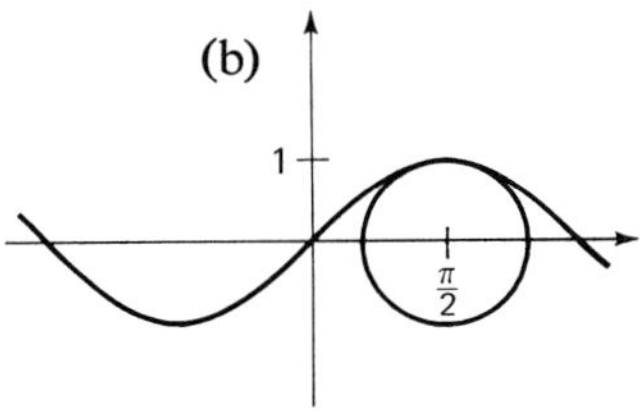

(c) $R = 145^{3/2}/12$, $(x+143)^2 + \left(y - \frac{241}{12}\right)^2 = \frac{145^3}{144}$;

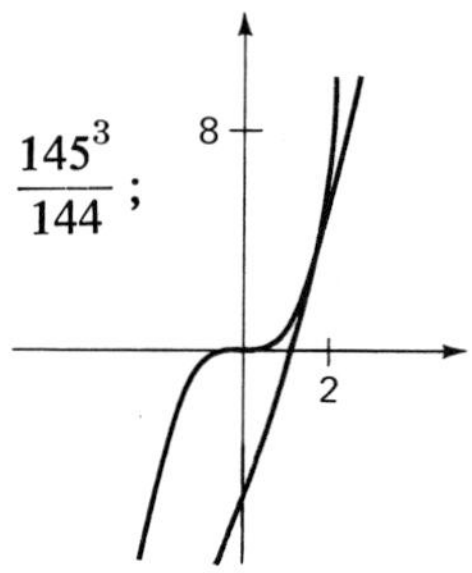

(d)

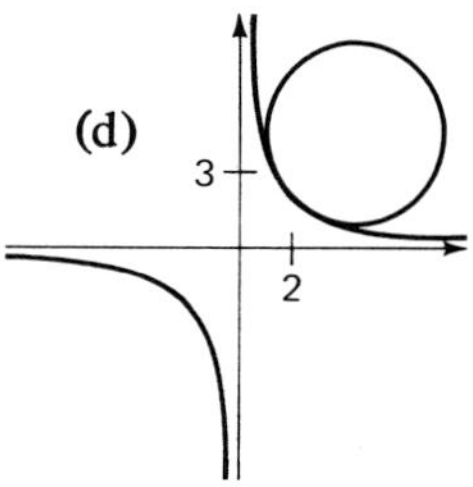

(e) $R = 5^{3/2}/2$, $(x-\frac{7}{2})^2 + (y+4)^2 = \frac{5^3}{4}$.

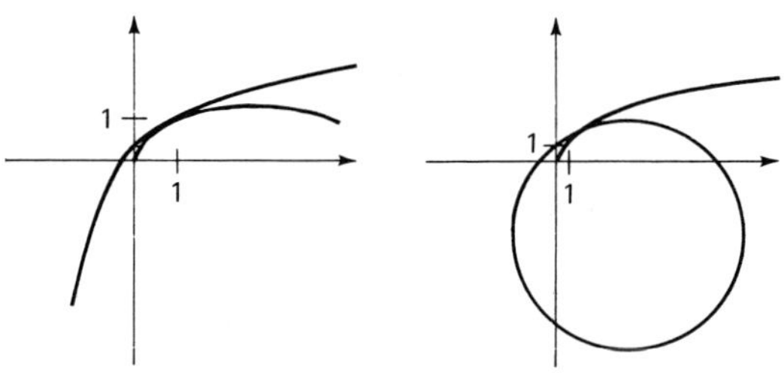

(f)

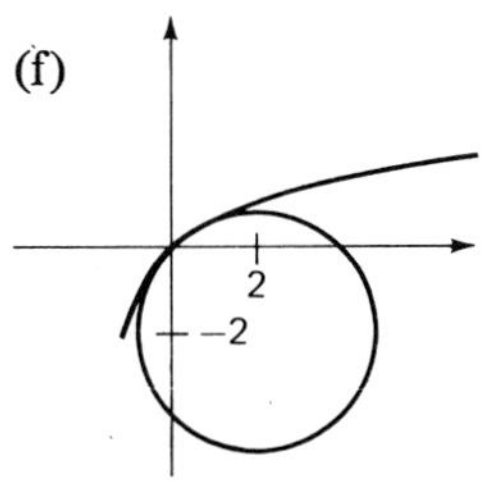

5 One can use the formulas in Exercise 3 directly, because the parameter may be eliminated ($y = x^2$). Thus, for each t, $x_0 = t$, $y_0 = t^2, f'(x_0) = 2x_0 = 2t, f''(x_0) = 2$. Hence

$$x_1(t) = t - 2t\frac{(1 + 4t^2)}{2} = -4t^3,$$

$$y_1(t) = t^2 + \frac{1 + 4t^2}{2} = 3t^2 + \tfrac{1}{2}$$

are parametric equations for the curve traced out.

7 Cartesian coordinates may always be regarded as parametric equations by using x as the parameter. Thus taking $y = f(x)$, $x = g(x)$, we have $g'(x) = 1$, $g''(x) = 0$. Substituting these values (at x_0) into the formula of the theorem gives

$$K = \frac{|1 \cdot f''(x_0) - f'(x_0) \cdot 0|}{(f'(x_0)^2 + 1)^{3/2}} = \frac{|f''(x_0)|}{(1 + f'(x_0)^2)^{3/2}}.$$

11 (6.5.10) From (6.5.8) and (6.5.9) we have

$$R' = \frac{1}{2}\frac{|PP'|^2}{|P'Q'|} = \frac{(x_1 - x_0)^2 + (y_1 - y_0)^2}{\left(\dfrac{|y_1 - y_0 - m(x_1 - x_0)|}{\sqrt{1 + m^2}}\right)}$$

$$= \tfrac{1}{2}\sqrt{1 + m^2}\left|\frac{(x_1 - x_0)^2 + (y_1 - y_0)^2}{y_1 - y_0 - m(x_1 - x_0)}\right|.$$

(6.5.11) From (6.5.10)

$$R' = \tfrac{1}{2}\sqrt{1 + (f'(t_0)/g'(t_0))^2}\left|\frac{[g(t) - g(t_0)]^2 + [f(t) - f(t_0)]^2}{f(t) - f(t_0) - (f'(t_0)/g'(t_0))(g(t) - g(t_0))}\right|$$

$$= \tfrac{1}{2}\sqrt{[g'(t_0)]^2 + [f'(t_0)]^2}\left|\frac{[g(t) - g(t_0)]^2 + [f(t) - f(t_0)]^2}{g'(t_0)(f(t) - f(t_0)) - f'(t_0)(g(t) - g(t_0))}\right|$$

13 Suppose the graph of f has zero curvature at each point. From the formula for the curvature given in Corollary 6.5.2, we see that $f''(x) = 0$ at every point. Thus $f'(x)$ is a constant, say A, and $f(x) = Ax + B$. This is the equation of a straight line.

---

page 509

6.6

1 $\langle 1, 6\rangle$. **3** $\langle -2, 1\rangle$. **5** $\langle 1, 11/4\rangle$. **7** $\langle 12, 7\rangle$.

9 $|\mathbf{A} + \mathbf{B}| = \sqrt{37}$, $|\mathbf{B} - \mathbf{D}| = \sqrt{74}$, $|2\mathbf{A} + 3\mathbf{C}| = \sqrt{34}$.

13 $\alpha = 2$ and $\beta = 4$.

14 Let $\mathbf{A} = \langle x, y \rangle$ and $\mathbf{B} = \langle r, s \rangle$. Then

$|\mathbf{A}|^2 = x^2 + y^2$, $|\mathbf{B}|^2 = r^2 + s^2$, $|\mathbf{A} + \mathbf{B}|^2 = (x + r)^2 + (y + s)^2 = x^2 + 2xr + r^2 + y^2 + 2ys + s^2$.

$$\begin{aligned} \text{Thus } |\mathbf{A} - \mathbf{B}|^2 &= (x - r)^2 + (y - s)^2 = x^2 - 2xr + r^2 + y^2 \\ &\quad - 2ys + s^2 \\ &= 2x^2 + 2r^2 + 2y^2 + 2s^2 \\ &\quad - (x^2 + 2xr + r^2 + y^2 + 2ys + s^2) \\ &= 2|\mathbf{A}|^2 + 2|\mathbf{B}|^2 - |\mathbf{A} + \mathbf{B}|^2 \\ &= 338 + 722 - 576 = 484. \end{aligned}$$

Hence $|\mathbf{A} - \mathbf{B}| = 22$.

16 $\langle 1, -10 \rangle$.

18 (a) $\langle -8, 6 \rangle$, 10; (c) $\langle -12, -5 \rangle$, 13.

---

page 514

6.7

1 $v(1) = \langle 1, 4 \rangle$, $a(1) = \langle 0, 4 \rangle$.

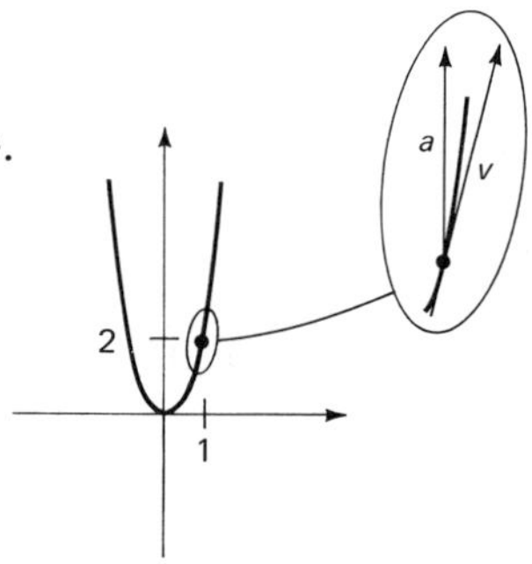

2

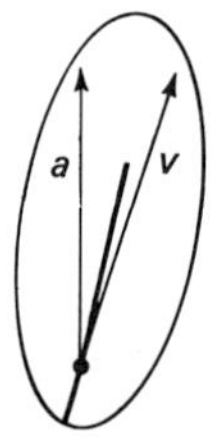

3 $v\left(\dfrac{3\pi}{4}\right) = \langle -\sqrt{2}, -\sqrt{2} \rangle$, $a\left(\dfrac{3\pi}{4}\right) = \langle \sqrt{2}, -\sqrt{2} \rangle$.

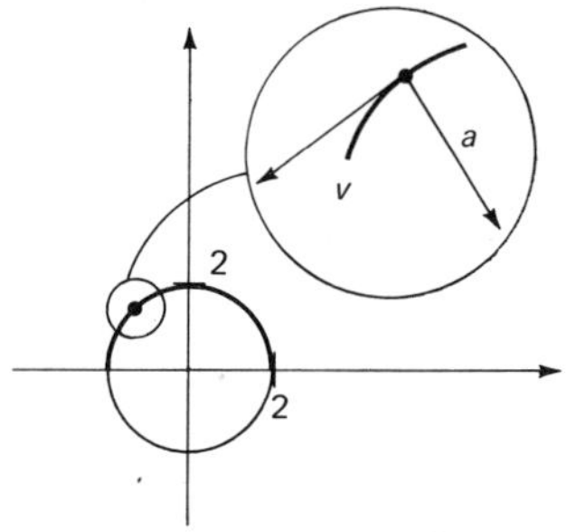

5 $v(0) = \langle 0, 1 \rangle$, $a(0) = \langle 2, 1 \rangle$.

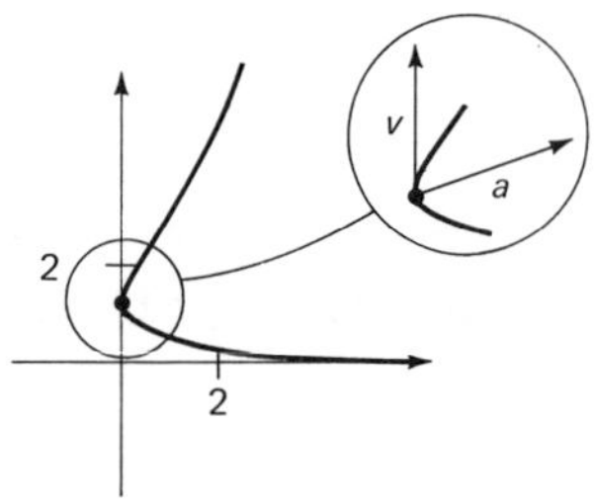

7 $v(-\pi) = \langle 0, 1\rangle$, $a(-\pi) = \langle 1, 0\rangle$.

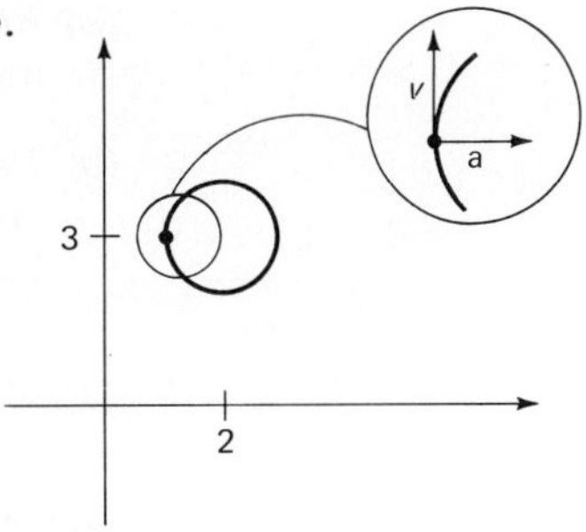

8

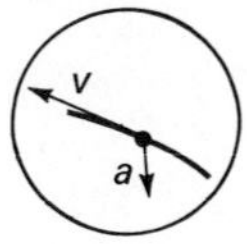

9 11,250 ft., $2{,}812\frac{1}{2}$ ft.

10 The maximum range is $(1/32)v_0^2$ feet.

11 The maximum height is reached when $t = (v_0 \sin\alpha)/32$. This maximum height is $(v_0 \sin\alpha)^2/64$ ft. The shell hits the ground when $(v_0 \sin\alpha)t - 16t^2 = 0$, or $t = (v_0 \sin\alpha)/16$. Its range is $(v_0 \cos\alpha) \times (v_0 \sin\alpha)/16 = \frac{1}{32}v_0^2 \sin 2\alpha$ ft.

---

chapter **7**

page 526

7.1

1

3

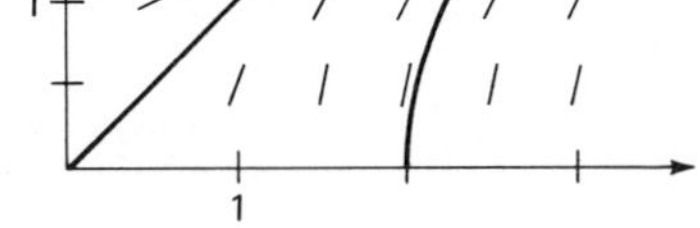

5 $f(x) = f'(x) + 1$.

7 $m(x, y) = -x/y$.

---

page 533

7.2

1 (a) second, (c) first.

3 (a) $\{(x, y) \mid y \neq 0\}$, (c) the entire xy-plane,
(e) $\{(x, y) \mid x^2 + y^2 \neq 4\}$.

4 (b) Any interval will do. (The unique solution $y = (x - 1)^3$ is valid for $-\infty < x < +\infty$.)

6 (a) The equation can be written $(x - c)^2 + y^2 = 1$. For fixed c the equation is the equation of a circle of radius 1 and center at $(c, 0)$.

7 (a) $x^2\left(1 - \dfrac{dy}{dx}\right) = 2xy$, (c) $\dfrac{dy}{dx} = \dfrac{2xy}{2x^2 + y}$.

9 (a) $x^2 \dfrac{d^2y}{dx^2} = -2x\dfrac{dy}{dx} + y = 0$, (c) $\dfrac{d^2y}{dx^2} = y$, (e) $\dfrac{d^2y}{dx^2} = -y$.

10 $(y')^2 + yy'' + 1 = 0$.

11 $dy/dx = -x/y$.

---

page 540

7.3

1 (a)

i	x_i	y_i
0	1.0	0.0000
1	1.2	0.2000
2	1.4	0.3667
3	1.6	0.5095
4	1.8	0.6345
5	2.0	0.7456

$\Delta x_i = 0.2$

(c)

i	x_i	y_i
0	0.0	−1.0000
1	−0.25	−1.0000
2	−0.5	−0.8750
3	−0.75	−0.6250
4	−1.00	−0.2500

$\Delta x_i = -0.25$

(e)

i	x_i	y_i
0	−2.0	2.0000
1	−1.9	1.2000
2	−1.8	0.9264
3	−1.7	0.7719
4	−1.6	0.6706
5	−1.5	0.5987

$\Delta x_i = 0.1$

3 (b) Let z_n denote the approximation to $y(1) = e$ obtained by taking $\Delta x_i = 1/n$. Then $\lim_{n\to\infty} z_n = \lim_{n\to\infty}\left(1 + \dfrac{1}{n}\right)^n = e$.

4 (c) Use the formula $y(x_{i+1}) = y(x_i) + y'(x_i)\,\Delta x_i + (y''(x_i)/2)(\Delta x_i)^2$.

(d) Note that $d^2y/dx^2 = dy/dx = y$ is obtained by differentiating the original differential equation. So in this case the formula in (c) gives

$$y_{i+1} = \left(1 + \Delta x_i + \frac{\Delta x_i^2}{2}\right) y_i = (1.105)y_i.$$

i	x_i	y_i
0	.0	1.0000
1	.1	1.1050
2	.2	1.2210
3	.3	1.3492
4	.4	1.4909
5	.5	1.6474
6	.6	1.8204
7	.7	2.0115
8	.8	2.2227
9	.9	2.4561
10	1.0	2.7140

5 (a) (*Hint:* According to Exercise 4, the error in our approximation is

$$\frac{y''(x_i)}{2}(\Delta x_i)^2 + \frac{y'''(x_i)}{3!}(\Delta x_i)^3 + \cdots + \frac{y^{(n)}(x_i)}{n!}(\Delta x_i)^n + \cdots.$$

For small values of Δx_i this error can be approximated well by its first term.)

6 (b) $y_{i+1} \approx y_i + \int_{x_i}^{x_{i+1}} \frac{1}{2}[m(x_i, y_i) + m(x_{i+1}, y_i + m(x_i, y_i)\,\Delta x_i)]\,dx$
$= y_i + \frac{1}{2}m(x_i, y_i)\,\Delta x_i + \frac{1}{2}m(x_{i+1}, y_i + m(x_i, y_i)\,\Delta x_i)\,\Delta x_i.$

---

page 546

7.4

1 $y = \frac{2}{3} + C(x^2 + 1)^{-3/2}.$ **3** $y = Ce^{-x} + \frac{1}{3}e^{2x}.$

5 $y = C'(x + 4)^{7/4}/x^{3/4}.$

7 $\frac{1}{2}x(3x^2 + 7)^{1/2} + \frac{7\sqrt{3}}{6}\log(x + \sqrt{x^2 + 7/3}) + (4 - 5y^2)^{1/2} - 2\left(\log\frac{2 + \sqrt{4 - 5y^2}}{y}\right) = C.$

9 $y = x^3 + \frac{C}{x^3}.$ **11** $e^y(y - 1) - \frac{1}{2}x^2 + \log x = C.$

13 $y(x) = 2\sin x + C/\sin^2 x.$

15 The equation is linear *in* x: $\frac{dx}{dy} - \frac{3}{y}x = y^4,$

$$x(y) = y^3\left[\int y^4 \cdot \frac{1}{y^3}\,dy + C\right] = \frac{1}{2}y^5 + Cy^3.$$

17 (*Hint:* If $y = vx$, then $dy = v\,dx + x\,dv$.) The solution is

$$x^2/\sqrt{4x^2 + y^2} = C.$$

---

page 552

7.5

1 $y = \left(\frac{x + C}{2}\right)^2.$

3 $y = \pm \sin(x + C)$. One obtains $\cos x$ by observing that $\cos x = \sin(x + \pi/2)$.

5 Let $x(t)$ denote the position of the particle at time t.

(a) The motion is governed by the equation $dx/dt = K(6 - x)$, which has as solutions $x(t) = 6 + Ce^{-Kt}$. Because $x(0) = 0$,

$$C = -6, \text{ and because } \left.\frac{dx}{dt}\right|_0 = 30, K = 5, \text{ and } x(t) = 6(1 - e^{-5t}).$$

7 (a) Let $x(t)$ denote the amount of salt in the tank at time t; $x(t) = 200(1 - e^{-t/20})$ lbs.

(b) $t = -20 \log(.95) \approx 1.02$ min.

9 Let $x(t)$ denote the amount of salt in tank A at time t. Then $200 - x(t)$ is the amount in tank B, and the conditions require

$$\frac{dx}{dt} = 5\left(\frac{200 - x(t)}{100}\right) - 5\left(\frac{x(t)}{100}\right) = \frac{100 - x}{10}.$$

The variables are separable, and

$$\log(x - 100) = -\frac{t}{10} + C,$$

or $x = 100 - C'e^{-t/10}$. Now $x(0) = 0$, so $C' = 100$ and $x(t) = 100(1 - e^{-t/10})$ lbs. is the amount of salt in tank A at time t. Then $100 + e^{-t/10}$ lbs. is the amount of salt in tank B at time t. Because $\lim_{t\to\infty} x(t) = 100$, there will be approximately 100 lbs. of salt in each tank after a long time.

10 (b) $v = a\left(\frac{C'e^{2akt/m} - 1}{C'e^{2akt/m} + 1}\right) = \sqrt{\frac{mg}{k}}\left(\frac{C'e^{2\sqrt{kg/m}t} - 1}{C'e^{2\sqrt{kg/m}t} + 1}\right).$

Note that $\lim_{t\to\infty} v(t) = \sqrt{mg/k}$.

11 (b) $v = (2gR^2/r + v_0^2 - 2gR)^{1/2}$.

(c) For the projectile to escape we must have $\lim_{r\to\infty} v(r) \geq 0$. Now, $\lim_{r\to\infty} v(r) = (v_0^2 - 2gR)^{1/2}$, so the least velocity which will allow the particle to escape is $v_e = \sqrt{2gR}$.

12 Approximately 82% of the colony remains at the end of 1 hour.

---

page 561

7.6

1 (a) $\sum_{i=0}^{\infty} u_i = \sum_{i=-2}^{\infty} u_{i+2} = \sum_{i=-1}^{\infty} u_{i+1},$

(c) $\sum_{i=1}^{\infty} i(i-1)x^{i-2} = \sum_{i=-1}^{\infty} (i+2)(i+1)x^i,$

(e) $x^2 \sum_{i=0}^{\infty} iu_i x^{i-1} = \sum_{i=0}^{\infty} iu_i x^{i+1} = \sum_{i=1}^{\infty} (i-1)u_{i-1}x^i,$

(g) $\sum_{i=1}^{\infty} u_i + \sum_{i=0}^{\infty} u_{i+2} = \sum_{i=1}^{\infty} u_i + \sum_{i=2}^{\infty} u_i = u_1 + 2\sum_{i=2}^{\infty} u_i.$

3 (a) We suppose a solution of the form

$$y(x) = u_0 + u_1 x + \cdots + u_i x^i + \cdots$$

exists. Then

$$y'(x) = u_1 + 2u_2 x^2 + \cdots + iu_i x^{i-1} + \cdots,$$

and the equation requires

$$(u_1 + 2u_2 x^2 + \cdots + (i+1)u_{i+1}x^i + \cdots) - (u_0 + u_1 x + \cdots + u_i x^i + \cdots) = 0,$$

or

$$(u_1 - u_0) + (2u_2 - u_1)x + (3u_3 - u_2)x^2 + \cdots + ((i+1)u_{i+1} - u_i)x^i + \cdots = 0.$$

This gives the recurrence relations $u_1 = u_0$, $u_2 = u_1/2, \cdots$, $u_{i+1} = u_i/(i+1), \cdots$. Thus if u_0 is arbitrarily chosen, the other coefficients are determined and

$$y(x) = u_0\left(1 + \frac{x^2}{2} + \frac{x^4}{8} + \cdots + \frac{x^{2i}}{2i(2i-2)\cdots 2} + \cdots\right).$$

(b) Choose $u_0 = 1$ in the expression in (a).

(c) Note that $2i(2i-2)\cdots 2 = 2^i i!$, so the solution can be expressed as

$$y(x) = u_0\left(\sum_{i=0}^{\infty} \frac{x^{2i}}{2^i i!}\right).$$

If we recall the power series expansion for e^x, we see that $y(x) = u_0 e^{x^2/2}$. Using separation of variables, we have $dy/y = x\,dx$. By integration, $\log y = \frac{1}{2}x^2 + C$, or $y = C'e^{x^2/2}$.

4 (a) $y(x) = (u_0 + 1)e^x - (1 + x)$.

5 $y(x) = u_0\left(1 + (x-1) + \frac{(x-1)^2}{2} + \cdots + \frac{(x-1)^i}{i!} + \cdots\right).$

This is of course the series expansion of $y(x) = Ce^x$ about $x_0 = 1$.

6 (a) $y(x) = u_0\left(1 - (x-1) + \frac{(x-1)^3}{3} - \frac{(x-1)^4}{3 \cdot 4} + \cdots\right).$

7 (a) $$y(x) = u_0\left(1 + \frac{x^3}{2\cdot 3} + \frac{x^6}{2\cdot 3\cdot 5\cdot 6} + \frac{x^9}{2\cdot 3\cdot 5\cdot 6\cdot 8\cdot 9} + \cdots + \frac{(3i-2)(3i-5)\cdots 7\cdot 4\cdot 1x^{3i}}{(3i)!} + \cdots\right)$$
$$+ u_1\left(x + \frac{x^4}{3\cdot 4} + \frac{x^7}{3\cdot 4\cdot 6\cdot 7} + \frac{x^{10}}{3\cdot 4\cdot 6\cdot 7\cdot 9\cdot 10} + \cdots + \frac{(3i-1)(3i-4)\cdots 8\cdot 5\cdot 2x^{3i+1}}{(3i+1)!} + \cdots\right).$$

(b) For $x < 0$ the equation is similar to $d^2y/dx^2 = -y$ (which has $\sin x$ and $\cos x$ for solutions), and for $x > 0$ the equation is similar to $d^2y/dx^2 = y$ (which has e^x and e^{-x} for solutions).

8 (a) The recurrence relation is $u_{i+2} = \dfrac{(1+i)u_i}{(i+2)(i+1)} = \dfrac{u_i}{i+2}$,
$$y(x) = u_0\left(1 + \frac{x^2}{2} + \frac{x^4}{2\cdot 4} + \frac{x^6}{2\cdot 4\cdot 6} + \cdots + \frac{x^{2i}}{2^i i!} + \cdots\right)$$
$$+ u_1\left(x + \frac{x^3}{3} + \frac{x^5}{3\cdot 5} + \frac{x^7}{3\cdot 5\cdot 7} + \cdots + \frac{2^i i! x^{2i+1}}{(2i+1)!} + \cdots\right).$$

9 $$y(x) = u_0\left(1 - \frac{x^2}{2!} - \frac{x^3}{3!} - \frac{x^4}{4!} + \cdots\right) + u_1\left(x - \frac{x^3}{3!} - \frac{x^4}{12} + \cdots\right).$$

10 $$y(x) = u_0\left(1 - \frac{n(n+1)}{2!}x^2 + \frac{n(n-2)(n+1)(n+3)}{4!}x^4 - \cdots + (-1)^i \frac{n(n-2)(n-4)\cdots(n-2i+2)\,(n+1)(n+3)\cdots(n+2i-1)}{(2i)!}x^{2i} + \cdots\right)$$
$$+ u_1\left(x - \frac{(n-1)(n+2)}{3!}x^3 + \frac{(n-1)(n-3)(n+2)(n+4)}{5!}x^5 - \cdots + (-1)^i \frac{(n-1)(n-3)\cdots(n-2i+1)(n+2)(n+4)\cdots(n+2i)}{(2i+1)!}x^{2i+1} + \cdots\right).$$

tables

Table 1 *Logarithms to Base 10*

x	0	1	2	3	4	5	6	7	8	9
1.0	.0000	.0043	.0086	.0128	.0170	.0212	.0253	.0294	.0334	.0374
1.1	.0414	.0453	.0492	.0531	.0569	.0607	.0645	.0682	.0719	.0755
1.2	.0792	.0828	.0864	.0899	.0934	.0969	.1004	.1038	.1072	.1106
1.3	.1139	.1173	.1206	.1239	.1271	.1303	.1335	.1367	.1399	.1430
1.4	.1461	.1492	.1523	.1553	.1584	.1614	.1644	.1673	.1703	.1732
1.5	.1761	.1790	.1818	.1847	.1875	.1903	.1931	.1959	.1987	.2014
1.6	.2041	.2068	.2095	.2122	.2148	.2175	.2201	.2227	.2253	.2279
1.7	.2304	.2330	.2355	.2380	.2405	.2430	.2455	.2480	.2504	.2529
1.8	.2553	.2577	.2601	.2625	.2648	.2672	.2695	.2718	.2742	.2765
1.9	.2788	.2810	.2833	.2856	.2878	.2900	.2923	.2945	.2967	.2989
2.0	.3010	.3032	.3054	.3075	.3096	.3118	.3139	.3160	.3181	.3201
2.1	.3222	.3243	.3263	.3284	.3304	.3324	.3345	.3365	.3385	.3404
2.2	.3424	.3444	.3464	.3483	.3502	.3522	.3541	.3560	.3579	.3598
2.3	.3617	.3636	.3655	.3674	.3692	.3711	.3729	.3747	.3766	.3784
2.4	.3802	.3820	.3838	.3856	.3874	.3892	.3909	.3927	.3945	.3962
2.5	.3979	.3997	.4014	.4031	.4048	.4065	.4082	.4099	.4116	.4133
2.6	.4150	.4166	.4183	.4200	.4216	.4232	.4249	.4265	.4281	.4298
2.7	.4314	.4330	.4346	.4362	.4378	.4393	.4409	.4425	.4440	.4456
2.8	.4472	.4487	.4502	.4518	.4533	.4548	.4564	.4579	.4594	.4609
2.9	.4624	.4639	.4654	.4669	.4683	.4698	.4713	.4728	.4742	.4757
3.0	.4771	.4786	.4800	.4814	.4829	.4843	.4857	.4871	.4886	.4900
3.1	.4914	.4928	.4942	.4955	.4969	.4983	.4997	.5011	.5024	.5038
3.2	.5051	.5065	.5079	.5092	.5105	.5119	.5132	.5145	.5159	.5172
3.3	.5185	.5198	.5211	.5224	.5237	.5250	.5263	.5276	.5289	.5302
3.4	.5315	.5328	.5340	.5353	.5366	.5378	.5391	.5403	.5416	.5428
3.5	.5441	.5453	.5465	.5478	.5490	.5502	.5514	.5527	.5539	.5551
3.6	.5563	.5575	.5587	.5599	.5611	.5623	.5635	.5647	.5658	.5670
3.7	.5682	.5694	.5705	.5717	.5729	.5740	.5752	.5763	.5775	.5786
3.8	.5798	.5809	.5821	.5832	.5843	.5855	.5866	.5877	.5888	.5899
3.9	.5911	.5922	.5933	.5944	.5955	.5966	.5977	.5988	.5999	.6010
4.0	.6021	.6031	.6042	.6053	.6064	.6075	.6085	.6096	.6107	.6117
4.1	.6128	.6138	.6149	.6160	.6170	.6180	.6191	.6201	.6212	.6222
4.2	.6232	.6243	.6253	.6263	.6274	.6284	.6294	.6304	.6314	.6325
4.3	.6335	.6345	.6355	.6365	.6375	.6385	.6395	.6405	.6415	.6425
4.4	.6435	.6444	.6454	.6464	.6474	.6484	.6493	.6503	.6513	.6522
4.5	.6532	.6542	.6551	.6561	.6571	.6580	.6590	.6599	.6609	.6618
4.6	.6628	.6637	.6646	.6656	.6665	.6675	.6684	.6693	.6702	.6712
4.7	.6721	.6730	.6739	.6749	.6758	.6767	.6776	.6785	.6794	.6803
4.8	.6812	.6821	.6830	.6839	.6848	.6857	.6866	.6875	.6884	.6893
4.9	.6902	.6911	.6920	.6928	.6937	.6946	.6955	.6964	.6972	.6981
5.0	.6990	.6998	.7007	.7016	.7024	.7033	.7042	.7050	.7059	.7067
5.1	.7076	.7084	.7093	.7101	.7110	.7118	.7126	.7135	.7143	.7152
5.2	.7160	.7168	.7177	.7185	.7193	.7202	.7210	.7218	.7226	.7235
5.3	.7243	.7251	.7259	.7267	.7275	.7284	.7292	.7300	.7308	.7316
5.4	.7324	.7332	.7340	.7348	.7356	.7364	.7372	.7380	.7388	.7396

x 0 1 2 3 4 5 6 7 8 9

Table 1 *Logarithms to Base 10*

x	0	1	2	3	4	5	6	7	8	9
5.5	.7404	.7412	.7419	.7427	.7435	.7443	.7451	.7459	.7466	.7474
5.6	.7482	.7490	.7497	.7505	.7513	.7520	.7528	.7536	.7543	.7551
5.7	.7559	.7566	.7574	.7582	.7589	.7597	.7604	.7612	.7619	.7627
5.8	.7634	.7642	.7649	.7657	.7664	.7672	.7679	.7686	.7694	.7701
5.9	.7709	.7716	.7723	.7731	.7738	.7745	.7752	.7760	.7767	.7774
6.0	.7782	.7789	.7796	.7803	.7810	.7818	.7825	.7832	.7839	.7846
6.1	.7853	.7860	.7868	.7875	.7882	.7889	.7896	.7903	.7910	.7917
6.2	.7924	.7931	.7938	.7945	.7952	.7959	.7966	.7973	.7980	.7987
6.3	.7993	.8000	.8007	.8014	.8021	.8028	.8035	.8041	.8048	.8055
6.4	.8062	.8069	.8075	.8082	.8089	.8096	.8102	.8109	.8116	.8122
6.5	.8129	.8136	.8142	.8149	.8156	.8162	.8169	.8176	.8182	.8189
6.6	.8195	.8202	.8209	.8215	.8222	.8228	.8235	.8241	.8248	.8254
6.7	.8261	.8267	.8274	.8280	.8287	.8293	.8299	.8306	.8312	.8319
6.8	.8325	.8331	.8338	.8344	.8351	.8357	.8363	.8370	.8376	.8382
6.9	.8388	.8395	.8401	.8407	.8414	.8420	.8426	.8432	.8439	.8445
7.0	.8451	.8457	.8463	.8470	.8476	.8482	.8488	.8494	.8500	.8506
7.1	.8513	.8519	.8525	.8531	.8537	.8543	.8549	.8555	.8561	.8567
7.2	.8573	.8579	.8585	.8591	.8597	.8603	.8609	.8615	.8621	.8627
7.3	.8633	.8639	.8645	.8651	.8657	.8663	.8669	.8675	.8681	.8686
7.4	.8692	.8698	.8704	.8710	.8716	.8722	.8727	.8733	.8739	.8745
7.5	.8751	.8756	.8762	.8768	.8774	.8779	.8785	.8791	.8797	.8802
7.6	.8808	.8814	.8820	.8825	.8831	.8837	.8842	.8848	.8854	.8859
7.7	.8865	.8871	.8876	.8882	.8887	.8893	.8899	.8904	.8910	.8915
7.8	.8921	.8927	.8932	.8938	.8943	.8949	.8954	.8960	.8965	.8971
7.9	.8976	.8982	.8987	.8993	.8998	.9004	.9009	.9015	.9020	.9025
8.0	.9031	.9036	.9042	.9047	.9053	.9058	.9063	.9069	.9074	.9079
8.1	.9085	.9090	.9096	.9101	.9106	.9112	.9117	.9122	.9128	.9133
8.2	.9138	.9143	.9149	.9154	.9159	.9165	.9170	.9175	.9180	.9186
8.3	.9191	.9196	.9201	.9206	.9212	.9217	.9222	.9227	.9232	.9238
8.4	.9243	.9248	.9253	.9258	.9263	.9269	.9274	.9279	.9284	.9289
8.5	.9294	.9299	.9304	.9309	.9315	.9320	.9325	.9330	.9335	.9340
8.6	.9345	.9350	.9355	.9360	.9365	.9370	.9375	.9380	.9385	.9390
8.7	.9395	.9400	.9405	.9410	.9415	.9420	.9425	.9430	.9435	.9440
8.8	.9445	.9450	.9455	.9460	.9465	.9469	.9474	.9479	.9484	.9489
8.9	.9494	.9499	.9504	.9509	.9513	.9518	.9523	.9528	.9533	.9538
9.0	.9542	.9547	.9552	.9557	.9562	.9566	.9571	.9576	.9581	.9586
9.1	.9590	.9595	.9600	.9605	.9609	.9614	.9619	.9624	.9628	.9633
9.2	.9638	.9643	.9647	.9652	.9657	.9661	.9666	.9671	.9675	.9680
9.3	.9685	.9689	.9694	.9699	.9703	.9708	.9713	.9717	.9722	.9727
9.4	.9731	.9736	.9741	.9745	.9750	.9754	.9759	.9763	.9768	.9773
9.5	.9777	.9782	.9786	.9791	.9795	.9800	.9805	.9809	.9814	.9818
9.6	.9823	.9827	.9832	.9836	.9841	.9845	.9850	.9854	.9859	.9863
9.7	.9868	.9872	.9877	.9881	.9886	.9890	.9894	.9899	.9903	.9908
9.8	.9912	.9917	.9921	.9926	.9930	.9934	.9939	.9943	.9948	.9952
9.9	.9956	.9961	.9965	.9969	.9974	.9978	.9983	.9987	.9991	.9996
x	0	1	2	3	4	5	6	7	8	9

Table 2 *Logarithms to Base e*

N	.0	.1	.2	.3	.4	.5	.6	.7	.8	.9
1	0.000	0.095	0.182	0.262	0.336	0.405	0.470	0.531	0.588	0.642
2	0.693	0.742	0.788	0.833	0.875	0.916	0.956	0.993	1.030	1.065
3	1.099	1.131	1.163	1.194	1.224	1.253	1.281	1.308	1.335	1.361
4	1.386	1.411	1.435	1.459	1.482	1.504	1.526	1.548	1.569	1.589
5	1.609	1.629	1.649	1.668	1.686	1.705	1.723	1.740	1.758	1.775
6	1.792	1.808	1.825	1.841	1.856	1.872	1.887	1.902	1.917	1.932
7	1.946	1.960	1.974	1.988	2.001	2.015	2.028	2.041	2.054	2.067
8	2.079	2.092	2.104	2.116	2.128	2.140	2.152	2.163	2.175	2.186
9	2.197	2.208	2.219	2.230	2.241	2.251	2.262	2.272	2.282	2.293
10	2.303	2.313	2.322	2.332	2.342	2.351	2.361	2.370	2.380	2.389

Table 3 *e^x and e^{-x}*

x	e^x	e^{-x}	x	e^x	e^{-x}
0.0	1.00	1.00	3.1	22.2	.045
0.1	1.11	.905	3.2	24.5	.041
0.2	1.22	.819	3.3	27.1	.037
0.3	1.35	.741	3.4	30.0	.033
0.4	1.49	.670	3.5	33.1	.030
0.5	1.65	.607	3.6	36.6	.027
0.6	1.82	.549	3.7	40.4	.025
0.7	2.01	.497	3.8	44.7	.022
0.8	2.23	.449	3.9	49.4	.020
0.9	2.46	.407	4.0	54.6	.018
1.0	2.72	.368	4.1	60.3	.017
1.1	3.00	.333	4.2	66.7	.015
1.2	3.32	.301	4.3	73.7	.014
1.3	3.67	.273	4.4	81.5	.012
1.4	4.06	.247	4.5	90.0	.011
1.5	4.48	.223	4.6	99.5	.010
1.6	4.95	.202	4.7	110	.0091
1.7	5.47	.183	4.8	122	.0082
1.8	6.05	.165	4.9	134	.0074
1.9	6.69	.150	5.0	148	.0067
2.0	7.39	.135	5.1	164	.0061
2.1	8.17	.122	5.2	181	.0055
2.2	9.02	.111	5.3	200	.0050
2.3	9.97	.100	5.4	221	.0045
2.4	11.0	.091	5.5	245	.0041
2.5	12.2	.082	5.6	270	.0037
2.6	13.5	.074	5.7	299	.0033
2.7	14.9	.067	5.8	330	.0030
2.8	16.4	.061	5.9	365	.0027
2.9	18.2	.055	6.0	403	.0025
3.0	20.1	.050			

Table 4 *Trigonometric Functions*

Degrees	Radians	Sin	Tan	Cot	Cos		
0°	.000	.000	.000		1.000	1.571	90°
1°	.017	.017	.017	57.29	1.000	1.553	89°
2°	.035	.035	.035	28.64	.999	1.536	88°
3°	.052	.052	.052	19.081	.999	1.518	87°
4°	.070	.070	.070	14.301	.998	1.501	86°
5°	.087	.087	.087	11.430	.996	1.484	85°
6°	.105	.105	.105	9.514	.995	1.466	84°
7°	.122	.122	.123	8.144	.993	1.449	83°
8°	.140	.139	.141	7.115	.990	1.431	82°
9°	.157	.156	.158	6.314	.988	1.414	81°
10°	.175	.174	.176	5.671	.985	1.396	80°
11°	.192	.191	.194	5.145	.982	1.379	79°
12°	.209	.208	.213	4.705	.978	1.361	78°
13°	.227	.225	.231	4.331	.974	1.344	77°
14°	.244	.242	.249	4.011	.970	1.326	76°
15°	.262	.259	.268	3.732	.966	1.309	75°
16°	.279	.276	.287	3.487	.961	1.292	74°
17°	.297	.292	.306	3.271	.956	1.274	73°
18°	.314	.309	.325	3.078	.951	1.257	72°
19°	.332	.326	.344	2.904	.946	1.239	71°
20°	.349	.342	.364	2.747	.940	1.222	70°
21°	.367	.358	.384	2.605	.934	1.204	69°
22°	.384	.375	.404	2.475	.927	1.187	68°
23°	.401	.391	.424	2.356	.921	1.169	67°
24°	.419	.407	.445	2.246	.914	1.152	66°
25°	.436	.423	.466	2.144	.906	1.134	65°
26°	.454	.438	.488	2.050	.899	1.117	64°
27°	.471	.454	.510	1.963	.891	1.100	63°
28°	.489	.469	.532	1.881	.883	1.082	62°
29°	.506	.485	.554	1.804	.875	1.065	61°
30°	.524	.500	.577	1.732	.866	1.047	60°
31°	.541	.515	.601	1.664	.857	1.030	59°
32°	.559	.530	.625	1.600	.848	1.012	58°
33°	.576	.545	.649	1.540	.839	.995	57°
34°	.593	.559	.675	1.483	.829	.977	56°
35°	.611	.574	.700	1.428	.819	.960	55°
36°	.628	.588	.727	1.376	.809	.942	54°
37°	.646	.602	.754	1.327	.799	.925	53°
38°	.663	.616	.781	1.280	.788	.908	52°
39°	.681	.629	.810	1.235	.777	.890	51°
40°	.698	.643	.839	1.192	.766	.873	50°
41°	.716	.656	.869	1.150	.755	.855	49°
42°	.733	.669	.900	1.111	.743	.838	48°
43°	.750	.682	.933	1.072	.731	.820	47°
44°	.768	.695	.966	1.036	.719	.803	46°
45°	.785	.707	1.000	1.000	.707	.785	45°
		Cos	Cot	Tan	Sin	Radians	Degrees

index